Switchable and Responsive Surfaces and Materials for Biomedical Applications

Related titles

Tissue engineering using ceramics and polymers: Second edition
(ISBN 978-0-85709-712-5)

Characterization of biomaterials
(ISBN 978-1-84569-810-2)

Bio-tribocorrosion in biomaterials and medical implants
(ISBN 978-0-85709-540-1)

Woodhead Publishing Series in Biomaterials:
Number 92

Switchable and Responsive Surfaces and Materials for Biomedical Applications

Edited by

Zheng Zhang

ELSEVIER

AMSTERDAM • BOSTON • CAMBRIDGE • HEIDELBERG
LONDON • NEW YORK • OXFORD • PARIS • SAN DIEGO
SAN FRANCISCO • SINGAPORE • SYDNEY • TOKYO
Woodhead Publishing is an imprint of Elsevier

WP WOODHEAD PUBLISHING

Woodhead Publishing is an imprint of Elsevier
80 High Street, Sawston, Cambridge, CB22 3HJ, UK
225 Wyman Street, Waltham, MA 02451, USA
Langford Lane, Kidlington, OX5 1GB, UK

British Library Cataloguing-in-Publication Data
A catalogue record for this book is available from the British Library

Library of Congress Control Number: 2014954008

ISBN: 978-0-85709-713-2 (print)
ISBN: 978-0-85709-717-0 (online)

For information on all Woodhead Publishing publications
visit our website at http://store.elsevier.com/

Typeset by TNQ Books and Journals
www.tnq.co.in

Printed in the United States of America

Contents

List of contributors ix
Woodhead Publishing Series in Biomaterials xi
Introduction xv

Part One Switchable and responsive materials and surfaces 1

1 Thermo-responsive polymers: structure and design of smart materials 3
A.K. Teotia, H. Sami, A. Kumar
1.1 Introduction 3
1.2 Thermo-responsive polymers and their characteristics 4
1.3 Types of thermo-responsive polymers 6
1.4 Physical forms of switchable materials and their applications 20
1.5 Summary 30
References 31

2 Environmentally responsive polyelectrolytes and zwitterionic polymers 45
M.T. Bernards
2.1 Introduction 45
2.2 Monomer subunits and polymerization approaches 46
2.3 General solution properties 49
2.4 Stimuli-induced changes and their applications 52
2.5 Future trends 58
2.6 Sources of further information 58
References 59

3 Peptide-based switchable and responsive surfaces 65
M. Zelzer
3.1 Introduction 65
3.2 Preparation of peptide surfaces 68
3.3 Responsive peptide surfaces 72
3.4 Peptides attached to responsive surfaces 77
3.5 Protein surfaces 81
3.6 Physical properties and characterisation 83

3.7 Emerging applications **88**
3.8 Future trends **89**
References **90**

4 Photonic sensitive switchable materials **93**
L. Jiang, L. Wen
4.1 Introduction **93**
4.2 Photonic sensitive switchable materials **94**
4.3 Potential applications **99**
4.4 Conclusions and future trends **111**
References **112**

5 Responsive polymer brushes for biomedical applications **119**
N. Akkilic, W.M. de Vos
5.1 Introduction **119**
5.2 Brush architecture **123**
5.3 Types of responsive polymer brushes **124**
5.4 Biomedical applications **131**
5.5 Summary and future trends **138**
References **141**

6 Preparation and analysis of switchable copolymers for biomedical application **147**
M. Nitschke, S. Gramm
6.1 Introduction **147**
6.2 Switchable copolymer coatings **147**
6.3 Advanced analytical techniques **153**
6.4 Future trends **161**
6.5 Sources of further information **161**
References **161**

Part Two Biological interactions and biomedical applications of switchable surfaces **165**

7 Interaction of switchable biomaterials surfaces with proteins **167**
Q. Yu, H. Chen
7.1 Introduction **167**
7.2 Protein adsorption on surfaces **168**
7.3 Protein adsorption on thermo-responsive surfaces **169**
7.4 Protein adsorption on pH and/or ionic strength-responsive surfaces **174**
7.5 Protein adsorption on other responsive surfaces **178**
7.6 Synergistic effect of surface chemistry and nanostructures on protein adsorption **180**

7.7 Aspects for future research **181**
Acknowledgments **182**
References **183**

8 Interaction of responsive/switchable surfaces with cells **189**
B.L. Banik, J.L. Brown
8.1 Introduction **189**
8.2 Controlling stem cell behavior **190**
8.3 Interfacial properties **191**
8.4 Physical behavior **194**
8.5 Future trends **198**
References **199**
Further reading **201**

9 Temperature-responsive polymers for cell culture and tissue engineering applications **203**
Y. Akiyama, T. Okano
9.1 Introduction **203**
9.2 Methods for preparing temperature-responsive cell culture surfaces (TRCSs) and their characteristics **204**
9.3 Cell sheet-based regenerative medicine **220**
9.4 Future application of temperature-responsive cell culture surface to support and promote regenerative medicine field **223**
References **225**

10 Stimuli-sensitive polymers for drug delivery and diagnostic systems interacting with biosurfaces **235**
T. Yoshida
10.1 Introduction **235**
10.2 Polymers sensitive to body's internal stimuli **236**
10.3 Polymers sensitive to the body's external stimuli **245**
10.4 Future trends **251**
References **252**

11 Application of responsive polymers in implantable medical devices and biosensors **259**
J. Li, M. Stachowski, Z. Zhang
11.1 Introduction **259**
11.2 Biocompatibility of medical devices and responsive polymers **260**
11.3 Cardiovascular devices **262**
11.4 Drug delivery/infection control coatings and devices **269**
11.5 Surgical devices **271**
11.6 Dental and orthopedic devices **274**
11.7 Ophthalmic devices **276**
11.8 Respiratory devices **278**
11.9 Gastrointestinal devices **279**

11.10 Urogenital devices/urinary devices **279**
11.11 Insulin delivery devices for diabetes management **280**
11.12 Biosensors **281**
11.13 Summary **287**
References **288**

Index **299**

List of contributors

Y. Akiyama Tokyo Women's Medical University, Tokyo, Japan

N. Akkilic University of Twente, Enschede, The Netherlands

B.L. Banik The Pennsylvania State University, University Park, PA, USA

M.T. Bernards University of Missouri, Columbia, MO, USA

J.L. Brown The Pennsylvania State University, University Park, PA, USA

H. Chen Soochow University, Suzhou, China

W.M. de Vos University of Twente, Enschede, The Netherlands

S. Gramm Max Bergmann Center of Biomaterials, Dresden, Germany

L. Jiang Chinese Academy of Sciences, Beijing, P.R. China

A. Kumar Indian Institute of Technology Kanpur, Kanpur, India

J. Li Teleflex Medical, Cambridge, MA, USA

M. Nitschke Max Bergmann Center of Biomaterials, Dresden, Germany

T. Okano Tokyo Women's Medical University, Tokyo, Japan

H. Sami Indian Institute of Technology Kanpur, Kanpur, India

M. Stachowski Teleflex Medical, Cambridge, MA, USA

A.K. Teotia Indian Institute of Technology Kanpur, Kanpur, India

L. Wen Chinese Academy of Sciences, Beijing, P.R. China

T. Yoshida Astellas Pharma Inc., Tsukuba-shi, Japan

Q. Yu Soochow University, Suzhou, China

M. Zelzer University of Nottingham, Nottingham, UK; National Physical Laboratory, Middlesex, UK

Z. Zhang Teleflex Medical, Cambridge, MA, USA

Woodhead Publishing Series in Biomaterials

1 **Sterilisation of tissues using ionising radiations**
Edited by J. F. Kennedy, G. O. Phillips and P. A. Williams
2 **Surfaces and interfaces for biomaterials**
Edited by P. Vadgama
3 **Molecular interfacial phenomena of polymers and biopolymers**
Edited by C. Chen
4 **Biomaterials, artificial organs and tissue engineering**
Edited by L. Hench and J. Jones
5 **Medical modelling**
R. Bibb
6 **Artificial cells, cell engineering and therapy**
Edited by S. Prakash
7 **Biomedical polymers**
Edited by M. Jenkins
8 **Tissue engineering using ceramics and polymers**
Edited by A. R. Boccaccini and J. Gough
9 **Bioceramics and their clinical applications**
Edited by T. Kokubo
10 **Dental biomaterials**
Edited by R. V. Curtis and T. F. Watson
11 **Joint replacement technology**
Edited by P. A. Revell
12 **Natural-based polymers for biomedical applications**
Edited by R. L. Reiss et al.
13 **Degradation rate of bioresorbable materials**
Edited by F. J. Buchanan
14 **Orthopaedic bone cements**
Edited by S. Deb
15 **Shape memory alloys for biomedical applications**
Edited by T. Yoneyama and S. Miyazaki
16 **Cellular response to biomaterials**
Edited by L. Di Silvio
17 **Biomaterials for treating skin loss**
Edited by D. P. Orgill and C. Blanco
18 **Biomaterials and tissue engineering in urology**
Edited by J. Denstedt and A. Atala
19 **Materials science for dentistry**
B. W. Darvell
20 **Bone repair biomaterials**
Edited by J. A. Planell, S. M. Best, D. Lacroix and A. Merolli

21 **Biomedical composites**
Edited by L. Ambrosio
22 **Drug–device combination products**
Edited by A. Lewis
23 **Biomaterials and regenerative medicine in ophthalmology**
Edited by T. V. Chirila
24 **Regenerative medicine and biomaterials for the repair of connective tissues**
Edited by C. Archer and J. Ralphs
25 **Metals for biomedical devices**
Edited by M. Ninomi
26 **Biointegration of medical implant materials: Science and design**
Edited by C. P. Sharma
27 **Biomaterials and devices for the circulatory system**
Edited by T. Gourlay and R. Black
28 **Surface modification of biomaterials: Methods analysis and applications**
Edited by R. Williams
29 **Biomaterials for artificial organs**
Edited by M. Lysaght and T. Webster
30 **Injectable biomaterials: Science and applications**
Edited by B. Vernon
31 **Biomedical hydrogels: Biochemistry, manufacture and medical applications**
Edited by S. Rimmer
32 **Preprosthetic and maxillofacial surgery: Biomaterials, bone grafting and tissue engineering**
Edited by J. Ferri and E. Hunziker
33 **Bioactive materials in medicine: Design and applications**
Edited by X. Zhao, J. M. Courtney and H. Qian
34 **Advanced wound repair therapies**
Edited by D. Farrar
35 **Electrospinning for tissue regeneration**
Edited by L. Bosworth and S. Downes
36 **Bioactive glasses: Materials, properties and applications**
Edited by H. O. Ylänen
37 **Coatings for biomedical applications**
Edited by M. Driver
38 **Progenitor and stem cell technologies and therapies**
Edited by A. Atala
39 **Biomaterials for spinal surgery**
Edited by L. Ambrosio and E. Tanner
40 **Minimized cardiopulmonary bypass techniques and technologies**
Edited by T. Gourlay and S. Gunaydin
41 **Wear of orthopaedic implants and artificial joints**
Edited by S. Affatato
42 **Biomaterials in plastic surgery: Breast implants**
Edited by W. Peters, H. Brandon, K. L. Jerina, C. Wolf and V. L. Young
43 **MEMS for biomedical applications**
Edited by S. Bhansali and A. Vasudev
44 **Durability and reliability of medical polymers**
Edited by M. Jenkins and A. Stamboulis
45 **Biosensors for medical applications**
Edited by S. Higson
46 **Sterilisation of biomaterials and medical devices**
Edited by S. Lerouge and A. Simmons

47 **The hip resurfacing handbook: A practical guide to the use and management of modern hip resurfacings**
Edited by K. De Smet, P. Campbell and C. Van Der Straeten
48 **Developments in tissue engineered and regenerative medicine products**
J. Basu and J. W. Ludlow
49 **Nanomedicine: Technologies and applications**
Edited by T. J. Webster
50 **Biocompatibility and performance of medical devices**
Edited by J.-P. Boutrand
51 **Medical robotics: Minimally invasive surgery**
Edited by P. Gomes
52 **Implantable sensor systems for medical applications**
Edited by A. Inmann and D. Hodgins
53 **Non-metallic biomaterials for tooth repair and replacement**
Edited by P. Vallittu
54 **Joining and assembly of medical materials and devices**
Edited by Y. (Norman) Zhou and M. D. Breyen
55 **Diamond-based materials for biomedical applications**
Edited by R. Narayan
56 **Nanomaterials in tissue engineering: Fabrication and applications**
Edited by A. K. Gaharwar, S. Sant, M. J. Hancock and S. A. Hacking
57 **Biomimetic biomaterials: Structure and applications**
Edited by A. J. Ruys
58 **Standardisation in cell and tissue engineering: Methods and protocols**
Edited by V. Salih
59 **Inhaler devices: Fundamentals, design and drug delivery**
Edited by P. Prokopovich
60 **Bio-tribocorrosion in biomaterials and medical implants**
Edited by Y. Yan
61 **Microfluidic devices for biomedical applications**
Edited by X.-J. James Li and Y. Zhou
62 **Decontamination in hospitals and healthcare**
Edited by J. T. Walker
63 **Biomedical imaging: Applications and advances**
Edited by P. Morris
64 **Characterization of biomaterials**
Edited by M. Jaffe, W. Hammond, P. Tolias and T. Arinzeh
65 **Biomaterials and medical tribology**
Edited by J. Paolo Davim
66 **Biomaterials for cancer therapeutics: Diagnosis, prevention and therapy**
Edited by K. Park
67 **New functional biomaterials for medicine and healthcare**
E. P. Ivanova, K. Bazaka and R. J. Crawford
68 **Porous silicon for biomedical applications**
Edited by H. A. Santos
69 **A practical approach to spinal trauma**
Edited by H. N. Bajaj and S. Katoch
70 **Rapid prototyping of biomaterials: Principles and applications**
Edited by R. Narayan
71 **Cardiac regeneration and repair Volume 1: Pathology and therapies**
Edited by R.-K. Li and R. D. Weisel
72 **Cardiac regeneration and repair Volume 2: Biomaterials and tissue engineering**
Edited by R.-K. Li and R. D. Weisel

73 **Semiconducting silicon nanowires for biomedical applications**
Edited by J. L. Coffer
74 **Silk biomaterials for tissue engineering and regenerative medicine**
Edited by S. Kundu
75 **Biomaterials for bone regeneration: Novel techniques and applications**
Edited by P. Dubruel and S. Van Vlierberghe
76 **Biomedical foams for tissue engineering applications**
Edited by P. Netti
77 **Precious metals for biomedical applications**
Edited by N. Baltzer and T. Copponnex
78 **Bone substitute biomaterials**
Edited by K. Mallick
79 **Regulatory affairs for biomaterials and medical devices**
Edited by S. F. Amato and R. Ezzell
80 **Joint replacement technology Second edition**
Edited by P. A. Revell
81 **Computational modelling of biomechanics and biotribology in the musculoskeletal system: Biomaterials and tissues**
Edited by Z. Jin
82 **Biophotonics for medical applications**
Edited by I. Meglinski
83 **Modelling degradation of bioresorbable polymeric medical devices**
Edited by J. Pan
84 **Perspectives in total hip arthroplasty: Advances in biomaterials and their tribological interactions**
S. Affatato
85 **Tissue engineering using ceramics and polymers Second edition**
Edited by A. R. Boccaccini and P. X. Ma
86 **Biomaterials and medical device-associated infections**
Edited by L. Barnes and I. R. Cooper
87 **Surgical techniques in total knee arthroplasty (TKA) and alternative procedures**
Edited by S. Affatato
88 **Lanthanide oxide nanoparticles for molecular imaging and therapeutics**
G. H. Lee
89 **Surface modification of magnesium and its alloys for biomedical applications Volume 1: Biological interactions, mechanical properties and testing**
Edited by T .S. N. Sankara Narayanan, I. S. Park and M. H. Lee
90 **Surface modification of magnesium and its alloys for biomedical applications Volume 2: Modification and coating techniques**
Edited by T .S. N. Sankara Narayanan, I. S. Park and M. H. Lee
91 **Medical modelling: The application of Advanced Design and Rapid Prototyping techniques in medicine Second Edition**
Edited by R. Bibb, D. Eggbeer and A. Paterson
92 **Switchable and responsive surfaces and materials for biomedical applications**
Edited by Z. Zhang
93 **Biomedical textiles for orthopaedic and surgical applications: Fundamentals, applications and tissue engineering**
Edited by T. Blair
94 **Surface coating and modification of metallic biomaterials**
Edited by C. Wen

Introduction

Z. Zhang[1], *R.S. Langer*[2]
[1]Teleflex Medical, Cambridge, MA, USA; [2]Massachusetts Institute of Technology, Cambridge, MA, USA

Materials or surfaces are said to be "responsive" if they display a pronounced response to an environmental stimulus, particularly a response that may be suitable for application. Some responses in the form of physical or phase changes can be "switchable" or "reversible." With the development of materials science, especially with the development of synthetic polymers and surface chemistry, these materials and surfaces have been designed for broad applications. "Smart" or "intelligent" has also been used to describe these materials since the 1980s. Shape-memory alloys and polymers, piezoelectric materials, and switchable glass are all good examples.

More recently responsive materials and surfaces have been used in various emerging biomedical areas such as drug delivery (Santini, Cima, & Langer, 1999), protein conjugates (Stayton et al., 1995), diagnostics (Hoffman, 2000), tissue engineering (Okano, Yamada, Sakai, & Sakurai, 1993), and medical devices (Lendlein & Langer, 2002). Moreover, materials and surfaces have been developed to direct specific cellular behaviors, or respond to specific cellular signals (Anderson, Burdick, & Langer, 2004). Over the past decade ambitious projects from both academia and industry have been deployed to mimic the smartness of biological systems such as tissues or organs. All these efforts tend to blur the boundary between traditional materials and living organisms, endowing a more ambitious meaning to the phrase "smart" or "intelligent materials."

The book provides an overview of recent advances in responsive surfaces and materials designed for biomedical applications. Both bulk responsive materials and surface modification techniques are included. The interactions of biomolecules and cells with responsive interfaces are specifically reviewed, with discussion of emerging applications that could change our lives. The book is divided into two parts with four interdisciplinary topics. The first part reviews switchable and responsive material technologies for biomedical applications (Chapters 1–6). The two topics of Part One are responsive materials and responsive surface modification respectively. Part Two (Chapters 7–11) explores two topics include the interaction of switchable surfaces with proteins and cells, as well as multidisplinary research toward different biomedical applications.

Responsive materials

The first section of Part One provides a detailed overview of switchable and responsive materials, exploring thermoresponsive polymers (Chapter 1), environmentally responsive polyelectrolytes and zwitterionic polymers (Chapter 2), peptide/protein-based natural responsive materials (Chapter 3), and photonic sensitive switchable materials (Chapter 4).

Responsive surface modification

Like a bulk responsive material, surface of a substrate can be responsive upon electrochemical, photo, temperature, pH, mechanical, or electrical stimuli (Lahann & Langer, 2005). Researchers are developing different ways to make these smart surfaces such as self-assembled monolayers (SAMs), polymer brushes, or copolymer coatings. A classic example is a switchable SAM surface in responsive to electric potential, which achieve hydrophobic–hydrophilic transition on a gold substrate. The switchable surfaces provide an idea platform to study surface-biological system interactions (Lahann et al., 2003). Polymer brushes and copolymers are more practical avenues to be applied with long-term performance.

Responsive surface-biological system interactions

The first section of Part Two discusses the biological interactions of responsive surfaces with proteins (Chapter 7) and cells (Chapter 8). Interactions with in vivo environment such as blood and tissue, as well as biocompatibility issues, are addressed in Chapter 11.

Multidisplinary research toward different biomedical applications

In this book, a few applications of responsive materials and surfaces are explored, which include cell culture and tissue engineering (Chapter 9), drug delivery and diagnostic systems (Chapter 10), implantable medical devices and biosensors (Chapter 11). These areas were chosen not only because they have huge economic value, but also involve challenges during the development toward applications.

Hopefully the book and chapters can stimulate further design and application of responsive materials and surfaces into the exciting biomedical area and improve patient health.

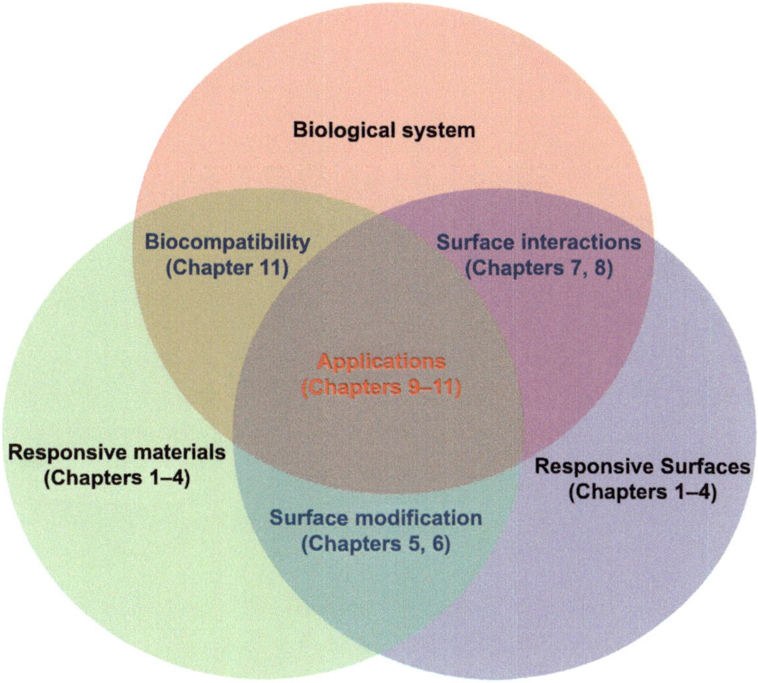

The book comprises interdisplinary overviews of responsive materials and surfaces, and their interaction with biological system, as well as their biomedical applications.

References

Anderson, D. G., Burdick, J. A., & Langer, R. (2004). Smart biomaterials. *Science*, *305*(5692), 1923–1924.

Hoffman, A. S. (2000). Bioconjugates of intelligent polymers and recognition proteins for use in diagnostics and affinity separations. *Clinical Chemistry*, *46*(9), 1478–1486.

Lahann, J., et al. (2003). A reversibly switching surface. *Science*, *299*(5605), 371–374.

Lahann, J., & Langer, R. (2005). Smart materials with dynamically controllable surfaces. *MRS Bulletin*, *30*(03), 185–188.

Lendlein, A., & Langer, R. (2002). Biodegradable, elastic shape-memory polymers for potential biomedical applications. *Science*, *296*(5573), 1673–1676.

Okano, T., Yamada, N., Sakai, H., & Sakurai, Y. (1993). A novel recovery system for cultured cells using plasma-treated polystyrene dishes grafted with poly(*N*-isopropylacrylamide). *Journal of Biomedical Materials Research*, *27*(10), 1243–1251.

Santini, J. T., Cima, M. J., & Langer, R. (1999). A controlled-release microchip. *Nature*, *397*(6717), 335–338.

Stayton, P. S., et al. (1995). Control of protein-ligand recognition using a stimuli-responsive polymer. *Nature*, *378*(6556), 472–474.

Part One

Switchable and responsive materials and surfaces

Thermo-responsive polymers: structure and design of smart materials

1

A.K. Teotia, H. Sami, A. Kumar
Indian Institute of Technology Kanpur, Kanpur, India

1.1 Introduction

"Smart or stimuli-responsive materials" is a class of materials that show response to external stimuli or change in their environment. These are one of the most rapidly evolving classes of materials and are classified based on the type of stimuli toward which they show responsive behavior, such as pH, temperature, variation in ionic strength, light, redox conditions, charged moieties in vicinity, electrical changes, magnetic field, and biological and chemical stimuli (Bae, Okano, & Kim, 1991a; Bae, Okano, & Kim, 1991b; Bromberg & Ron, 1998; Brownlee & Cerami, 1979; Firestone & Siegel, 1991; Hassan, Doyle, & Peppas, 1997; Ito, Casolaro, Kono, & Imanishi, 1989; Kabanov, 1994; Kwon, Bae, & Kim, 1991; Kwon, Bae, Okano, & Kim, 1991; Mamada, Tanaka, Kungwatchakun, & Irie, 1990; Okano, Bae, Jacobs, & Kim, 1990; Sawahata, Hara, Yasunaga, & Osada, 1990; Suzuki & Tanaka, 1990). Responsive polymer systems are polymeric materials falling under smart material category, which show reversible transitions in properties such as shape, solubility, surface characteristics, and molecular assembly in response to a stimulus. Stimuli such as pH, temperature, mechanical, electric, or magnetic fields alter the levels of various interaction energies, thus altering molecular interactions and hydrophilic–hydrophobic balance in these materials, leading to reversible changes in their microstructure (Taylor & Cerankowski, 1975). The ability to control properties of polymer systems by an external stimulus is highly desirable for fabricating smart materials that can be actuated to perform functions on demand, by altering external environmental conditions. Owing to their unique behavior of changing properties in response to a trigger, these polymer systems are finding increasing applications in diverse areas such as drug delivery, tissue engineering, gene delivery, and sensors. This chapter focuses on thermo-responsive polymers, their different classes, and the structural, chemical, and physical properties responsible for their responsive behavior, along with strategies to control their temperature responsiveness and to design switchable materials.

Switchable and Responsive Surfaces and Materials for Biomedical Applications. http://dx.doi.org/10.1016/B978-0-85709-713-2.00001-8

1.2 Thermo-responsive polymers and their characteristics

1.2.1 Phase transition and critical solution temperature

Thermo-responsive polymers are the most comprehensively studied responsive polymers because of their unique property of sol–gel transition above certain temperature. Some of them also show phase transition near to physiological human body temperature. Moreover, these polymers can be modified to show sol–gel transition at desired temperature. Based on their response to change in temperature, these polymers are categorized in two classes; first, polymers that become insoluble above a critical temperature called the lower critical solution temperature (LCST), and second, polymers that precipitate and undergo phase change below a critical temperature called as the upper critical solution temperature (UCST) (Figure 1.1). When polymers exhibiting LCST are dissolved in an aqueous system, they are completely miscible at normal temperature, but their solubility decreases with increase in temperature and above a critical value, i.e., LCST, they show phase separation (Boutris, Chatzi, & Kiparissides, 1997; Yan, Zhu, & Kenkare, 2000). These types of materials are also called "negative temperature–sensitive polymers," that is, poly(*N*-isopropylacrylamide) (PNIPAAm). The second class of polymers, which show UCST, are also referred to as "positive temperature–sensitive polymers"; that is, the temperature above which these polymers remain miscible in solution and as temperature of the solution falls below critical value, phase separation occurs, for example, poly(acrylic acid) (PAA), polyacrylamide (PAAm), and poly(acrylamide-*co*-butyl methacrylate).

For a polymer in aqueous solution, LCST is the point in the phase diagram at which entropy of the water in the system increases due to less ordered arrangement of water molecules and becomes more than enthalpy of water hydrogen bonded to the polymer (Kumar, Srivastava, Galaev, & Mattiasson, 2007), therefore entropy of system governs LCST, and enthalpy of system governs UCST (Southall, Dill, & Haymet, 2002).

When temperature of an aqueous solution of thermo-responsive polymer exhibiting LCST behavior is increased, the polymer chains show coil-to-globule-to-aggregate

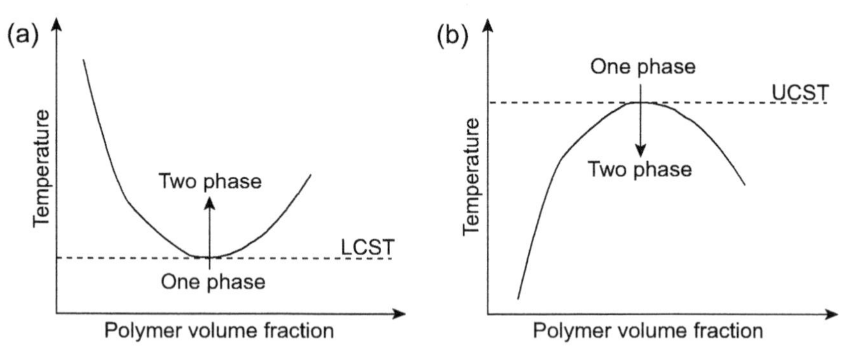

Figure 1.1 Curves showing phase transition phenomenon. (a) Lower critical solution temperature (LCST) and (b) upper critical solution temperature (UCST) phase transition behaviors of thermo-responsive polymers in solution.

transition (Figure 1.2). This conformational transition has been ascribed to different reasons based on different studies. Studies conducted on transition enthalpies of aqueous solutions of polymers showed interruption in polymer–water hydrogen bonding interactions that are involved in phase separation (Schild & Tirrell, 1990). Other studies have shown that during phase transition, there is loss of ordered water molecule arrangement around the hydrophobic polymer chain, which leads to chain collapse (Hwang & Hogenesch, 1995; Shibayama, Suetoh, & Nomura, 1996). Recent studies have shown that both hydrogen bonding and hydrophobic interactions in polymer solvent system are responsible for phase transition and hydrated random coil to hydrophobic globule transition above critical solution temperatures (Lin, Chen, & Liang, 1999; Volpert, Selb, & Candau, 1998). In PNIPAAm, as the temperature of polymer solution approaches to transition point, the hydrogen bonds formed between N–H or C=O groups of the polymer and water around the polymer chain (Qiu & Park, 2001) breaks, thus collapsing the polymer molecule and forming a hydrophobic globule (Markvicheva et al., 1991). As the temperature further increases, the remaining hydrogen bonds also break, leading to an increase in hydrophobic interactions between polymer chains, which causes aggregation and separation of polymer out of solution (Boutris et al., 1997; Fujishige, Kubota, & Ando, 1989).

1.2.2 Thermodynamics of polymer in solution

For a process to be thermodynamically feasible, it is required that Gibbs free energy (ΔG) ((ΔG) = $\Delta H - T \cdot \Delta S$, where T = reaction temperature) for the process be negative at the reaction conditions. Dissolution of polymer in solvent involves a variety of interactions between the polymer and solvent. Thus, the overall outcome of all of these interactions should result in negative values for enthalpy (ΔH) and entropy (ΔS) for spontaneous dissolution of the polymer in solvent, resulting in a homogeneous system. If the value for (ΔG) turns out to be positive for a polymer solvent system at certain temperature, the polymer will not be miscible with that solvent at those conditions, and two different phases of polymer and solvent will separately co-exist.

For polymers showing LCST behavior, Gibbs free energy is negative at normal temperatures, and they tend to dissolve in aqueous systems. The water molecules from the solvent form a thin layer surrounding the hydrophilic part of the polymer, and this hydration process gives negative enthalpy to the dissolution process. Furthermore, there is formation of a structured arrangement of water molecules around

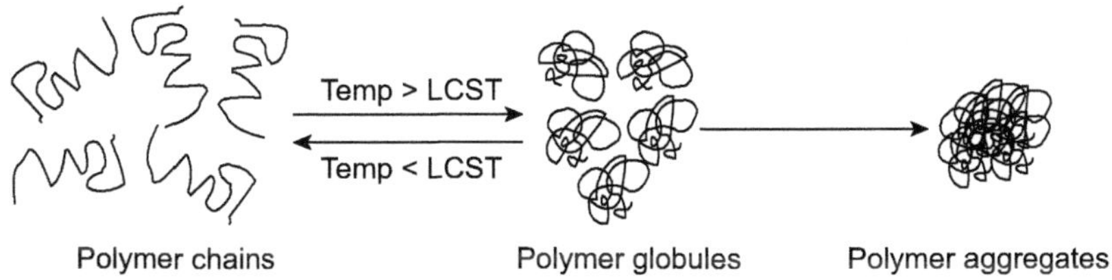

Figure 1.2 Temperature-induced phase transitions in thermo-responsive polymer chains. Extended chains-to-globules-to-aggregate transition is shown.

the hydrophobic part of the polymer, where there is formation of extensive hydrogen bonding among the surrounding water molecules, which provides negative value to entropy (ΔS) of the reaction (Horne, Almeida, Day, & Yu, 1971). When temperature of the aqueous polymer system is increased, entropy (ΔS) of the system increases; as a result, hydrogen bonds break, leading to collapse of the hydration shell formed by the water molecules. These are excluded into the bulk, which leads to an increase in entropy (ΔS), the overall entropy of system becomes greater than the negative enthalpy (ΔH) of the reaction, giving positive value to Gibbs free energy (ΔG). Breakage of hydrogen bonds and loss of hydration shell lead to increased hydrophobic interactions between hydrophobic parts of the same or different polymer molecules. This leads to chain collapse and inter-molecular aggregation and thereby separation into two phases, which ultimately leads to suppression in conformation changes of polymer chains decreasing ΔS and ΔH (Cheon, Jeong, & Cho, 1999; Idziak, Avoce, Lessard, Gravel, & Zhu, 1999; Schild, 1992; Zeng, Tong, & Feng, 1997). For polymers that show UCST behavior, both enthalpy (ΔH) and entropy (ΔS) are positive, which decreases with increase in temperature, showing opposite behavior to that shown by LCST polymers.

1.3 Types of thermo-responsive polymers

Based on their origin, thermo-responsive polymers can be broadly classified as synthetic and natural polymers. The synthetic thermo-responsive polymers are of different types, for example, poly(*N*-alkyl substituted acrylamides), poly(*N*-vinyl-alkyl-amides), poly(ethylene glycol)-poly(propylene glycol)-poly(ethylene glycol) copolymer (PEG–PPG–PEG), poly(ethylene glycol)-poly(D,L-lactic acid)-poly(ethylene glycol) copolymer (PEG–PLLA/PDLA–PEG), and biodegradable poly(ethylene glycol)-poly(lactic-*co*-glycolic acid) (PEG–PLGA) (Figure 1.3). A new class of synthetic thermo-responsive polymers based on a polypeptide-like elastin (elastin-like polypeptides (ELPs)) has also been reported (Figure 1.3). Combinations of the above synthetic polymers have also been synthesized as block copolymers (Figure 1.4). Natural thermo-responsive polymers and their derivatives include agarose, gelatin, agar, and pectin, as well as chitosan/glycerol phosphate polymer system. This chapter focuses on polymer materials with thermo-responsive behavior due to chemical interactions in their molecules.

1.3.1 Synthetic thermo-responsive polymers

1.3.1.1 N-substituted acrylamide polymers

N-substituted acrylamides are an extensive class of thermo-responsive polymers that have attracted a lot of interest due to their unique thermo-responsive behavior in aqueous media. Among this class, poly(*N*-isopropylacrylamide) (PNIPAAm) is the most extensively studied polymer for structure activity relationships (Bae, Okano, & Kim, 1990; Feil, Bae, Feijen, & Kim, 1992) and potential thermo-responsive applications.

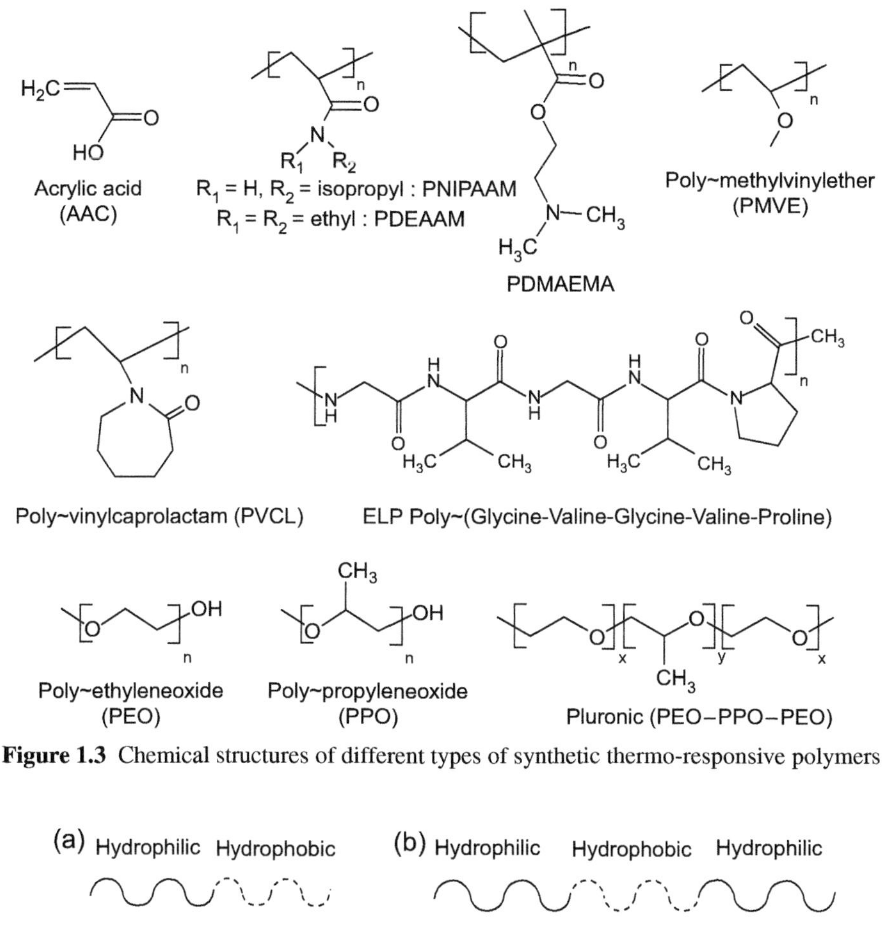

Figure 1.3 Chemical structures of different types of synthetic thermo-responsive polymers.

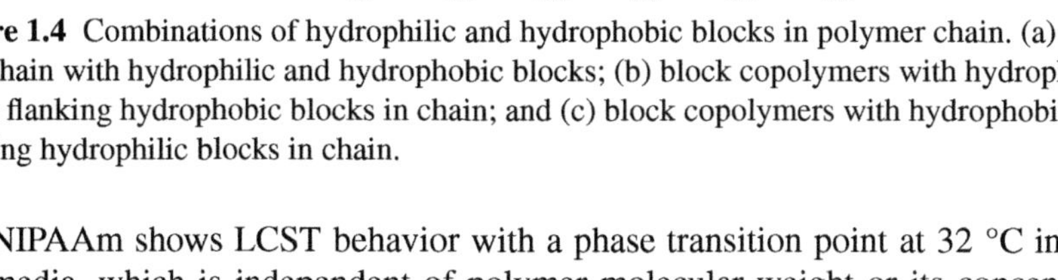

Figure 1.4 Combinations of hydrophilic and hydrophobic blocks in polymer chain. (a) Polymer chain with hydrophilic and hydrophobic blocks; (b) block copolymers with hydrophilic block flanking hydrophobic blocks in chain; and (c) block copolymers with hydrophobic block flanking hydrophilic blocks in chain.

PNIPAAm shows LCST behavior with a phase transition point at 32 °C in aqueous media, which is independent of polymer molecular weight or its concentration in solution (Fujishige et al., 1989). Below LCST, it shows extended chain conformation because of hydrogen bonding between hydrophilic groups in polymer and water molecules. Above transition temperature, hydrophobic interactions in polymer chains dominate because of dehydration of hydrophobic isopropyl group, which leads to phase separation and precipitation of the polymer, showing coil-to-globule to aggregate transition (Fujishige et al., 1989). The phase separation temperatures

Table 1.1 Effect of side chain lengths and properties on phase transition temperature (LCST) of poly(*N*-substituted acrylamide) polymers

R	Poly(*N*-alkylacrylamide)	LCST (°C)
NH_2	Poly(acrylamide)	No LCST
$NH–CH_3$	Poly(*N*-methylacrylamide)	No LCST
$NH–CH_2–CH_3$	Poly(*N*-ethylacrylamide)	82
$NH–CH_2–CH_2–CH_3$	Poly(*N*-n-propylacrylamide)	22
$NH–CH_2–(CH_3)_2$	Poly(*N*-isopropylacrylamide)	32–34
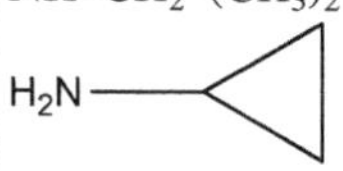	Poly(*N*-cyclopropylacrylamide)	47
$NH–CH_2–CH_2–CH_2–CH_3$	Poly(*N*-n-butylacrylamide)	No LCST
$N–C–(CH_3)_3$	Poly(*N*-*t*-butylacrylamide)	No LCST
$N–(CH_3)_2$	Poly(*N*,*N*-dimethylacrylamide)	No LCST
$N–(CH_2–CH_3)_2$	Poly(*N*,*N*-diethylacrylamide)	32–42

LCST, lower critical solution temperature.
Reproduced with permission from Galaev and Mattiasson (1993).

of these polymers show variation depending on the chemical structure of their side chain (Galaev & Mattiasson, 1993) (Table 1.1). Some of simple analogs (polyacrylamide and PNIPAAm) are soluble in water at room temperature, but others with more hydrophobic groups are insoluble in water, for example, poly(*N*-n-butylacrylamide) and poly(*N*-tert-butylacrylamide). Few of these that are soluble in water at room temperature show phase transition at increased temperatures (e.g., PNIPAAm). The thermo-responsiveness of these polymers can be tailored such that the incorporation of a hydrophobic co-monomer into a hydrophilic monomer can render it to possess a critical solution temperature and show phase transitions (Taylor & Cerankowski, 1975).

The LCST of this class of polymers can be easily altered by modifying side chains or by incorporating co-monomers having different hydrophilic/hydrophobic values (Kuckling, Adler, Arndt, Ling, & Habicher, 2000; Principi, Goh, Liu, & Winnik, 2000; Wang, Qiu, & Wu, 1998). Polymerization with a hydrophobic co-monomer will further lower the LCST, and addition of a hydrophilic co-monomer will increase the LCST. For example, copolymerization of NIPAAm with acrylamide (AAm) increased LCST from 32 °C to 45 °C, whereas incorporation of *N*-tert butylacrylamide (*N*-tBAAm) decreased the LCST by 10 °C (Hoffman et al., 2000). The copolymerization of hydrophilic and hydrophobic monomers changes polymer chain segmental mobility. These chains are easily permeated by water even at temperatures above critical solution temperature (CST), with enhanced segmental chain mobility yielding more expanded and flexible chains. Alternatively, the LCST of a polymer is also influenced by its concentration in solution and molecular weight of the molecule. In the case of PNIPAAm and poly(*N*-vinylcaprolactam) (PVCL), an increase in polymer chain length or its concentration in solution decreases LCST value (Chilkoti, Dreher, Meyer, &

Raucher, 2002; Kirsh, 1998; Schild & Tirrell, 1990). Incorporation of hydrophilic or hydrophobic co-monomers with these polymers has also been reported to influence the LCST (Eeckman, Amighi, & Moes, 2001; Eeckman, Moes, & Amighi, 2002; Mikheeva et al., 1997). Moreover, the LCST of PNIPAAm and PVCL polymers were decreased in the presence of salts (Eeckman et al., 2001, 2002; Mikheeva et al., 1997; Schild & Tirrell, 1990) and other polar solvents such as alcohol (Kirsh, 1998), acetone, and dioxane (Costa & Freitas, 2002; Schild, Muthukumar, & Tirrell, 1991; Wu, Liu, Heng, & Yang, 2005). The incorporation of proteins in polymer solution of PNIPAAm led to increased hydrophobic interactions and increased the LCST values.

To understand their responsive behavior, extensive experimental and computational studies have been performed to elucidate various molecular interactions occurring among or within the chains of polymers; most of these were carried out using PNIPAAm as a model thermo-responsive polymer. Initial studies concentrated on chain conformation studies of PNIPAAm in aqueous solution for analyzing bond formation and interaction between polymer groups and water molecules in solution. These studies showed that, with an increase in temperature, there is an extended coil-to-globule transition due to intra-molecular segment aggregation, leading to sharp conformational changes and chain collapse. When the temperature is decreased, the extended coil conformation is restored. These studies were performed using surfactant as co-solvent, which interacts with the isopropyl group of polymer due to hydrophobic interactions. At lower concentrations, surfactant assists in intermolecular solubilization and formation of isolated collapsed polymer globules, whereas on further increasing surfactant concentration, the isolated globules were converted to flexible expanded coils (Ricka, Meewes, Nyffenegger, & Binkert, 1990). Other studies have revealed that coil-to-globule state transition involves two additional thermodynamically stable "crumped coil" and "molten globule" intermediate states. On reversing the conditions, a hysteresis effect was observed in which a time lapse occurs when there is globule-to-coil transition, which was more profound on rapid thermal transitions. Due to the hysteresis effect, reversion time required for globule-to-coil transitions is more than required for coil-to-globule transitions (Wu & Wang, 1998).

The initial studies conducted were carried out using aqueous solutions of thermo-responsive (PNIPAAm) polymer. In pure water there is a well-distributed hydrogen bond network present between water molecules. On addition of polymer to water, there occurs a perturbation in the existing inter-molecular H_2O–H_2O hydrogen bond network, leading to restructuring of water molecules. As a result, the water gets redistributed around hydrophobic groups and hydrophilic groups of polymer, as well as bulk water molecules. Based on water molecule distribution and type of functional groups present in the vicinity, formation of new hydrogen bonds takes place between (1) polar groups in polymer ((N–)H–O(=C)) pair (polymer–polymer); (2) polymer polar groups and water ((N–)H–O(water)) (polymer–water); and (3) water oxygen and polymer oxygen groups ((water)O–H–O(polymer)).

Simulation studies have been used by researchers to analyze hydrogen bond formation dynamics and molecular distribution in solution. There are reports carried out with systems that have large numbers of polymer chains in solution, and in systems that have smaller numbers of isolated polymer linear chains in solution. It has been shown that there is a stabilization of water–water hydrogen bonds around hydrophobic regions of polymer due to hydrophobic stabilization, which occurs much more in presence of larger hydrophobic groups in the polymer chain and/or at lower water concentrations. This is due to generated constraints on orientation of water and polar groups of polymer (Tamai, Tanaka, & Nakanishi, 1996a). The polymer chains with small hydrophobic groups form less compact structures with larger flexibility, enhancing molecular interactions and slowing the rate of aggregation in comparison to chains with larger hydrophobic groups (Gangemi et al., 2008). The hydrogen bonds formed between polymer polar groups and water molecules are in a dynamic state and are categorized into two types based on their lifetimes; one, with a shorter lifetime, present in the majority; and second, with a longer lifetime and relatively few in number. The presence of these hydrogen bonds greatly reduces mobility of water molecules around polar groups of polymer chain (Tamai, Tanaka, & Nakanishi, 1996b).

To more precisely elucidate the interactions involved in volume phase transitions at higher temperatures, the effect of concentration and degree of crosslinking on phase transition behavior and polymer–polymer and polymer–solvent interactions, molecular simulation studies on multiple and isolated single polymer chains in aqueous systems were carried out. These studies revealed that water–polymer hydrogen bonds around polar groups in polymer chains (–NH in PNIPAAm) play a vital role in polymer–polymer and polymer–water interactions and in the hydration of polymer chains. With an increase in temperature, there is a decrease in number of intermolecular hydrogen bonds between polymer and water molecules, leading to dehydration of polymer chains. As a result, the intramolecular and interchain hydrogen bonds among polymers increases. Moreover, these inter- and intrachain crosslinkings occur much more at higher polymer concentrations in solution. The reduction in polymer–water hydrogen bonds above LCST leads to the formation of more collapsed and compact-folded globular structures (Alaghemandi & Spohr, 2012; Baltes, Garret-Flaudy, & Freitag, 1999; Deshmukh, Mooney, Mcdermott, Kulkarni, & Macelroy, 2009; Galaev & Mattiasson, 1993).

Poly(acrylamides) such as poly(*N*-isopropylacrylamide) (PNIPAAm) and poly(*N*,*N*-diethylacrylamide) (PDEAAm), which both have LCST of approximately 32–33 °C, are the most widely studied thermo-responsive polymers. However, because of issues such as nonbiodegradability of the polymers, very sharp phase transition effects, burst release of the entrapped molecules, and poor mechanical properties of gels formed (Bae, Okano, Hsu, & Kim, 1987), there is a constant need to improve their properties to suit the application. For these reasons, several block copolymers of NIPAAm with hydrophobic co-monomer have been synthesized and studied, for example, PNIPAAm-*b*-poly(butylmethacrylate) (PNIPAAm-PBMA) (Chung et al., 1999), PNIPAAm-polystyrene (PNIPAAm-PSt) (Cammas et al., 1997; Chung, Yokoyama, & Okano, 2000), and PNIPAAm-*b*-polymethyl methacrylate (PNIPAAm-*b*-PMMA) (Wei, Zhang, Zhou, Cheng, & Zhuo, 2006). Copolymers of *N*-isopropylacrylamide grafted with oligo-NIPAAm (Yoshida et al., 1995), PNIPAAm with poly(acrylic acid) (PAAC) and poly-ethyleneglycol (PEG) (Kaneko et al., 1998), and Pluronics and poly(*N*-isopropylacrylamide-*co*-*N*-hydroxylmethylacrylamide) (Liu, Pramoda, Yang, Chow, & He, 2004) have also been synthesized. These polymers showed rapid changes in response to temperature fluctuations, because of rapid dehydration and rehydration effects due to oligo-NIPAAm and PEG, respectively, in PNIPAAm-PEG block copolymer systems. Recently PEG-*b*-PNIPAAm-*b*-PEG triblock copolymers (Topp, Dijkstra, Talsma, & Feijen, 1997) and star block copolymers of NIPAAm and PEG (Lin & Cheng, 2001) were synthesized and showed reversible gelation in the temperature range of 26–33 °C. In another study, an NIPAAm-acrylic acid (AAC) copolymer was synthesized, which, in aqueous solutions, showed reversible gelation at 32 °C without any hysteresis effects (Han & Bae, 1998).

1.3.1.2 Poly(N-vinylamides)

Poly(*N*-vinylamides) are another class of synthetic thermo-responsive polymers. Two of the most common polymers of this class are poly(*N*-vinylpyrrolidone) (PVP) and poly(*N*-vinylcaprolactam) (PVCL) (Figure 1.3). PVCL is a water-soluble nonionic amphiphilic polymer with its basic unit comprising a seven-membered cyclic amide with a polar hydrophilic carboxyl group and an amide group connected directly to a hydrophobic vinyl chain.

PVCL is soluble in water at normal temperature, but shows phase transition at ~32 °C, similar to that of PNIPAAm (Kirsh, 1998), whereas PVP is insoluble in water. Although PVCL has LCST similar to PNIPAAm, its phase transition thermodynamics and mechanism of phase separation are quite different (Laukkanen, Valtola, Winnik, & Tenhu, 2004; Makhaeva, Tenhu, & Khokhlov, 1998). In PVCL, the lactam ring is directly connected to carbon backbone chain via amide linkage, which makes it very stable to hydrolysis. Even on hydrolysis occurring at relatively higher acidic values, PVCL does not form any toxic small amide compounds as are observed on hydrolysis of PNIPAAm. This property makes PVCL a more preferred polymer for biological applications in comparison to PNIPAAm and its derivatives (Vihola, Laukkanen, Valtola, Tenhu, & Hirvonen, 2005). This observation is also substantiated by studies that have shown better tolerance and no cytotoxic effect of PVCL on cells (Vihola et al., 2005). However, PVCL has the drawbacks of not being biodegradable and of absorbing various organic compounds and charged surfactants from its environment (Kirsh, 1998; Makhaeva, Tenhu, & Khokhlov, 2000). In the presence of surfactants, it behaves like a polyelectrolyte because of the presence of adsorbed counter ions inside the polymer matrix along with surfactants, leading to an increase in osmotic pressure triggering extensive swelling of polymer chains (Makhaeva, Thanh, Starodoubtsev, & Khokhlov, 1996). Also, within biological systems, at physiological temperature it undergoes coagulation, and, during this process, it entraps proteins and other small molecules. Additionally there is increased adsorption of proteins to the polymer surface due to its increased hydrophobicity at LCST, which thus affect its properties (Kirsh, 1998).

The thermo-responsive behavior of PVCL has been extensively studied to elucidate the mechanism of its phase transition behavior. Studies with PVCL in aqueous systems below its LCST have revealed that the seven-membered ring of the polymer attains a chair-like conformation, with the carbon backbone chain reconfiguring and attaining syndiotactic configuration (Kirsh, Yanul, & Kalninsh, 1999; Sun & Wu, 2011).

This generates a specific favorable conformation in which two molecules of water are interacting with each monomer unit, leading to formation of a highly hydrated system below LCST. These conformational changes and hydrogen bonding prevent any hydrophobic interactions between the hydrophobic groups that are involved in aggregation and precipitation of polymer chains. When the temperature of the aqueous polymer solution is raised, the molecular motion of water molecules bound to PVCL increases. At higher temperatures, the molecular motion becomes too high, leading to displacement of water molecules from their bound state with polymer chains. Displacement of water molecules causes partial dehydration of PVCL macromolecules with the generation of compact globules followed by precipitation of polymer chains. Similarly, in swollen gel networks of PVCL, a rise in temperature above LCST excludes water molecules from the gel network, causing shrinkage of the gel.

Factors governing the LCST of PVCL include molecular weight of polymer and concentration of polymer molecules in the solution (Lau & Wu, 1999). It has been observed that PVCL polymer has two transition temperatures. It follows a biphasic transition via two different transition states, one at 31.5 °C and another at approximately 37.5 °C. The first of these is a micro-segregated state due to dehydration of the polymer chains, followed by second state in which there is volume collapse because of displacement of water molecules and hydrophobic effects that are presumed to be responsible for phase behavior of polymers with hydrophobic side group and amphiphilic backbone chains (Hirokawa & Tanaka, 1984; Khokhlov, Starodubtzev, & Vasilevskaya, 1993; Mikheeva et al., 1997; Otake, Inomata, Konno, & Saito, 1989). At lower temperatures, a large number of water molecules become highly bound to the PVCL chains, which even shields hydrophobic groups present in the molecule chain. With a rise in temperature, these interactions between hydrophobic groups and water molecules become thermodynamically less stable compared to interactions among hydrophobic groups, resulting in conformational transitions in which hydrophobic groups are shielded by hydrophilic groups of polymer chain that got exposed to water. These interactions causes micromicelle formation at first transition state; however, due to topological constraints, even these hydrophobic group conformational transitions do not completely shield all hydrophobic groups into the micromicelle core, and some of the hydrophobic groups remain exposed on surface. Thus the surface of the micromicelle is still partly hydrophobic and leads to further precipitation of polymer and to the development of the collapsed phase at higher temperatures. Both transition temperatures decrease in the presence of salt (NaCl) and surfactant (sodium dodecyl sulphate (SDS)). Addition of NaCl molecule interferes with interactions between water molecules and hydrophilic groups, whereas the addition of SDS decreases surface tension favoring micelle formation (Mikheeva et al., 1997).

Interesting phase transition behavior was observed in copolymers of PVCL, which include PVCL–PEO (poly(ethylene oxide); hydrophilic) block copolymer, and copolymers of PVCL with (*N,N*-dimethyl amino) ethyl methacrylate (DMAEMA), hydrophilic co-monomers like ethylacrylamide, polyacrylamide-PVA, or polyacrylamide-PEO copolymers. All of these combinations showed decrease in LCST of the PVCL copolymer system. The LCST behavior observed was not in accordance with conventional theory of increase/decrease in LCST with copolymerization with hydrophilic/hydrophobic co-monomers, respectively, depicting a number of factors governing these transitions (Shin, Jhon, Lee, & Yuk, 1998; Yuk, Cho, & Lee, 1997). In case of PVCL–PEO polymer systems, this anomalous behavior was observed due to competition between PVCL and PEO molecules to interact with water molecules because of their differential affinity toward water. This results in weakening of interactions between PVCL and water molecules in the vicinity. In crosslinked PVCL–PEO copolymer systems, intermolecular interactions take place among bulk water molecules, soluble PVCL–PEO–H_2O domains, and water-insoluble domains at crosslinked positions distributed in the network. Depending on crosslinking density, the PEO–H_2O molecule interactions also vary. A decreased number of PEO–H_2O molecule interactions were observed with increase in degree of crosslinking. Steric hindrances around crosslinks obstruct hydration of PVCL–PEO segments, leading to generation of bulk water domains, and low-water PVCL–PEO–H_2O domains, which leads to partial dehydration, as less water is available to hydrate polymer chains. The dehydration of polymer chains at PVCL–PEO domains leads to decreased flexibility of these domains. All of these interactions cause higher transition temperatures in crosslinked systems when compared to polymer in solutions with equal water contents and no intermolecular crosslinks present (Shin et al., 1998; Yuk et al., 1997).

1.3.1.3 Poly(ethylene oxide)–poly(propylene oxide)–poly(ethylene oxide) (PEO–PPO–PEO) block copolymers (poloxamers/pluronics)

Structurally, these are block copolymers with a poly(propylene oxide) (PPO) middle block flanked by poly(ethylene oxide) (PEO) block on either side (Figure 1.3). A great variety of these polymers have been synthesized by varying the ratio of individual block lengths. Owing to this structural organization of individual hydrophobic (PPO) and hydrophilic (PEO) blocks, the overall polymer is amphiphilic in nature and possesses surfactant-like properties. These polymers are used as nonionic surfactants in aqueous solutions at low temperatures. At concentrations below critical micelle concentration (CMC), polymer chains exist as individual unimer coils, whereas on further increasing the polymer concentration in aqueous solution above CMC or when solution temperature is raised above critical micellization temperature (CMT), there is formation of stable micelles. Micelle formation at room temperature was not observed in solutions of less hydrophobic polymers (Alexandridis, Holzwarth, & Hatton, 1994) but with an increase in temperature micelle formation occurs, which could be due to decreased solubility at higher temperatures. It has also been shown that the CMC or CMT for a polymer depends on its molecular weight (PEO/PPO block length) and PEO (hydrophilic)–PPO (hydrophobic) content ratio (Deng, Yu, Price, & Booth, 1992; Price, 1983). These polymers fall in the category of block copolymers as shown in Figure 1.4. By altering these parameters, the polymer properties can be finely tuned. Polymers in this class have been designed to have phase separation temperatures ranging from 10 °C (polymers with low PEO contents) to 100 °C (polymers with high PEO contents) (Alexandridis & Hatton, 1995). Within the micelle structure, PPO chains occupy hydrophobic cores, and PEO chains arrange themselves to form hydrophilic shells (Alexandridis & Hatton, 1995). With an increase in concentration or at higher temperatures, micelles come closer to each other and become entangled due to interactions in hydrophilic PEO molecules present in the outer shell of the micelle structure. These interactions lead to structural arrangement of micelles into cubic quasi-crystalline lattices with face-centered or body-centered cubic lattice arrangements (Jiang et al., 2007; Mortensen, Batsberg, & Hvidt, 2008). This structural organization macroscopically leads to gelation (Meznarich, Juggernauth, Batzli, & Love, 2011; Meznarich & Love, 2011).

Studies using Pluronic P103 solutions have demonstrated two sol–gel transition states, one at lower temperature (polymer concentrations above 29 wt%) called as hard gel-1 and second at higher temperature called as hard gel-2. Hard gel-1 state involves those structural arrangements, where the micelles are packed into close structures due to interactions in micellar PEO molecules (Alexandridis & Hatton, 1995). With increase in temperature, the PEO molecules become dehydrated due to displacement of interacting water molecules. These decreased water–polymer interactions induce hydrophobic interactions of PEO molecules with hydrophobic core PPO molecules, leading to phase mixing and interconnecting all the micellar clusters formed (Alexandridis & Hatton, 1995). At this point (temperature above 43 °C), another opaque gel state was observed for all gels (even with low polymer concentrations) called as hard gel-2 (Park & Char, 2002). The formation of hard gel-1 occurs in solution of Pluronics with concentration above ~29 wt% polymer concentration, the formed hard gel-1 de-gelates when temperature of polymer solution is increased. As the temperature rises to about 43 °C the polymer solution becomes turbid which is observed at all polymer concentration solutions. In solutions with 14–32 wt% polymer concentration when temperature increases above ~55 °C there is formation of hard gel-2.

When temperature of polymer solutions is raised the unimeric polymer chains aggregate to give rise to micelles, the concentration of micelles in solution increase with rise in temperature. Aggregation and close packing of these micelles gives rise to formation of hard gel-1 (Kabanov et al., 1995; Liu, Chen, & Huang, 1998). As the temperature of solution further increases, it leads to enhanced hydrophobic interactions between core PPO blocks. At these conditions, there is also an increase in hydrophobic behavior of PEO chains and they get partially dehydrated. The increase in hydrophobicity of PEO chains results in phase mixing with PPO chains (Hvidt, Jorgensen, Brown, & Schillen, 1994). At this point, all of the micellar clusters interconnect, leading to formation of hard gel-2. This type of gel formation was observed in Pluronics with short hydrophilic end blocks, for example, Pluronic P103, as short PEO chains cannot act as strong steric layers and there is occurrence of macroscopic liquid–liquid phase separation. The hard gel-1 formed was reversible; that is, it re-forms upon cooling to the first gelation temperature, but re-formation of hard gel-2 was not observed upon cooling the polymer solution (King, Heenan, Cloke, & Washington, 1997; Li et al., 1997; Liu et al., 1998; Park & Char, 2002).

NMR studies with Pluronics have revealed that their gelation mechanism at higher temperatures involves dehydration of methyl groups of PPO, as indicated by conformational changes in methyl groups (Rassing, Mckenna, Bandyopadhyay, & Eyring, 1984). Change in entropy of the system because of water molecule rearrangement has also been suggested to cause sol–gel transition (Vadnere, Amidon, Lindenbaum, & Haslam, 1984).

Pluronics have gained interest for use in a variety of biological applications such as drug delivery, tissue engineering, and cell culture because of their unique sol–gel transition behavior at normal temperatures. However, there are issues with their shelf lives, biocompatibility, and elimination profiles from biological systems. Additionally, all of these limitations are prominent with high-molecular-weight PEO/PPO block systems. Thus, this class of polymers needs attention with respect to the above-mentioned issues before they can be extensively used in biological and biotechnological applications.

1.3.1.4 Poly(ethylene glycol)–poly L-lactic acid (PEG–PLLA) and poly(ethylene glycol)–poly lactic acid-co-glycolic acid (PEG–PLGA) block copolymers

Another promising class of smart materials includes block copolymers based on PEG, lactides, and glycolides. This class of thermo-responsive polymers form physically crosslinked gels and have advantages of biocompatibility and biodegradability as opposed to pluronics, as discussed before. These polymers have been designed and studied in different block combinations and orientations, for example, poly(ethylene glycol)-poly-L-lactic acid-*co*-glycolic acid (PEG–PLGA), poly(ethylene glycol)-poly-D-lactic acid-*co*-glycolic acid (PEG–PDGA), PLLA–PEG–PLLA, PDLA–PEG–PDLA, PEG–PLLA, and PEG–PDLA.

PEG–PLLA–PEG is the most extensively studied block polymer system from this class, which exhibits sol–gel transition in aqueous solutions at body temperature. Structurally it is an amphiphilic polymer system with a hydrophobic PLLA middle segment flanked by hydrophilic PEG segments on either side (Figure 1.4(b)). It has been demonstrated that its sol–gel transition temperature is dependent on block lengths, block ratio in polymer, and polymer concentration. Keeping the PEG chain length constant with variations in only the PLLA chain length led to variation in gelation temperature of the copolymer (Jeong, Bae, Lee, & Kim, 1997; Jeong, Kim, & Bae, 2002). In systems containing oligo-LLA or oligo-DLA, stereocomplexation in the polymer chains occur, which causes crosslinking among oppositely oriented chains, leading to gelation. In the PEG–PLLA–PEG polymer system, it was observed that with an increase in temperature, hydrophilic interactions with water become less favorable and become more abrupt at transition temperatures, leading to gel dehydration. It further leads to increased polymer–polymer interaction relative to polymer–water interactions, which causes micellar growth and intermicellar interactions, affecting sol-to-gel transition. It has also been observed that these interactions are much more favorable at higher polymer concentrations and higher temperatures (Chung, Lee, & Park, 2008). Similar to other block copolymers, their properties are also influenced by chain block lengths and polymer block ratios. An increase in overall hydrophobicity of triblock copolymer leads to decrease in sol–gel transition temperature and critical gelation concentration (Jeong, Bae, & Kim, 1999). When lengths of flanking PEG chains in block copolymer were increased keeping the oligo-lactic acid chain length constant, a shift in gelation temperature toward higher values was observed. Alternatively, when the ratio of lactic acid to glycolic acid in PEG–PLGA–PEG polymer chain was decreased, a decrease in sol-to-gel temperature was observed. In these systems, micelle formation and intermicellar interactions have been shown to be responsible for their gelation behavior. The gels formed by this class of polymers are stable and possess more mechanical strength than gels formed by Pluronics. The rate of degradation of these polymeric gels has been shown to be very slow and to exhibit linear mass erosion rates (Jeong et al., 1997, 2002; Mukose et al., 2004).

In another type of system when block arrangements were shuffled to form PLLA–PEG–PLLA/PDLA–PEG–PDLA system where PEG is positioned in the middle position in a triblock copolymer, giving a hydrophobic–hydrophilic–hydrophobic arrangement (Figure 1.4(c)), it was observed that solutions of individual triblock copolymers did not show any gelation behavior at any temperature or concentration values. On the other hand, when the solution of these enantiomeric systems were mixed with each other, it led to a system in which spontaneous gelation occurred at around 37 °C. This phenomenon was explained by occurrence of interchain interactions between these two different copolymers due to stereocomplexation between polylactic acid (PLA) blocks of different orientations. In aqueous solutions of polylactic acids (PLLA/PDLA), the PLLA/PDLA segments interact with each other, leading to formation of hydrophobic core, and PEG chain blocks fold to form outer hydrophilic shell surrounding the hydrophobic core, giving rise to a micellar structure. On heating, the interactions between the PLLA/PLDA core and PEG shell weakens, which causes opening of micellar structure and mixing of the polymer chains with each other. Due to opposite orientations of chains in enantiomeric mixtures, they form stereocomplexed interchain crosslinking with each other, leading to gelation. The gelation in PLLA/PDLA systems was observed to be irreversible; that is, after gelation there is no gel-to-sol conversion (Fujiwara & Kimura, 2002; Fujiwara, Miyamoto, & Kimura, 2000; Fujiwara, Miyamoto, Kimura, Iwata, & Doi, 2001; Fujiwara, Mukose, et al., 2001; Jeong et al., 1997).

In another system in which enantiomeric mixtures of PEG–PLLA–PEG and PEG–PDLA–PEG were mixed, these underwent gelation due to stereo- and hydrophobic interactions between oppositely oriented L- and D-chains (Jeong et al., 1997). Within these systems, the PEG chains flanking the PLLA/PDLA chains attain conformational orientation similar to middle block polylactic acid, and undergo a conformational change. In the aqueous state, both of these PEG–PLLA–PEG and PEG–PDLA–PEG systems forms micellar structures in which the hydrophobic polylactic acid chain folds to form a hydrophobic internal core part and the hydrophilic PEG chains folds to form an outer shell. In enantiomeric systems, the chains of opposite enantiomers attain opposite orientations, when these PEG–PLLA–PEG and PEG–PDLA–PEG copolymers are mixed with each other, the oppositely oriented chains interact and intercross with each other. Due to this arrangement, they crosslink micelles with each other, giving rise to network formation and ultimately gelation at 30–40 °C. The PEG–PLLA–PEG and PEG–PDLA–PEG gels maintained their consistency to approximately 75 °C, and showed reversible gelation, as hydrophilic PEG chains always formed outer shell part of the micelles, and were always present on the surface to interlink with oppositely oriented chains (Jeong et al., 1997).

PEG–PLGA–PEG triblock copolymers in aqueous medium exhibit sol-to-gel transition similar to PEO–PPO–PEO block copolymer systems at higher temperature (between 20 and 70 °C), depending on block length, composition, and polymer concentration in solution (Jeong et al., 1999). These polymers showed dual transitions, sol-to-gel at lower and gel-to-sol transition at higher temperatures (Jeong et al., 1999). In aqueous solutions, these polymer chains arrange to generate PLGA (hydrophobic) core and PEG (hydrophilic) shell micellar structures. With an increase in temperature, the interactions between PEG chains and PLGA chains increase. Further increase in temperature causes increased hydrophobic interactions between chains, thus enhancing the intermicellar interactions. These interactions cause enhanced intermicellar PEG chain interactions, leading to intermicellar PEG chain interpenetration. Intermicellar interactions and crosslinking gives rise to an extensively packed network. Their sol-to-gel transition temperature depends on individual block lengths of the copolymer chains. An increase in PLGA to PEG block length ratio decreased gelling concentration and transition temperature. Salts enhance dehydration effects in polymer chains and lead to sol–gel transition at lower temperatures (Jeong et al., 1999).

Similarly, copolymers having a PLGA–PEG–PLGA triblock also showed sol-to-gel transition with increasing temperature due to an increase in micelle aggregation, thereby causing gelation (Jeong, Kim, & Park, 2004). These block copolymers have a middle hydrophilic block (PEG) flanked by a hydrophobic block (PLGA). PLGA–PEG–PLGA triblock copolymers showed sol–gel transitions with an increase in temperature above a certain gelation concentration. In aqueous solutions, these polymer chains form micelles that have PLGA chains in their hydrophobic core and hydrophilic PEG chains forming an outer shell. With an increase in temperature, dehydration of PEG chains takes place and causes micelles to shrink, thus bringing PEG and PLGA chains closer, which enhances interactions between PEG chains and hydrophobic PLGA chains. At higher temperatures, the PLGA blocks can diffuse from one micelle to another, forming a bridged structure between micelles. Further increase in temperature causes the number of micelles and interconnecting bridge to increase, and finally leads to aggregation of micelles and thereby gel formation (Ghahremankhani, Dorkoosh, & Dinarvand, 2008; Jeong et al., 2004; Lee et al., 2001).

The di-block copolymers of hydrophobic–hydrophilic PLGA–PEG also showed thermo-responsive behavior with sol-to-gel transition on increase in temperature (Choi et al., 1999). These polymers formed micelles, with a core of hydrophobic PLGA and an outer shell composed of hydrophilic PEG blocks. There is formation of bridged micelles due to interactions between PEG chains of adjacent micelles. The bridge density increases with increase in temperature, leading to aggregation and gelation.

The di-block system of PLLA–PEG/PDLA–PEG enantiomers also showed thermo-responsive sol–gel transition. However, the solution of individual polymers does not show any gelation with temperature variation. In an aqueous solution, individual copolymers of PLLA–PEG or PDLA–PEG form micelles with a core of PLLA/PDLA and a shell of PEG chains; on mixing with each other, these lead to hexagonal crystal formation of polymer chains and gelation at room temperature. With a rise in temperature, hydrophobic interactions increase, which enhances micellar aggregation. These polymers exhibit irreversible gel-to-sol conversion at 75 °C. At higher temperatures there is formation of stereocomplexed crystals, whereas hexagonal crystals break. These hexagonal crystals fail to re-form when the temperature is lowered again (Fujiwara & Kimura, 2002; Fujiwara, Miyamoto, et al., 2001; Kister, Cassanas, & Vert, 1998).

1.3.1.5 Elastin-like oligo- and polypeptides (ELPs)

Elastin is a connective tissue protein that confers an elastic nature to tissues to maintain and regain their shape after stretching. The presence of elastin in skin and blood vessels enables them to be of a highly elastic nature. Elastin is made up of repeating units of valine (V), glycine (G), alanine (A), and proline (P).

Pentameric synthetic polypeptides with elastin-like repeating units of amino acids valine–proline–glycine–valine–glycine (VPGVG) have been reported that exhibit inverse transition temperature. Almost all of the elastin-like polypeptides (ELPs) (Figure 1.3) are derived from the sequence Val/Ile–Pro–Xaa–Yaa–Gly, where Xaa is generally glycine, and Yaa can be any other amino acid except proline (VPG'X'G). The transition temperature of ELPs depends on the molecular weight and amino acid composition of the chains, which generally vary at the fourth position of the block peptide chain. Incorporation of a more polar residue increases transition temperature, whereas incorporation of a hydrophobic residue decreases transition temperature. Increase in hydrophobic interactions with a rise in temperature in polymer chains makes them more structurally ordered and closely packed, resulting in polymer contraction (Banta, Wheeldon, & Blenner, 2010; Urry, 1997). Below this critical temperature, polymer chains remain in expanded soluble state in aqueous solution, whereas above it, an increase in hydrophobic interactions in chain leads to folding of the chain into a dynamic, nonrandom β-spiral structure, involving type-II β-turn conformation with three pentamer units per turn. These are stabilized by inter-turn, intra- and inter-spiral hydrophobic contacts (Urry, Trapane, & Prasad, 1985; Urry, Peng, Parker, Gowda, & Harris, 1993), attaining a β-spiral arrangement that grows to give rise to fibrillar particles several nanometers long (Reiersen, Clarke, & Rees, 1998).

Below the transition temperature, the ELP chains in aqueous solution are in extended disordered conformation and are soluble in aqueous system, whereas above transition temperatures, dehydration of ELP chains leads to conformational changes in polypeptide chains, generating β-turns and attaining β-spiral structural conformation. This structuring of water molecules and peptide chains increases entropy of the system. The dehydrated chains get much more structurally aligned because of hydrophobic interactions in nonpolar amino acids in their chains, and interlink with each other (Chang, Venkatachalam, Prasad, & Urry, 1989; Schmidt, Dybal, Rodriguez-Cabello, & Reboto, 2005; Urry, Trapane, Iqbal, Venkatachalam, & Prasad, 1985). This crosslinking between chains gives rise to a physically crosslinked gel network. With a decrease in temperature, the opposite occurs, and ELP chains again solubilize and go into solution. The transition temperature for ELPs decreased with an increase in hydrophobicity of the system, an increase in hydrophobic amino acids, variations in pH, or variations in ionic strength (Reiersen et al., 1998; Urry, 1992; Urry, Trapane, & Prasad, 1985).

These ELPs can be synthesized using standard peptide synthesis methods such as solid phase synthesis. Recently ELPs were synthesized using recombinant DNA techniques in *Escherichia coli*, where peptides having 1000–1500 amino acids can be easily synthesized (Chilkoti, Dreher, & Meyer, 2002) without any errors and variations in peptide chains. ELP-based materials have found application in peptides (Massodi, Bidwell, & Raucher, 2005), drugs (Dreher et al., 2003; Mcdaniel, Callahan, & Chilkoti, 2010), and gene delivery (Dash et al., 2011). Janorkar et al. have generated by expressing in *E. coli* poly$(VPGVG)_{120}$ elastin-like peptides having a molecular weight of 51 KDa; the generated ELP was used in synthesizing ELP–collagen composite material. The generated ELP had a transition temperature of 32 °C (Amruthwar & Janorkar, 2013; Amruthwar, Puckett, & Janorkar, 2013). ELP–collagen composite material was characterized for physical and mechanical properties. The material was used in the synthesis of scaffolds by freezing the ELP–collagen gel at −20 °C and then lyophilizing for 12 h. The mechanical characterization of scaffolds revealed that incorporation of ELPs in collagen scaffolds decreased their strain-bearing capacity (Amruthwar et al., 2013). The collagen and elastin are two major proteins in tissue extracellular matrix; in this direction, scaffolds made of ELP–collagen composite material have been evaluated for tissue engineering applications (Amruthwar & Janorkar, 2013; Nettles, Chilkoti, & Setton, 2010).

ELPs are promising thermo-responsive materials for *in vivo* tissue engineering and drug delivery applications because of their high biocompatibility, as their degradation products are naturally occurring amino acids (Banta et al., 2010; Chilkoti, Dreher, Meyer, & Raucher, 2002).

1.3.2 Natural thermo-responsive polymers and their derivatives

Natural thermo-responsive biopolymer gels have been extensively used in the food and pharmaceutical industries and are finding increasing applications in biomaterials and tissue engineering. Some of most commonly used materials are agarose, gelatin, amylose, amylopectin, carageenans, gellan, xyloglucan, and cellulose derivatives. These are isolated from biological sources and are used either as such or after purification or are modified to endow them with certain properties. The polymers like gelatin and agarose, when dissolved in water, are in the sol state at higher temperatures; but, on decreasing the temperature, show gel phase transition known as physical gelation (San Biagio, Bulone, Martorana, Dowd, & Newman, 2001). Another natural polymer is cellulose, which is insoluble in water because of limited access to C_6–OH groups in cellulose structure, which are responsible for formation of intermolecular hydrogen bonds with water, enabling solubilization (Itagaki, Tokai, & Kondo, 1997; Kondo, 1997); however, its derivatives that have hydrophilic groups are soluble at normal temperature and are known to exhibit reverse thermo-gelation at higher temperatures (Bekturov & Bimendina, 1981).

A reverse thermo-gelation property system was generated using chitosan and glycerol phosphate di-sodium salt (Chenite et al., 2000). The aqueous solution of this mixture behaved as sol at normal temperatures and showed gel phase transition when the temperature was increased to 37 °C. The structuring effect of glycerol on water molecules surrounding the chitosan polymer at higher temperatures leads to increased hydrophobic interactions between chitosan molecules, leading in turn to gelation in which the gelation temperature is influenced by the degree of de-acetylation of chitosan chains, making them neutral.

1.4 Physical forms of switchable materials and their applications

Thermo-responsive switchable materials have been synthesized in a variety of formats for diverse applications such as bioseparation, drug delivery, tissue engineering, and sensing. The responsive component of smart materials is generally the thermosensitive polymer, which has been formulated into the desired physical form, and performs the actuator function in response to temperature change. Although thermo-responsive polymers have been designed into a variety of structures, they can be broadly classified into the following physical forms: free linear chains, cross-linked gels, surface-grafted forms, and nano/micro-sized particle assemblies (Jeong & Gutowska, 2002; Kumar, Srivastava, Galaev, & Mattiasson, 2007). Free linear polymeric chains have been synthesized in solutions, in which they switch between a precipitated and miscible form in response to an external trigger. On the other hand, gels are covalently or physically crosslinked polymeric networks that exhibit environmentally triggered reversible swelling behavior.

These thermo-responsive gels possess a crosslinked polymeric network that can be formed by both physical and chemical crosslinking and that have been fabricated in both microscopic (e.g., microgels) and macroscopic dimensions (e.g., hydrogels). Smart thermo-responsive surfaces possess surface-grafted polymer chains that can reversibly swell or collapse in response to temperature change, thereby endowing the surface with alternating hydrophilic and hydrophobic properties, respectively. Nano/micro-sized particle assemblies include micelles and polymersomes, which are made of temperature-sensitive polymers and exhibit temperature-dependent changes in their properties. This section of the chapter highlights the designing of thermo-responsive materials in different formats, with particular emphasis on PNIPAAm-based materials and their applications.

1.4.1 Free polymer chains in solution

In aqueous solution, the thermo-responsive polymers are in extended conformations and form a homogeneous phase below LCST. In this form, these polymeric systems have found use in number of biological applications because of their unique responsive nature. In addition, these polymers also have no denaturing effects toward biological substances such as proteins, cells, and other low-molecular-weight biological substances (Diamond & Hsu, 1992; Rito-Palomares, 2004). One area in which thermo-responsive polymers have been used is separating and purifying biomolecules using the solubility or affinity properties of these substances. Two of most commonly used systems are the aqueous two phase system (ATPS) and affinity precipitation.

In the ATPS, the thermo-responsive polymer interacts with proteins and acts as a ligand carrier, forming polymer–ligand complexes. In ATPS with thermo-responsive polymers, the polymer–ligand complex becomes partitioned and separated into separate phases when environmental conditions are changed by increasing temperature above the polymer LCST. The separated complex can be easily recovered from the precipitated phase. In thermo-responsive polymer ATPS, polymers such as PNIPAAm, PVCL, ethyl(hydroxyethyl) cellulose (EHEC), and block and random copolymer of ethylene oxide–propylene oxide (EOPO) have been used (Persson, Johansson, Galaev, Mattiasson, & Tjerneld, 2000). Polymer systems of EOPO and poly(*N*-vinylcaprolactam-*co*-vinyl imidazole) forming two-phase systems with dextran have also been used in the purification of proteins (Farkas, Stalbrand, & Tjerneld, 1996; Franco et al., 1997; Lu, Albertsson, Johansson, & Tjerneld, 1996).

Using thermo-responsive polymers in ATPS provides an advantage in that they can be easily recovered by inducing temperature triggered phase separation. The recovered polymers can be recycled and reused in purification cycle (Persson, Johansson, & Tjerneld, 1999). EOPO (50:50) random copolymer and copolymer of EOPO after hydrophobic modification of EOPO (HMEOPO) by adding hydrophobic chains on both ends of the polymer molecule have also been used in ATPS for purifying proteins (Persson et al., 1999). In aqueous solutions when the temperature is raised above critical temperature, the phase separation occurs, generating two distinct phases, the upper one composed of water and the lower one composed of polymer. The EOPO and HMEOPO polymer systems have been extensively studied and well characterized for separation and partitioning behavior using proteins such as bovine serum albumin (BSA), lysozyme, and apolipoprotein A-1 as model systems (Persson et al., 1999). ATPS have shown to be capable of purifying enzymes such as endo-polygalacturonase from *Kluyveromyces marxiamus* (Pereira, Wu, Venâncio, & Teixeira, 2003). ATPS are also used in extractive bioconversion of starch using α-amylase and amyloglucosidase (Li, Kim, & Peeples, 2002). One of the area in which ATPS have been shown to be effective is in separating and isolating specific type of animal cells by coupling specific antibodies against that specific cell type with a thermo-responsive smart polymer. This type of ATPS has been used in separating CD-34–positive human acute myeloid leukemia cells (KG-1) from human T-lymphoma cells using anti–CD-34 antibody conjugated to PNIPAAm in ATPS (Kumar, Kamihira, Galaev, Mattiasson, & Iijima, 2001).

Affinity precipitation of protein is another system in which thermo-responsive smart polymers have been used (Kumar, Galaev, & Mattiasson, 1998; Kumar, Wahlund, Kepka, Galaev, & Mattiasson, 2003; Mattiasson, Kumar, Ivanov, & Galaev, 2007). Protein precipitation is one commonly used technique in protein purification. Generally protein precipitation is done using salts, organic solvents, and polymers such as polyethylene oxide, but these techniques did not provide any selectivity in precipitation, as precipitation is dependent on surface properties of protein molecules. To increase the selectivity of protein precipitation method, smart polymers provide good alternative technique. For affinity precipitation, two approaches can be employed using either a homo-bifunctional ligand or a hetero-bifunctional ligand. We will discuss only hetero-bifunctional approach here, as it involves responsive polymers. In a hetero-bifunctional approach, an affinity ligand is coupled covalently to a smart polymer to generate an affinity micro ligand (AML) (Kumar et al., 1998, 2007). The polymer to be used for generation of AML should possess low interactions with ligand and other impurities, should have reactive groups for ligand coupling, and should show complete phase separation for maximum recovery during the precipitation and solubilization cycle.

In affinity precipitation, the target protein gets bound to affinity ligand attached to the thermo-responsive smart polymer, the complex can be phase separated into insoluble state by triggering change in microenvironment. The target protein can then be eluted from precipitated complex by reversing the conditions. After removal of the protein, the AML recovers and is again available for reuse in purification cycle. In some cases, the polymer itself can act as an AML; polymers such as chitosan, Eudragit-S100, PNIPAAm, and PVCL have been used either as such or coupled to a ligand (Kumar et al., 2007; Mattiasson et al., 2007). In metal chelate affinity precipitation smart polymers conjugated with metal ions have been synthesized and used (Galaev, Kumar, & Mattiasson, 1999). AML with different affinity ligand system has been used in purification of proteins depending on their affinity toward, for example, concanavalin A, wheat germ lectin, and metal ions covalently attached to the smart polymer (Pan & Chien, 2003). The recombinant, six-histidine tagged, single Fv-antibody fragments from *E. coli* cell culture have been separated by an affinity precipitation system (Kumar et al., 2003). The affinity precipitation system has also been used in the purification of RNA and DNA (Balan et al., 2003) by the interaction of metal ions with exposed aromatic nitrogen in purines in single-stranded nucleic acids (Murphy, Jewell, White, Fox, & Willson, 2003). Double-stranded plasmid DNA was purified on the basis of a triple helix affinity interaction (Costioli, Fisch, Garret–Flaudy, Hilbrig, & Freitag, 2003). Recently ELPs have also been used in affinity precipitation of recombinant proteins from cell cultures; these provide easily tunable systems that are highly stable and show no denaturing effects toward biological molecules and cells (Kostal, Mulchandani, & Chen, 2001; Meyer & Chilkoti, 1999; Shimazu, Mulchandani, & Chen, 2003; Stiborova, Kostal, Mulchandani, & Chen, 2003; Urry, Luan, Harris, & Parker, 1997).

1.4.2 Macroscopic hydrogels and microgels

Gels are crosslinked networks of polymers that are insoluble in their dispersion medium. In case of hydrogels, the dispersion medium is water and thus hydrogels are defined as crosslinked polymeric porous networks that can absorb huge quantities of water into their interstitial pores (Hoffman, 2002). For fabrication of temperature-sensitive hydrogels, polymer materials have been processed into both three-dimensional macroscopic materials called as bulk hydrogels and microscopic materials called microgels and used for relevant applications.

Hydrogels have been synthesized based on both chemical (covalent bonds between polymer chains) and physical crosslinking (e.g., polymer chain entanglements, ionic interactions, and hydrogen bonds) methods (Hoffman, 2002) (Figure 1.5). Covalently crosslinked thermo-responsive hydrogels exhibit reversible swelling of the crosslinked

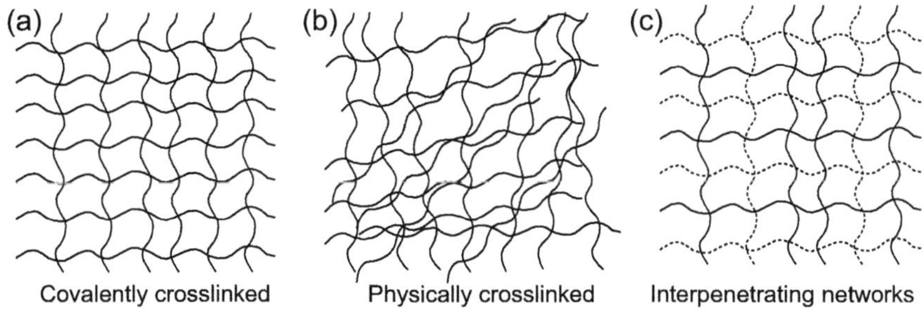

Figure 1.5 Different types of polymeric gel networks. (a) Covalently crosslinked networks; (b) physically crosslinked polymer networks; and (c) interpenetrating polymer networks (subclass of covalently crosslinked networks).

polymeric network as a function of temperature stimulus (Ward & Georgiou, 2011). A subclass of chemically crosslinked hydrogels is interpenetrating polymer networks (IPN), in which a second hydrogel network is formed within an already polymerized hydrogel network either in the absence of a cross-linker (semi-IPN) or in the presence of a cross-linker (full-IPN). The de-swelling behavior of thermo-responsive hydrogels at higher temperatures is due to increased hydrophobic interactions at higher temperatures, which leads to solvent expulsion (Hirotsu, 1988; Kumar et al., 2007). Thermo-responsive hydrogels has been used for a variety of applications such as drug delivery (Bromberg & Ron, 1998; Coughlan, Quilty, & Corrigan, 2004; Liu, Shao, & Lü, 2006; Na, Park, et al., 2006; Qiu & Park, 2001), actuators (Eddington & Beebe, 2004), and immobilized catalysis (Arica, Öktem, Öktem, & Tuncel, 1999).

Within the domain of drug delivery applications, thermo-responsive hydrogel materials have been extensively used for the delivery of peptides and proteins (Bromberg & Ron, 1998) and a variety of drugs (Coughlan et al., 2004; Na, Park, et al., 2006; Okano et al., 1990). For drug release, hydrogels based on PNIPAAm have been used as on–off drug release systems as a function of the temperature of the external environment (Dinarvand & D'Emanuele, 1995; Okano et al., 1990). Devices for on–off release of drugs exploit the thermo-responsive reversible swelling behavior of these polymeric networks. PNIPAAm-based copolymers can be designed into hydrogels and interpenetrating networks for fabricating such on–off drug release systems. Okano and his group have reported fabrication of crosslinked hydrogels of poly(*N*-isopropylacrylamide (NIPAAm)-*co*-butylmethacrylate (BMA)) and interpenetrating polymer networks of PNIPAAm and poly tetramethylene ether glycol (PTMEG) (Okano et al., 1990) with controlled release properties in response to temperature. Both of these thermo-responsive matrices were characterized for on–off release of indomethacin as a function of temperature because of their temperature dependent swelling properties (Okano et al., 1990). In another study, an interpenetrating hydrogel network of poly(*N*-isopropylacrylamide) (PNIPAAm) and hydrophobic poly(ethyl acrylate) (PEA) was used for studying the controlled release of daidzein as model drug (Liu et al., 2006). Similarly, thermosensitive hydrogels based on PNIPAAm copolymerized with butyl methacrylate (BMA; hydrophobic) or acrylic acid (AAc; hydrophilic) have been reported for release of heparin (Gutowska, Bae, Feijen, & Kim, 1992). This study demonstrated that the loading and release of heparin was affected by gel composition, temperature, and swelling characteristics of the gel network. Thus, the properties of the polymeric network affect the release profile of the encapsulated drug. However, properties of the drug can also affect the swelling behavior of the polymeric network and thereby the drug release, as demonstrated by an important study performed by Coughlan and coworkers. Drugs with different physicochemical properties were loaded into a thermo-responsive hydrogel, and it was demonstrated that drug size, chemical nature, and solubility affect drug release and hydrogel swelling (Coughlan et al., 2004). In an interesting study, Park and coworkers reported delivery of drug, growth factor, and cells by using poly(*N*-isopropylacrylamide-*co*-acrylic acid) (PNIPAAm-*co*-AAc)–based injectable hydrogels (Na, Park, et al., 2006).

Switchable swelling behavior of thermo-responsive hydrogels has also been exploited for applications in the designing of actuators, in which their stimulus-triggered volume change can be used for controlling flow in microfluidic drug delivery devices (Eddington & Beebe, 2004). PNIPAAm-based monolithic hydrogels cross-linked with *N*,*N*-methylenebisacrylamide were used as nonmechanical valves for controlling flow within analytical microfluidic chips (Yu et al., 2003). Thermo-responsiveness of the hydrogels involves rapid de-swelling and swelling of the monolith, which was exploited to open or close the microfluidic channel and exhibited closing and opening time of 5 and 3.5 s, respectively (Yu et al., 2003). In another, related study, PNIPAAm-based hydrogels were successfully used as thermally controlled valves in distribution and metering components of microfluidic analytical devices and closing of a polymerase chain reaction (PCR) reactor chamber (Wang et al., 2005). In an extensive study, PNIPAAm-based thermo-responsive hydrogels were evaluated in detail for their use as thermally controlled actuators in lab-on-chip and microfluidic devices (Harmon, Tang, & Frank, 2003). Frank and coworkers fabricated PNIPAAm hydrogel actuators in micrometer dimensions and studied their anisotropic swelling and kinetics of the volume phase change (Harmon et al., 2003). It was also demonstrated that the rate of hydrogel response can be tuned by using semi-interpenetrating hydrogel networks (Harmon et al., 2003). Applications such as flow control in microfluidic PCR chambers need valves to be operated at higher temperatures, and thus these thermo-responsive hydrogel-based actuators need to be designed for inclusion of a broader temperature-responsive range. Toward this goal, Luo and coworkers have reported a thermally controlled actuator based on photocopolymerized NIPAAm and *N*-(ethylacrylamide) hydrogel, which exhibits adjustable LCST within a broad temperature range (Luo, Mutlu, Gianchandani, Svec, & Fréchet, 2003). Moreover, these thermally actuated, hydrogel-based valves exhibited continued performance even after repeated open–close cycles (Luo et al., 2003; Yu et al., 2003).

Microgels are crosslinked polymer particles synthesized in the size range of tens of nanometers to a few micrometers (Guan & Zhang, 2011; Saunders & Vincent, 1999). Microgels have been used for potential applications in drug delivery and regenerative medicine owing to their promising properties, as they combine virtues of both hydrogels and nanoparticles within one structure (Saunders et al., 2009). Due to their small size, the rate of response of microgels to stimulus is faster than that of bulk hydrogels, and even smaller gels in nanometer dimensions show quicker volume–phase transitions (Reese, Mikhonin, Kamenjicki, Tikhonov, & Asher, 2004; Tanaka & Fillmore, 1979). Thermo-responsive PNIPAAm microgels have been extensively investigated and characterized (Pelton, 2000) for a variety of uses and in different formats, for example, dispersed particles, microgel-assembled thin films, and 3D assemblies (Guan & Zhang, 2011). PNIPAAm microgels have been largely employed for drug delivery applications, where they have been used for delivering protein-based molecules and small drugs, as discussed further. Lyon and coworkers have reported synthesis of PNIPAAm microgels for temperature sensitive release of insulin. By using NMR studies, the authors demonstrated enhanced loading of insulin via a swelling technique in comparison to an equilibrium partitioning method (Nolan, Gelbaum, & Lyon, 2006). In another study, Peppas and coworkers reported fabrication of thermo-responsive insulin-loaded nanospheres based on poly(*N*-isopropylacrylamide), poly(ethylene glycol) dimethacrylate, and poly(ethylene glycol) methacrylate, and showed efficient protection of loaded insulin against temperature and shear stress (Leobandung, Ichikawa, Fukumori, & Peppas, 2002). Loading of water-soluble drugs within PNIPAAm-based hydrogels has been extensively examined in a study that investigated interactions of a variety of drugs (having different properties) with PNIPAAm-based microgels exhibiting differential distribution of functionalities, thus elucidating the factors that affect drug loading and release (Hoare & Pelton, 2008). Thermo-responsive microgels also offer the advantage of surface modification with bio-moieties for potential targeting drug delivery applications. Nayak et al. synthesized thermo-responsive microgel surface functionalized with folic acid for targeting of cancer cells and showed efficient receptor-mediated endocytosis of particles by cells (Nayak, Lee, Chmielewski, & Lyon, 2004).

1.4.3 Surface-grafted polymers

Fabrication of surfaces with switchable properties as a function of an external trigger has attracted a great deal of interest because of their potential use in a variety of applications such as cell sheet engineering and bioseparation (Brun-Graeppi, Richard, Bessodes, Scherman, & Merten, 2010; Nagase, Kobayashi, & Okano, 2009). In this format, the thermo-responsive polymer is present on the surface as polymer brushes and crosslinked polymeric networks. As the polymer is linked to the surface, its degree of freedom is restricted, and thus on temperature change there is no aggregation or phase change of the polymer. However, the polymeric chains grafted to the surface swell/collapse as a function of temperature and allow switchable hydrophilic/hydrophobic properties to the surface (Kumar et al., 2007). When the polymer is in the swollen state, the surface becomes hydrophilic and, on collapsing of the polymer chains the surface, becomes relatively hydrophobic, thus offering a way of thermally controlling the behavior of surfaces. Thermo-responsive surfaces have been fabricated by attaching temperature-sensitive polymers onto the substrate surface by a variety of techniques depending upon the nature of substrate and the polymer being used. Thermo-responsive polymers have been grafted onto surfaces by using radiation-induced polymerization approaches such as electron beam irradiation (Akiyama, Kikuchi, Yamato, & Okano, 2004; Hirose, Kwon, Yamato, Kikuchi, & Okano, 2000), UV-induced polymerization (Sugawara &

Matsuda, 1994; Von Recum et al., 1998), and plasma polymerization (Cheng et al., 2005; Pan, Wesley, Luginbuhl, Denton, & Ratner, 2001). UV-induced polymerization grafting is advantageous, as it is inexpensive, faster, and the grafting is near the surface without affecting the properties of bulk materials (Brun-Graeppi et al., 2010). Other grafting methods include "grafting to" and "grafting from" approaches. "Grafting to" methods involve attachment of pre-formed polymer onto the surface by exploiting the chemical functionalities of both the surface and the polymer (Cunliffe, De Las Heras Alarcón, Peters, Smith, & Alexander, 2003; Kanazawa et al., 1996). However, this method offers low grafting densities on the surface because of steric restrictions of the chemical groups on the surface (Nagase et al., 2009). On the other hand, the "grafting from" approach has been used for grafting larger amounts of polymer on the surfaces. The "grafting from" method uses pre-attached initiators/precursors on the surface, which are further used for polymerization from the surface and thus allow better control of grafting density of the polymer (Suzuki et al., 2000; Yakushiji et al., 1999). An interesting approach to grafting polymers in a more controlled way is by using atom transfer radical polymerization, which enables surface grafting of dense polymer brushes with more controlled molecular weights and better properties as compared to conventional "grafting from" approaches (Idota et al., 2005; Nagase et al., 2007, 2009). PNIPAAm-based thermo-responsive surfaces have been extensively used for applications such as cell sheet engineering because of the properties of these surfaces to be hydrophobic above their LCST and hydrophilic below the LCST of the grafted polymer (Brun-Graeppi et al., 2010; Nagase et al., 2009). The goal of cell sheet engineering is to grow and harvest mammalian cells in the form of a sheet so that they can be further manipulated/used for tissue engineering applications. Formation of a cell sheet has been achieved by first growing mammalian cells on these surfaces at temperatures when the surface is hydrophobic and thus allows cells to attach and grow (Yamada et al., 1990). When the temperature of these surfaces is reduced, the surface becomes hydrophilic and the cells are detached as a sheet from the substrate (Yamada et al., 1990). This approach of using thermo-responsive surfaces for growing cell sheets has been used for culture of kidney epithelial cells (Kushida et al., 2000), co-culture of lung cells (Nandkumar et al., 2002), and fabrication of cell sheets of designed shapes (Hirose et al., 2000). Apart from cell sheet engineering applications, temperature-controlled hydrophobicity of these surfaces has also been used for chromatographic separation of substances based on hydrophobic interactions, and there are extensive reviews dealing with this aspect (Kikuchi & Okano, 2002; Kumar et al., 2007; Nagase et al., 2009).

1.4.4 Polymeric micelles

Surfactant molecules having a hydrophilic head and hydrophobic tail portion when present in aqueous solution generate micellar structures above critical micelle concentration (CMC) by noncovalent aggregation of individual surfactant molecules. The micelles generated can be either spherical or irregular in shape. In micelles, surfactant molecules form walls of spherical structure in such a manner that the hydrophilic head portion faces outward and hydrophobic tail is toward the interior of the micelles, thus generating a hydrophobic core (Figure 1.6). Surfactants have been used for micellar solubilization of hydrophobic drugs. The micellar structures generated using surfactant molecules have been used in delivering poorly soluble or hydrophobic drugs. These micellar structures can be loaded with hydrophobic drugs during their assembly and can be used to release the drugs on application of a temperature trigger, as shown in Figure 1.7. The micellar drug delivery systems provide advantages such as

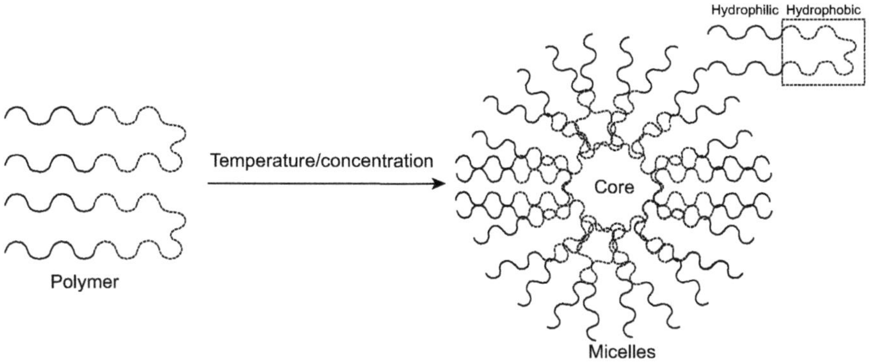

Figure 1.6 Micelles generated by different types of thermo-responsive polymers in aqueous solutions. Block copolymers with hydrophobic blocks flanked by hydrophilic blocks.

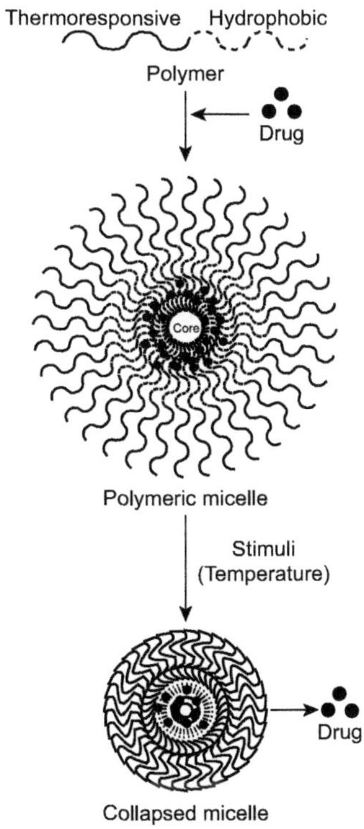

Figure 1.7 Micelles of thermo-responsive polymers used as controlled drug delivery vehicle.

the following: less drug degradation and drug loss; optimized drug delivery; increased drug stability and bioavailability (Rangel-Yagui, Pessoa, & Tavares, 2005); fewer chances of drug side effects and toxic effects; and localized drug delivery to diseased and damaged body parts (Torchilin, 2001).

Drug solubilization capacity of a surfactant depends on factors such as chemical structure of surfactant and drug molecule, temperature, pH, ionic strength, molecular nature, and size of the drug molecule (Torchilin, 2001). Having low CMC values, the nonionic surfactants are better solubilizing agents for hydrophobic drugs.

Polymeric molecules such as PEO, PPO, and Pluronics have amphiphilic structures with properties of nonionic surfactants. Furthermore, certain polymers and block copolymers in solution generate polymeric micellar structures. The polymers provide higher numbers of uniform size self-assembled micelles (Gadelle, Koros, & Schechter, 1995). The polymeric micelles can also be functionalized by attaching specific ligands to the surface of these micelles. PEO coating on the micelles increases their blood circulation life, preventing their uptake and removal by macrophages; this allows their affective distribution and delivery to a desired location in the body (Kwon & Kataoka, 1995; Kwon & Okano, 1996; Lee, Lee, & Andrade, 1995) and can also be chemically crosslinked to make them more stable. The polymeric micelles provide the advantage of ease of sterilization by filtration (Kwon & Okano, 1996). The aqueous propylene glycol nonionic surfactant polymeric micellar systems were used in solubilization of hydrophobic corticosteroids used in anti-inflammatory therapy (Ong & Manoukian, 1988).

The use of responsive smart polymers for generating polymeric micellar structures provides the advantage of incorporating responsive properties in micelles, and for providing controlled and optimized release of drug molecules. The drug loaded to these systems can be released in a controlled and optimized manner under internal microenvironment changes or by providing measured amounts of external stimuli which destabilize the polymeric micellar system in a controlled manner, precipitating the release of payload. Responsive polymeric micelles generated using PEO, PPO, Pluronics, PNIPAAm, PVCL, PLA, PLGA, and other block copolymers have been studied for their application in drug delivery and have been used in generating micellar thermo-responsive drug delivery systems (Batrakova & Kabanov, 2008; Chung et al., 1999; Kabanov & Alakhov, 2002; Liang, Kozlovskaya, Chen, Zavgorodnya, & Kharlampieva, 2012; Neradovic, Hinrichs, Kettenes-Van Den Bosch, Van Nostrum, & Hennink, 2001; Prabaharan, Grailer, Steeber, & Gong, 2008, Prabaharan, Grailer, 2009; Ward & Georgiou, 2011).

Micelles of PEO-*b*-PLLA have been used in generating hydrolysis/enzymatic degradation–triggered drug release systems (Adams, Lavasanifar, & Kwon, 2003). Other stimuli that can be used for triggered drug release are pH, glutathione presence (reduction), enzymes, and physical stimuli such as light, sound, vibration/pressure, and magnetic fields (Katayama, Sonoda, & Maeda, 2001; Miyata et al., 2005; Oishi, Nagasaki, Itaka, Nishiyama, & Kataoka, 2005; Rangel-Yagui et al., 2005).

Use of thermo-responsive polymeric micellar systems for drug delivery application has received significant attention (Bae et al., 1987; Bae, Choi, Park, Lee, & Park, 2006; Cohn, Sosnik, & Levy, 2003; Na, Lee, Lee, & Bae, 2006). Polymeric systems generated using PEG-PNIPAAm copolymers have been used to generate micelles that incorporated hydrophobic drug in them above LCST, and released it at a localized site upon cooling the site (Neradovic, Soga, Van Nostrum, & Hennink, 2004). Micelles of block copolymers with poly(*N*-isopropyl acrylamide)-*N*,*N*-dimethyl acrylamide (PNIPAAm-DMAAm) as thermo-responsive block and poly(D,L-lactide-*co*-ε-caprolactone) hydrophobic block in P(NIPAAm-*co*-DMAAm)-*b*-poly(LA-*co*-CL) block copolymer were synthesized and showed temperature-triggered release of doxorubicin at 42.5 °C (Nakayama et al., 2006). Another hydrolytically sensitive thermo-responsive block copolymer with hydrophilic PEG and (NIPAAm)-poly[*N*(2-hydroxypropyl) methacrylamide-oligolactide] (PNIPAAm-HPMA-PLA) was synthesized; in this system, upon hydrolysis of lactic acid, the hydrophobicity of micelle core increases, causing dissociation of micellar architecture and drug release (Rijcken et al., 2005). Other systems were also generated based on similar methodology (Neradovic et al., 2004; Soga et al., 2005). A reversibly swelling- and shrinking-based responsive delivery system was generated using Pluronics and crosslinking them using gold nanoparticles (Bae et al., 2006).

The higher stability of polymeric micelles, as well as nano-range size distribution of generated micelles, provides more penetration into body tissues, thereby enhancing bioavailability, along with earlier discussed advantages associated with polymeric micellar systems over other colloidal systems; these advantages make polymeric micellar systems more promising drug delivery systems.

1.4.5 Polymersomes

Polymersomes or polymer vesicles are hollow vesicular polymeric structures that possess a hydrophilic core surrounded by hydrophobic membrane and a hydrophilic outer surface, as shown in the schematic representation in Figure 1.8 (Discher & Eisenberg, 2002). Polymersomes are nanometer-sized vesicles that are formed as a result of self-assembly of amphiphilic block copolymers (Du & O'Reilly, 2009). These hollow vehicles have generated considerable interest for their application in biomedical and material science because of their attractive properties. Polymersomes can be loaded with both hydrophobic and hydrophilic cargoes simultaneously, along with

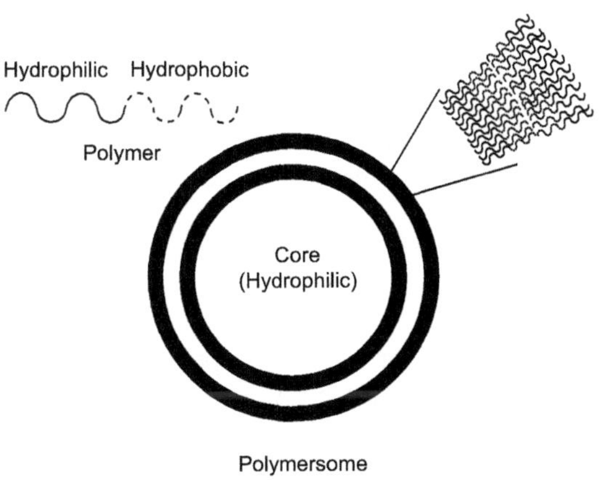

Figure 1.8 Polymersomes generated by thermo-responsive polymers having hydrophobic and hydrophilic blocks in their chains.

the prospect of surface modification for biological application, and are more robust structures because of their polymeric origin as opposed to liposomes, which are based on low-molecular-weight lipids (Du and O'Reilly, 2009). Endowment of polymersomes with stimuli-responsive properties is highly desirable for smart drug delivery applications and has been achieved for temperature, pH, and redox responsive stimuli (Bellomo, Wyrsta, Pakstis, Pochan, & Deming, 2004; Meng, Zhong, & Feijen, 2009). Thermo-responsive polymersomes have been synthesized by using temperature-sensitive polymers that exhibit temperature-dependent changes in their hydrophobicity/hydrophilicity and thus lead to their self-assembly into polymersomes at particular temperatures as discussed further. Thermally induced selfassembly of responsive di-block copolymers of poly[*N*-(3-aminopropyl)-methacrylamide hydrochloride]-*b*-PNIPAAm into polymersomes was demonstrated by McCormick and coworkers (Li, Lokitz, & Mccormick, 2006). The authors used reversible addition fragmentation chain-transfer polymerization for the synthesis of the polymer, which assembled into polymersomes at temperatures above the LCST of the PNIPAAm block. It was also demonstrated that these thermo-responsive polymersomes can be structurally "locked" or crosslinked by using a polyelectrolyte of opposite charge (Li et al., 2006). In another interesting study, Qin and coworkers demonstrated employment of temperature not only for the assembly of polymersomes but also for the release of encapsulated drug from these polymeric vesicles. It was demonstrated that above 32 °C, the block copolymer of poly(ethylene oxide)-*b*-poly(*N*-isopropylacrylamide) assembled into polymersomes, and these self-assembled vesicles could also encapsulate hydrophobic dye within their membranes and doxorubicin in their hydrophilic core (Qin, Geng, Discher, & Yang, 2006). Temperature-triggered release of the encapsulated molecules was observed when these polymersomes were exposed to temperature below 32 °C (Qin et al., 2006). Shell-crosslinked polymersomes based on PNIPAAm with other copolymers have also been reported. Block copolymers of polylactide-*b*-PNIPAAm were shown to be aggregated into vesicles in aqueous solutions, which were further crosslinked by using hexamethylene diacrylate for the chain extension of PNIPAAm (Hales, Barner-Kowollik, Davis, & Stenzel, 2004). Chen et al. demonstrated photo-crosslinking of the shells of thermosensitive polymersomes based on block copolymer of poly(2-cinnamoylethyl methacrylate)-*b*-poly(*N*-isopropylacrylamide), where thermosensitive release of 4-aminopyridine from these polymersomes was also shown (Chen, Ding, Zheng, & Peng, 2006). Polymersomes provide the capability of loading both hydrophobic and hydrophilic drugs into same delivery vehicle; thus, dual drugs with different properties and of different amounts can be delivered to the same target using a single delivery system. This system can provide a great advantage in multi-drug therapy regimens.

1.5 Summary

Thermo-responsive materials have huge potential for a suite of applications, owing to their responsiveness to an external trigger and the ability to be used in different formats. There are a variety of thermo-responsive polymers that differ in their mechanism of temperature sensitivity and thereby have different properties. These temperature-sensitive polymers have been designed into different formats for fabrication of a variety of switchable materials such as drug delivery vehicles, materials for bioseparation, and tissue engineering scaffolds, among other applications. These materials are increasingly finding applications in different research areas especially in biomedical applications for site-specific triggered delivery, where their temperature responsiveness can be exploited to control drug release. However, use of thermo-responsive release within *in vivo* conditions needs a high degree of fine-tuning of the temperature-sensitive character and is the current challenge with respect to these switchable materials. Another property of hydrogel-based thermo-responsive materials that needs improvement is the timescale of their response to stimuli for practical applications. Further studies also need to concentrate on the aspects of biocompatibility and biodegradability of these materials, with detailed *in vivo* studies for establishing their true potential and development of smart biomedical technologies.

References

Adams, M. L., Lavasanifar, A., & Kwon, G. S. (2003). Amphiphilic block copolymers for drug delivery. *Journal of Pharmaceutical Sciences*, *92*, 1343–1355.

Akiyama, Y., Kikuchi, A., Yamato, M., & Okano, T. (2004). Ultrathin poly(*N*-isopropylacrylamide) grafted layer on polystyrene surfaces for cell adhesion/detachment control. *Langmuir*, *20*, 5506–5511.

Alaghemandi, M., & Spohr, E. (2012). Molecular dynamics investigation of the thermo-responsive polymer poly(*N*-isopropylacrylamide). *Macromolecular Theory and Simulations*, *21*, 106–112.

Alexandridis, P., & Hatton, T. A. (1995). Poly(ethylene oxide)-poly(propylene oxide)-poly(ethylene oxide) block-copolymer surfactants in aqueous-solutions and at interfaces - thermodynamics, structure, dynamics, and modeling. *Colloids and Surfaces A-Physicochemical and Engineering Aspects*, *96*, 1–46.

Alexandridis, P., Holzwarth, J. F., & Hatton, T. A. (1994). Micellization of poly(ethylene oxide)-poly(propylene oxide)-poly(ethylene oxide) triblock copolymers in aqueous-solutions–thermodynamics of copolymer association. *Macromolecules*, *27*, 2414–2425.

Amruthwar, S. S., & Janorkar, A. V. (2013). In vitro evaluation of elastin-like polypeptide-collagen composite scaffold for bone tissue engineering. *Dental Materials*, *29*, 211–220.

Amruthwar, S. S., Puckett, A. D., & Janorkar, A. V. (2013). Preparation and characterization of novel elastin-like polypeptide-collagen composites. *Journal of Biomedical Materials Research Part A*, *101*, 2383–2391.

Arica, M. Y., Öktem, H. A., Öktem, Z., & Tuncel, S. A. (1999). Immobilization of catalase in poly(isopropylacrylamide-*co*-hydroxyethylmethacrylate) thermally reversible hydrogels. *Polymer International*, *48*, 879–884.

Bae, K. H., Choi, S. H., Park, S. Y., Lee, Y., & Park, T. G. (2006). Thermosensitive pluronic micelles stabilized by shell cross-linking with gold nanoparticles. *Langmuir*, *22*, 6380–6384.

Bae, Y. H., Okano, T., Hsu, R., & Kim, S. W. (1987). Thermosensitive polymers as on–off switches for drug release. *Makromolekulare Chemie-Rapid Communications*, *8*, 481–485.

Bae, Y. H., Okano, T., & Kim, S. W. (1990). Temperature-dependence of swelling of cross-linked poly(*N,N'*-alkyl substituted acrylamides) in water. *Journal of Polymer Science Part B-Polymer Physics*, *28*, 923–936.

Bae, Y. H., Okano, T., & Kim, S. W. (1991a). On–off thermocontrol of solute transport. 1. Temperature-dependence of swelling of *N*-isopropylacrylamide networks modified with hydrophobic components in water. *Pharmaceutical Research*, *8*, 531–537.

Bae, Y. H., Okano, T., & Kim, S. W. (1991b). On–off thermocontrol of solute transport. 2. Solute release from thermosensitive hydrogels. *Pharmaceutical Research*, *8*, 624–628.

Balan, S., Murphy, J., Galaev, I., Kumar, A., Fox, G. E., Mattiasson, B., et al. (2003). Metal chelate affinity precipitation of RNA and purification of plasmid DNA. *Biotechnology Letters*, *25*, 1111–1116.

Baltes, T., Garret-Flaudy, F., & Freitag, R. (1999). Investigation of the LCST of polyacrylamides as a function of molecular parameters and the solvent composition. *Journal of Polymer Science Part A-Polymer Chemistry*, *37*, 2977–2989.

Banta, S., Wheeldon, I. R., & Blenner, M. (2010). Protein engineering in the development of functional hydrogels. *Annual Review of Biomedical Engineering*, *12*(12), 167–186.

Batrakova, E. V., & Kabanov, A. V. (2008). Pluronic block copolymers: evolution of drug delivery concept from inert nanocarriers to biological response modifiers. *Journal of Controlled Release*, *130*, 98–106.

Bekturov, E. A., & Bimendina, L. A. (1981). Interpolymer complexes. *Advances in Polymer Science*, *41*, 99–147.

Bellomo, E. G., Wyrsta, M. D., Pakstis, L., Pochan, D. J., & Deming, T. J. (2004). Stimuli-responsive polypeptide vesicles by conformation-specific assembly. *Nature Materials*, *3*, 244–248.

Boutris, C., Chatzi, E. G., & Kiparissides, C. (1997). Characterization of the LCST behavior of aqueous poly(*N*-isopropylacrylamide) solutions by thermal and cloud point techniques. *Polymer*, *38*, 2567–2570.

Bromberg, L. E., & Ron, E. S. (1998). Temperature-responsive gels and thermogelling polymer matrices for protein and peptide delivery. *Advanced Drug Delivery Reviews*, *31*, 197–221.

Brownlee, M., & Cerami, A. (1979). Glucose-controlled insulin-delivery system - semi-synthetic insulin bound to lectin. *Science*, *206*, 1190–1191.

Brun-Graeppi, A. K. A. S., Richard, C., Bessodes, M., Scherman, D., & Merten, O.-W. (2010). Thermoresponsive surfaces for cell culture and enzyme-free cell detachment. *Progress in Polymer Science*, *35*, 1311–1324.

Cammas, S., Suzuki, K., Sone, C., Sakurai, Y., Kataoka, K., & Okano, T. (1997). Thermo-responsive polymer nanoparticles with a core-shell micelle structure as site-specific drug carriers. *Journal of Controlled Release*, *48*, 157–164.

Chang, D. K., Venkatachalam, C. M., Prasad, K. U., & Urry, D. W. (1989). Nuclear overhauser effect and computational characterization of the beta-spiral of the polypentapeptide of elastin. *Journal of Biomolecular Structure & Dynamics*, *6*, 851–858.

Chen, X., Ding, X., Zheng, Z., & Peng, Y. (2006). Thermosensitive cross-linked polymer vesicles for controlled release system. *New Journal of Chemistry*, *30*, 577–582.

Cheng, X., Canavan, H. E., Stein, M. J., Hull, J. R., Kweskin, S. J., Wagner, M. S., et al. (2005). Surface chemical and mechanical properties of plasma-polymerized *N*-isopropylacrylamide. *Langmuir*, *21*, 7833–7841.

Chenite, A., Chaput, C., Wang, D., Combes, C., Buschmann, M. D., Hoemann, C. D., et al. (2000). Novel injectable neutral solutions of chitosan form biodegradable gels in situ. *Biomaterials*, *21*, 2155–2161.

Cheon, J. B., Jeong, Y. I., & Cho, C. S. (1999). Effects of temperature on diblock copolymer micelle composed of poly(gamma-benzyl L-glutamate) and poly(*N*-isopropylacrylamide). *Polymer*, *40*, 2041–2050.

Chilkoti, A., Dreher, M. R., & Meyer, D. E. (2002). Design of thermally responsive, recombinant polypeptide carriers for targeted drug delivery. *Advanced Drug Delivery Reviews*, *54*, 1093–1111.

Chilkoti, A., Dreher, M. R., Meyer, D. E., & Raucher, D. (2002). Targeted drug delivery by thermally responsive polymers. *Advanced Drug Delivery Reviews, 54*, 613–630.

Choi, S. W., Choi, S. Y., Jeong, B., Kim, S. W., & Lee, D. S. (1999). Thermoreversible gelation of poly(ethylene oxide) biodegradable polyester block copolymers, II. *Journal of Polymer Science Part A-Polymer Chemistry, 37*, 2207–2218.

Chung, H. J., Lee, Y. H., & Park, T. G. (2008). Thermo-sensitive and biodegradable hydrogels based on stereocomplexed Pluronic multi-block copolymers for controlled protein delivery. *Journal of Controlled Release, 127*, 22–30.

Chung, J. E., Yokoyama, M., & Okano, T. (2000). Inner core segment design for drug delivery control of thermo-responsive polymeric micelles. *Journal of Controlled Release, 65*, 93–103.

Chung, J., Yokoyama, M., Yamato, M., Aoyagi, T., Sakurai, Y., & Okano, T. (1999). Thermo-responsive drug delivery from polymeric micelles constructed using block copolymers of poly(*N*-isopropylacrylamide) and poly(butylmethacrylate). *Journal of controlled release, 62*, 115–127.

Cohn, D., Sosnik, A., & Levy, A. (2003). Improved reverse thermo-responsive polymeric systems. *Biomaterials, 24*, 3707–3714.

Costa, R. O. R., & Freitas, R. F. S. (2002). Phase behavior of poly(*N*-isopropylacrylamide) in binary aqueous solutions. *Polymer, 43*, 5879–5885.

Costioli, M. D., Fisch, I., Garret–Flaudy, F., Hilbrig, F., & Freitag, R. (2003). DNA purification by triple–helix affinity precipitation. *Biotechnology and Bioengineering, 81*, 535–545.

Coughlan, D. C., Quilty, F. P., & Corrigan, O. I. (2004). Effect of drug physicochemical properties on swelling/deswelling kinetics and pulsatile drug release from thermoresponsive poly(*N*-isopropylacrylamide) hydrogels. *Journal of Controlled Release, 98*, 97–114.

Cunliffe, D., De Las Heras Alarcón, C., Peters, V., Smith, J. R., & Alexander, C. (2003). Thermoresponsive surface-grafted poly(N-isopropylacrylamide) copolymers: surface-properties and effect of phase transitions on protein and bacterial attachment. *Langmuir, 19*, 2888–2899.

Dash, B. C., Mahor, S., Carroll, O., Mathew, A., Wang, W., Woodhouse, K. A., et al. (2011). Tunable elastin-like polypeptide hollow sphere as a high payload and controlled delivery gene depot. *Journal of controlled release, 152*, 382–392.

Deng, Y., Yu, G. E., Price, C., & Booth, C. (1992). Thermodynamics of micellization and gelation of oxyethylene oxypropylene diblock copolymers in aqueous-solution studied by light-scattering and differential scanning calorimetry. *Journal of the Chemical Society-Faraday Transactions, 88*, 1441–1446.

Deshmukh, S., Mooney, D. A., Mcdermott, T., Kulkarni, S., & Macelroy, J. M. D. (2009). Molecular modeling of thermo-responsive hydrogels: observation of lower critical solution temperature. *Soft Matter, 5*, 1514–1521.

Diamond, A., & Hsu, J. (1992). *Aqueous two-phase systems for biomolecule separation. Bioseparation*. Springer.

Dinarvand, R., & D'emanuele, A. (1995). The use of thermoresponsive hydrogels for on–off release of molecules. *Journal of Controlled Release, 36*, 221–227.

Discher, D. E., & Eisenberg, A. (2002). Polymer vesicles. *Science, 297*, 967–973.

Dreher, M. R., Raucher, D., Balu, N., Colvin, O. M., Ludeman, S. M., & Chilkoti, A. (2003). Evaluation of an elastin-like polypeptide–doxorubicin conjugate for cancer therapy. *Journal of Controlled Release, 91*, 31–43.

Du, J., & O'reilly, R. K. (2009). Advances and challenges in smart and functional polymer vesicles. *Soft Matter, 5*, 3544–3561.

Eddington, D. T., & Beebe, D. J. (2004). Flow control with hydrogels. *Advanced Drug Delivery Reviews, 56*, 199–210.

Eeckman, F., Amighi, K., & Moes, A. J. (2001). Effect of some physiological and non-physiological compounds on the phase transition temperature of thermoresponsive polymers intended for oral controlled-drug delivery. *International Journal of Pharmaceutics, 222*, 259–270.

Eeckman, F., Moes, A. J., & Amighi, K. (2002). Evaluation of a new controlled-drug delivery concept based on the use of thermoresponsive polymers. *International Journal of Pharmaceutics, 241*, 113–125.

Farkas, T., Stalbrand, H., & Tjerneld, F. (1996). Partitioning of beta-mannanase and alpha-galactosidase from *Aspergillus niger* in ucon/reppal aqueous two-phase systems and using temperature-induced phase separation. *Bioseparation, 6*, 147–157.

Feil, H., Bae, Y. H., Feijen, J., & Kim, S. W. (1992). Mutual influence of pH and temperature on the swelling of ionizable and thermosensitive hydrogels. *Macromolecules, 25*, 5528–5530.

Firestone, B. A., & Siegel, R. A. (1991). Kinetics and mechanisms of water sorption in hydrophobic, ionizable copolymer gels. *Journal of Applied Polymer Science, 43*, 901–914.

Franco, T. T., Galaev, I. Y., Hatti-Kaul, R., Holmberg, N., Bu, L., & Mattiasson, B. (1997). Aqueous two-phase system formed by thermoreactive vinyl imidazole/vinyl caprolactam copolymer and dextran for partitioning of a protein with a polyhistidine tail. *Biotechnology Techniques, 11*, 231–236.

Fujishige, S., Kubota, K., & Ando, I. (1989). Phase-transition of aqueous-solutions of poly(*N*-isopropylacrylamide) and poly(*N*-isopropylmethacrylamide). *Journal of Physical Chemistry, 93*, 3311–3313.

Fujiwara, T., & Kimura, Y. (2002). Macromolecular organization of poly(L-lactide)-block-polyoxyethylene into bio-inspired nano-architectures. *Macromolecular Bioscience, 2*, 11–23.

Fujiwara, T., Miyamoto, M., & Kimura, Y. (2000). Crystallization-induced morphological changes of a poly(L-lactide)/poly(oxyethylene) diblock copolymer from sphere to band via disk: a novel macromolecular self-organization process from core-shell nanoparticles on surface. *Macromolecules, 33*, 2782–2785.

Fujiwara, T., Miyamoto, M., Kimura, Y., Iwata, T., & Doi, Y. (2001). Self-organization of diblock and triblock copolymers of poly(L-lactide) and poly(oxyethylene) into nanostructured bands and their network system. Proposition of a doubly twisted chain conformation of poly(L-lactide). *Macromolecules, 34*, 4043–4050.

Fujiwara, T., Mukose, T., Yamaoka, T., Yamane, H., Sakurai, S., & Kimura, Y. (2001). Novel thermo-responsive formation of a hydrogel by stereo-complexation between PLLA–PEG–PLLA and PDLA–PEG–PDLA block copolymers. *Macromolecular Bioscience, 1*, 204–208.

Gadelle, F., Koros, W. J., & Schechter, R. S. (1995). Solubilization of aromatic solutes in block copolymers. *Macromolecules, 28*, 4883–4892.

Galaev, I. Y., Kumar, A., & Mattiasson, B. (1999). Metal-copolymer complexes of *N*-isopropylacrylamide for affinity precipitation of proteins. *Journal of Macromolecular Science—Pure and Applied Chemistry, 36*, 1093–1105.

Galaev, I. Y., & Mattiasson, B. (1993). Thermoreactive water-soluble polymers, nonionic surfactants, and hydrogels as reagents in biotechnology. *Enzyme and Microbial Technology, 15*, 354–366.

Gangemi, F., Longhi, G., Abbate, S., Lebon, F., Cordone, R., Ghilardi, G. P., et al. (2008). Molecular dynamics simulation of aqueous solutions of 26-unit segments of p(NIPAAm) and of p(NIPAAm) "Doped" with amino acid based comonomers. *Journal of Physical Chemistry B, 112*, 11896–11906.

Ghahremankhani, A. A., Dorkoosh, F., & Dinarvand, R. (2008). PLGA–PEG–PLGA tri-block copolymers as in situ gel-forming peptide delivery system: effect of formulation properties on peptide release. *Pharmaceutical Development and Technology, 13*, 49–55.

Guan, Y., & Zhang, Y. (2011). PNIPAM microgels for biomedical applications: from dispersed particles to 3D assemblies. *Soft Matter*, *7*, 6375–6384.
Gutowska, A., Bae, Y. H., Feijen, J., & Kim, S. W. (1992). Heparin release from thermosensitive hydrogels. *Journal of Controlled Release*, *22*, 95–104.
Hales, M., Barner-Kowollik, C., Davis, T. P., & Stenzel, M. H. (2004). Shell-cross-linked vesicles synthesized from block copolymers of poly(D,L-lactide) and poly(*N*-isopropyl acrylamide) as thermoresponsive nanocontainers. *Langmuir*, *20*, 10809–10817.
Han, C. K., & Bae, Y. H. (1998). Inverse thermally-reversible gelation of aqueous *N*-isopropylacrylamide copolymer solutions. *Polymer*, *39*, 2809–2814.
Harmon, M. E., Tang, M., & Frank, C. W. (2003). A microfluidic actuator based on thermoresponsive hydrogels. *Polymer*, *44*, 4547–4556.
Hassan, C. M., Doyle, F. J., & Peppas, N. A. (1997). Dynamic behavior of glucose-responsive poly(methacrylic acid-g-ethylene glycol) hydrogels. *Macromolecules*, *30*, 6166–6173.
Hirokawa, Y., & Tanaka, T. (1984). Volume phase-transition in a nonionic gel. *Journal of Chemical Physics*, *81*, 6379–6380.
Hirose, M., Kwon, O. H., Yamato, M., Kikuchi, A., & Okano, T. (2000). Creation of designed shape cell sheets that are noninvasively harvested and moved onto another surface. *Biomacromolecules*, *1*, 377–381.
Hirotsu, S. (1988). Critical points of the volume phase transition in *N*-isopropylacrylamide gels. *The Journal of Chemical Physics*, *88*, 427–431.
Hoare, T., & Pelton, R. (2008). Impact of microgel morphology on functionalized Microgelâ˜'Drug interactions. *Langmuir*, *24*, 1005–1012.
Hoffman, A. S. (2002). Hydrogels for biomedical applications. *Advanced Drug Delivery Reviews*, *54*, 3–12.
Hoffman, A. S., Stayton, P. S., Bulmus, V., Chen, G. H., Chen, J. P., Cheung, C., et al. (2000). Really smart bioconjugates of smart polymers and receptor proteins. *Journal of Biomedical Materials Research Part A*, *52*, 577–586.
Horne, R. A., Almeida, J. P., Day, A. F., & Yu, N. T. (1971). Macromolecule hydration and effect of solutes on cloud point of aqueous solutions of polyvinyl methyl ether - a possible model for protein denaturation and temperature control in homeothermic animals. *Journal of Colloid and Interface Science*, *35*, 77.
Hvidt, S., Jorgensen, E. B., Brown, W., & Schillen, K. (1994). Micellization and gelation of aqueous-solutions of a triblock copolymer studied by rheological techniques and scanning calorimetry. *Journal of Physical Chemistry*, *98*, 12320–12328.
Hwang, F. S., & Hogenesch, T. E. (1995). Effects of water-soluble spacers on the hydrophobic association of fluorocarbon-modified poly(acrylamide). *Macromolecules*, *28*, 3328–3335.
Idota, N., Kikuchi, A., Kobayashi, J., Akiyama, Y., Sakai, K., & Okano, T. (2005). Thermal modulated interaction of aqueous steroids using polymer-grafted capillaries. *Langmuir*, *22*, 425–430.
Idziak, I., Avoce, D., Lessard, D., Gravel, D., & Zhu, X. X. (1999). Thermosensitivity of aqueous solutions of poly(*N,N*-diethylacrylamide). *Macromolecules*, *32*, 1260–1263.
Itagaki, H., Tokai, M., & Kondo, T. (1997). Physical gelation process for cellulose whose hydroxyl groups are regioselectively substituted by fluorescent groups. *Polymer*, *38*, 4201–4205.
Ito, Y., Casolaro, M., Kono, K., & Imanishi, Y. (1989). An insulin-releasing system that is responsive to glucose. *Journal of Controlled Release*, *10*, 195–203.
Jeong, B., Bae, Y. H., & Kim, S. W. (1999). Thermoreversible gelation of PEG–PLGA–PEG triblock copolymer aqueous solutions. *Macromolecules*, *32*, 7064–7069.

Jeong, B., Bae, Y. H., Lee, D. S., & Kim, S. W. (1997). Biodegradable block copolymers as injectable drug-delivery systems. *Nature, 388*, 860–862.

Jeong, B., & Gutowska, A. (2002). Lessons from nature: stimuli-responsive polymers and their biomedical applications. *Trends in Biotechnology, 20*, 305–311.

Jeong, B., Kim, S. W., & Bae, Y. H. (2002). Thermosensitive sol–gel reversible hydrogels. *Advanced Drug Delivery Reviews, 54*, 37–51.

Jeong, J. H., Kim, S. W., & Park, T. G. (2004). Biodegradable triblock copolymer of PLGA–PEG–PLGA enhances gene transfection efficiency. *Pharmaceutical Research, 21*, 50–54.

Jiang, J., Burger, C., Li, C. H., Li, J., Lin, M. Y., Colby, R. H., et al. (2007). Shear-induced layered structure of polymeric micelles by SANS. *Macromolecules, 40*, 4016–4022.

Kabanov, V. A. (1994). Physicochemical basis and the prospects of using soluble interpolyelectrolyte complexes. *Vysokomolekulyarnye Soedineniya Seriya A & Seriya B, 36*, 183–197.

Kabanov, A. V., & Alakhov, V. Y. (2002). Pluronic® block copolymers in drug delivery: from micellar nanocontainers to biological response modifiers. *Critical Reviews in Therapeutic Drug Carrier Systems, 19*, 1–72.

Kabanov, A. V., Nazarova, I. R., Astafieva, I. V., Batrakova, E. V., Alakhov, V. Y., Yaroslavov, A. A., et al. (1995). Micelle formation and solubilization of fluorescent probes in poly (oxyethylene-*b*-oxypropylene-*b*-oxyethylene) solutions. *Macromolecules, 28*, 2303–2314.

Kanazawa, H., Yamamoto, K., Matsushima, Y., Takai, N., Kikuchi, A., Sakurai, Y., et al. (1996). Temperature-responsive chromatography using poly(*N*-isopropylacrylamide)-modified silica. *Analytical Chemistry, 68*, 100–105.

Kaneko, Y., Nakamura, S., Sakai, K., Aoyagi, T., Kikuchi, A., Sakurai, Y., et al. (1998). Rapid deswelling response of poly(*N*-isopropylacrylamide) hydrogels by the formation of water release channels using poly(ethylene oxide) graft chains. *Macromolecules, 31*, 6099–6105.

Katayama, Y., Sonoda, T., & Maeda, M. (2001). A polymer micelle responding to the protein kinase A signal. *Macromolecules, 34*, 8569–8573.

Khokhlov, A. R., Starodubtzev, S. G., & Vasilevskaya, V. V. (1993). Conformational transitions in polymer gels - theory and Experiment. *Advances in Polymer Science, 109*, 123–175.

Kikuchi, A., & Okano, T. (2002). Intelligent thermoresponsive polymeric stationary phases for aqueous chromatography of biological compounds. *Progress in Polymer Science, 27*, 1165–1193.

King, S. M., Heenan, R. K., Cloke, V. M., & Washington, C. (1997). Neutron scattering from a poly(oxyethylene)-poly(oxypropylene)-poly(oxyethylene) copolymer in dilute aqueous solution under shear flow. *Macromolecules, 30*, 6215–6222.

Kirsh, Y. E. (1998). *Water soluble poly-N-vinylamides: synthesis and physicochemical properties*. Chichester, U.K.: John Wiley & Sons, Inc.

Kirsh, Y. E., Yanul, N., & Kalninsh, K. (1999). Structural transformations and water associate interactions in poly-*N*-vinylcaprolactam–water system. *European Polymer Journal, 35*, 305–316.

Kister, G., Cassanas, G., & Vert, M. (1998). Effects of morphology, conformation and configuration on the IR and Raman spectra of various poly(lactic acid)s. *Polymer, 39*, 267–273.

Kondo, T. (1997). The relationship between intramolecular hydrogen bonds and certain physical properties of regioselectively substituted cellulose derivatives. *Journal of Polymer Science Part B-Polymer Physics, 35*, 717–723.

Kostal, J., Mulchandani, A., & Chen, W. (2001). Tunable biopolymers for heavy metal removal. *Macromolecules, 34*, 2257–2261.

Kuckling, D., Adler, H. J. P., Arndt, K. F., Ling, L., & Habicher, W. D. (2000). Temperature and pH dependent solubility of novel poly(*N*-isopropylacrylamide) copolymers. *Macromolecular Chemistry and Physics, 201*, 273–280.

Kumar, A., Galaev, I. Y., & Mattiasson, B. (1998). Metal chelate affinity precipitation: a new approach to protein purification. *Bioseparation*, *7*, 185–194.

Kumar, A., Kamihira, M., Galaev, I. Y., Mattiasson, B., & Iijima, S. (2001). Type-specific separation of animal cells in aqueous two-phase systems using antibody conjugates with temperature-sensitive polymers. *Biotechnology and Bioengineering*, *75*, 570–580.

Kumar, A., Srivastava, A., Galaev, I. Y., & Mattiasson, B. (2007). Smart polymers: physical forms and bioengineering applications. *Progress in Polymer Science*, *32*, 1205–1237.

Kumar, A., Wahlund, P. O., Kepka, C., Galaev, I. Y., & Mattiasson, B. (2003). Purification of histidine–tagged single–chain Fv–antibody fragments by metal chelate affinity precipitation using thermoresponsive copolymers. *Biotechnology and Bioengineering*, *84*, 494–503.

Kushida, A., Yamato, M., Konno, C., Kikuchi, A., Sakurai, Y., & Okano, T. (2000). Temperature-responsive culture dishes allow nonenzymatic harvest of differentiated madin-darby canine kidney (MDCK) cell sheets. *Journal of Biomedical Materials Research*, *51*, 216–223.

Kwon, I. C., Bae, Y. H., & Kim, S. W. (1991). Electrically erodible polymer gel for controlled release of drugs. *Nature*, *354*, 291–293.

Kwon, I. C., Bae, Y. H., Okano, T., & Kim, S. W. (1991). Drug release from electric-current sensitive polymers. *Journal of Controlled Release*, *17*, 149–156.

Kwon, G. S., & Kataoka, K. (1995). Block copolymer micelles as long-circulating drug vehicles. *Advanced Drug Delivery Reviews*, *16*, 295–309.

Kwon, G. S., & Okano, T. (1996). Polymeric micelles as new drug carriers. *Advanced Drug Delivery Reviews*, *21*, 107–116.

Laukkanen, A., Valtola, L., Winnik, F. M., & Tenhu, H. (2004). Formation of colloidally stable phase separated poly(*N*-vinylcaprolactam) in water: a study by dynamic light scattering, microcalorimetry, and pressure perturbation calorimetry. *Macromolecules*, *37*, 2268–2274.

Lau, A. C. W., & Wu, C. (1999). Thermally sensitive and biocompatible poly(*N*-vinylcaprolactam): synthesis and characterization of high molar mass linear chains. *Macromolecules*, *32*, 581–584.

Lee, H., Lee, J., & Andrade, J. D. (1995). Blood compatibility of polyethylene oxide surfaces. *Progress in Polymer Science (UK)*, *20*, 1043–1079.

Lee, D. S., Shim, M. S., Kim, S. W., Lee, H., Park, I., & Chang, T. Y. (2001). Novel thermoreversible gelation of biodegradable PLGA-*block*-PEO-*block*-PLGA triblock copolymers in aqueous solution. *Macromolecular Rapid Communications*, *22*, 587–592.

Leobandung, W., Ichikawa, H., Fukumori, Y., & Peppas, N. A. (2002). Preparation of stable insulin-loaded nanospheres of poly(ethylene glycol) macromers and *N*-isopropyl acrylamide. *Journal of Controlled Release*, *80*, 357–363.

Liang, X., Kozlovskaya, V., Chen, Y., Zavgorodnya, O., & Kharlampieva, E. (2012). Thermosensitive multilayer hydrogels of poly(*N*-vinylcaprolactam) as nanothin films and shaped capsules. *Chemistry of Materials*, *24*, 3707–3719.

Li, M., Kim, J.-W., & Peeples, T. L. (2002). Amylase partitioning and extractive bioconversion of starch using thermoseparating aqueous two-phase systems. *Journal of Biotechnology*, *93*, 15–26.

Li, Y., Lokitz, B. S., & Mccormick, C. L. (2006). Thermally responsive vesicles and their structural "Locking" through polyelectrolyte complex formation. *Angewandte Chemie International Edition*, *45*, 5792–5795.

Lin, H. H., & Cheng, Y. L. (2001). In-situ thermoreversible gelation of block and star copolymers of poly(ethylene glycol) and poly(*N*-isopropylacrylamide) of varying architectures. *Macromolecules*, *34*, 3710–3715.

Lin, S. Y., Chen, K. S., & Liang, R. C. (1999). Thermal micro ATR/FT-IR spectroscopic system for quantitative study of the molecular structure of poly(*N*-isopropylacrylamide) in water. *Polymer*, *40*, 2619–2624.

Liu, Y., Chen, S.-H., & Huang, J. S. (1998). Small-angle neutron scattering analysis of the structure and interaction of triblock copolymer micelles in aqueous solution. *Macromolecules*, *31*, 2236–2244.

Liu, X. M., Pramoda, K. P., Yang, Y. Y., Chow, S. Y., & He, C. B. (2004). Cholesteryl-grafted functional amphiphilic poly(*N*-isopropylacrylamide-*co*-*N*-hydroxylmethylacrylamide): synthesis, temperature sensitivity, self-assembly and encapsulation of a hydrophobic agent. *Biomaterials*, *25*, 2619–2628.

Liu, Y.-Y., Shao, Y.-H., & Lü, J. (2006). Preparation, properties and controlled release behaviors of pH-induced thermosensitive amphiphilic gels. *Biomaterials*, *27*, 4016–4024.

Li, H., Yu, G. E., Price, C., Booth, C., Hecht, E., & Hoffmann, H. (1997). Concentrated aqueous micellar solutions of diblock copoly(oxyethylene/oxybutylene) E(41)B(8): a study of phase behavior. *Macromolecules*, *30*, 1347–1354.

Lu, M., Albertsson, P. A., Johansson, G., & Tjerneld, F. (1996). Ucon-benzoyl dextran aqueous two-phase systems: protein purification with phase component recycling. *Journal of Chromatography B-Biomedical Applications*, *680*, 65–70.

Luo, Q., Mutlu, S., Gianchandani, Y. B., Svec, F., & Fréchet, J. M. J. (2003). Monolithic valves for microfluidic chips based on thermoresponsive polymer gels. *Electrophoresis*, *24*, 3694–3702.

Makhaeva, E. E., Tenhu, H., & Khokhlov, A. R. (1998). Conformational changes of poly(vinylcaprolactam) macromolecules and their complexes with ionic surfactants in aqueous solution. *Macromolecules*, *31*, 6112–6118.

Makhaeva, E. E., Tenhu, H., & Khokhlov, A. R. (2000). Behavior of poly(*N*-vinylcaprolactam) macromolecules in the presence of organic compounds in aqueous solution. *Polymer*, *41*, 9139–9145.

Makhaeva, E. E., Thanh, L. T. M., Starodoubtsev, S. G., & Khokhlov, A. R. (1996). Thermoshrinking behavior of poly(vinylcaprolactam) gels in aqueous solution. *Macromolecular Chemistry and Physics*, *197*, 1973–1982.

Mamada, A., Tanaka, T., Kungwatchakun, D., & Irie, M. (1990). Photoinduced phase-transition of gels. *Macromolecules*, *23*, 1517–1519.

Markvicheva, E. A., Kuzkina, I. F., Pashkin, I. I., Plechko, T. N., Kirsh, Y. E., & Zubov, V. P. (1991). A novel technique for entrapment of hybridoma cells in synthetic thermally reversible polymers. *Biotechnology Techniques*, *5*, 223–226.

Massodi, I., Bidwell, G. L., III, & Raucher, D. (2005). Evaluation of cell penetrating peptides fused to elastin-like polypeptide for drug delivery. *Journal of Controlled Release*, *108*, 396–408.

Mattiasson, B., Kumar, A., Ivanov, A. E., & Galaev, I. Y. (2007). Metal-chelate affinity precipitation of proteins using responsive polymers. *Nature Protocols*, *2*, 213–220.

Mcdaniel, J. R., Callahan, D. J., & Chilkoti, A. (2010). Drug delivery to solid tumors by elastin-like polypeptides. *Advanced Drug Delivery Reviews*, *62*, 1456–1467.

Meng, F., Zhong, Z., & Feijen, J. (2009). Stimuli-responsive polymersomes for programmed drug delivery. *Biomacromolecules*, *10*, 197–209.

Meyer, D. E., & Chilkoti, A. (1999). Purification of recombinant proteins by fusion with thermally-responsive polypeptides. *Nature biotechnology*, *17*, 1112–1115.

Meznarich, N. A. K., Juggernauth, K. A., Batzli, K. M., & Love, B. J. (2011). Structural changes in PEO–PPO–PEO gels induced by methylparaben and dexamethasone observed using time-resolved SAXS. *Macromolecules*, *44*, 7792–7798.

Meznarich, N. A. K., & Love, B. J. (2011). The kinetics of gel formation for PEO–PPO–PEO triblock copolymer solutions and the effects of added methylparaben. *Macromolecules, 44*, 3548–3555.

Mikheeva, L. M., Grinberg, N. V., Mashkevich, A. Y., Grinberg, V. Y., Thanh, L. T. M., Makhaeva, E. E., et al. (1997). Microcalorimetric study of thermal cooperative transitions in poly(*N*-vinylcaprolactam) hydrogels. *Macromolecules, 30*, 2693–2699.

Miyata, K., Kakizawa, Y., Nishiyama, N., Yamasaki, Y., Watanabe, T., Kohara, M., et al. (2005). Practically applicable cross-linked polyplex micelle with high tolerability against freeze-drying for in vivo gene delivery. *Journal of Controlled Release, 109*, 15–23.

Mortensen, K., Batsberg, W., & Hvidt, S. (2008). Effects of PEO–PPO diblock impurities on the cubic structure of aqueous PEO–PPO–PEO pluronics micelles: fcc and bcc ordered structures in F127. *Macromolecules, 41*, 1720–1727.

Mukose, T., Fujiwara, T., Nakano, J., Taniguchi, I., Miyamoto, M., Kimura, Y., et al. (2004). Hydrogel formation between enantiomeric B–A–B–type block copolymers of polylactides (PLLA or PDLA: A) and polyoxyethylene (PEG: B); PEG–PLLA–PEG and PEG–PDLA–PEG. *Macromolecular Bioscience, 4*, 361–367.

Murphy, J. C., Jewell, D. L., White, K. I., Fox, G. E., & Willson, R. C. (2003). Nucleic acid separations utilizing immobilized metal affinity chromatography. *Biotechnology Progress, 19*, 982–986.

Nagase, K., Kobayashi, J., Kikuchi, A., Akiyama, Y., Kanazawa, H., & Okano, T. (2007). Interfacial property modulation of thermoresponsive polymer brush surfaces and their interaction with biomolecules. *Langmuir, 23*, 9409–9415.

Nagase, K., Kobayashi, J., & Okano, T. (2009). Temperature-responsive intelligent interfaces for biomolecular separation and cell sheet engineering. *Journal of the Royal Society Interface, 6*, S293–S309.

Nakayama, M., Okano, T., Miyazaki, T., Kohori, F., Sakai, K., & Yokoyama, M. (2006). Molecular design of biodegradable polymeric micelles for temperature-responsive drug release. *Journal of Controlled Release, 115*, 46–56.

Na, K., Lee, K. H., Lee, D. H., & Bae, Y. H. (2006). Biodegradable thermo-sensitive nanoparticles from poly(L-lactic acid)/poly(ethylene glycol) alternating multi-block copolymer for potential anti-cancer drug carrier. *European Journal of Pharmaceutical Sciences, 27*, 115–122.

Na, K., Park, J. H., Kim, S. W., Sun, B. K., Woo, D. G., Chung, H.-M., et al. (2006). Delivery of dexamethasone, ascorbate, and growth factor (TGF Î²-3) in thermo-reversible hydrogel constructs embedded with rabbit chondrocytes. *Biomaterials, 27*, 5951–5957.

Nandkumar, M. A., Yamato, M., Kushida, A., Konno, C., Hirose, M., Kikuchi, A., et al. (2002). Two-dimensional cell sheet manipulation of heterotypically co-cultured lung cells utilizing temperature-responsive culture dishes results in long-term maintenance of differentiated epithelial cell functions. *Biomaterials, 23*, 1121–1130.

Nayak, S., Lee, H., Chmielewski, J., & Lyon, L. A. (2004). Folate-mediated cell targeting and cytotoxicity using thermoresponsive microgels. *Journal of the American Chemical Society, 126*, 10258–10259.

Neradovic, D., Hinrichs, W. L. J., Kettenes-Van Den Bosch, J. J., Van Nostrum, C. F., & Hennink, W. E. (2001). Poly(*N*-isopropylacrylamide) with hydrolyzable lactic acid ester side groups: a new type of thermosensitive polymer suitable for drug delivery purposes. *Journal of Controlled Release, 72*, 252–253.

Neradovic, D., Soga, O., Van Nostrum, C., & Hennink, W. (2004). The effect of the processing and formulation parameters on the size of nanoparticles based on block copolymers of poly(ethylene glycol) and poly(*N*-isopropylacrylamide) with and without hydrolytically sensitive groups. *Biomaterials, 25*, 2409.

Nettles, D. L., Chilkoti, A., & Setton, L. A. (2010). Applications of elastin-like polypeptides in tissue engineering. *Advanced Drug Delivery Reviews*, *62*, 1479–1485.

Nolan, C. M., Gelbaum, L. T., & Lyon, L. A. (2006). 1H NMR investigation of thermally triggered insulin release from poly(*N*-isopropylacrylamide) microgels. *Biomacromolecules*, *7*, 2918–2922.

Oishi, M., Nagasaki, Y., Itaka, K., Nishiyama, N., & Kataoka, K. (2005). Lactosylated poly(ethylene glycol)-siRNA conjugate through acid-labile β-thiopropionate linkage to construct pH-sensitive polyion complex micelles achieving enhanced gene silencing in hepatoma cells. *Journal of the American Chemical Society*, *127*, 1624–1625.

Okano, T., Bae, Y. H., Jacobs, H., & Kim, S. W. (1990). Thermally on off switching polymers for drug permeation and release. *Journal of Controlled Release*, *11*, 255–265.

Ong, J. T., & Manoukian, E. (1988). Micellar solubilization of timobesone acetate in aqueous and aqueous propylene glycol solutions of nonionic surfactants. *Pharmaceutical Research*, *5*, 704–708.

Otake, K., Inomata, H., Konno, M., & Saito, S. (1989). A new model for the thermally induced volume phase-transition of gels. *Journal of Chemical Physics*, *91*, 1345–1350.

Pan, L.-C., & Chien, C.-C. (2003). A novel application of thermo-responsive polymer to affinity precipitation of polysaccharide. *Journal of Biochemical and Biophysical Methods*, *55*, 87–94.

Pan, Y. V., Wesley, R. A., Luginbuhl, R., Denton, D. D., & Ratner, B. D. (2001). Plasma polymerized *N*-isopropylacrylamide: synthesis and characterization of a smart thermally responsive coating. *Biomacromolecules*, *2*, 32–36.

Park, M. J., & Char, K. (2002). Two gel states of a PEO–PPO–PEO triblock copolymer formed by different mechanisms. *Macromolecular Rapid Communications*, *23*, 688–692.

Pelton, R. (2000). Temperature-sensitive aqueous microgels. *Advances in Colloid and Interface Science*, *85*, 1–33.

Pereira, M., Wu, Y.-T., Venâncio, A., & Teixeira, J. (2003). Aqueous two-phase extraction using thermoseparating polymer: a new system for the separation of endo-polygalacturonase. *Biochemical Engineering Journal*, *15*, 131–138.

Persson, J., Johansson, H. O., Galaev, I., Mattiasson, B., & Tjerneld, F. (2000). Aqueous polymer two-phase systems formed by new thermoseparating polymers. *Bioseparation*, *9*, 105–116.

Persson, J., Johansson, H. O., & Tjerneld, F. (1999). Purification of protein and recycling of polymers in a new aqueous two-phase system using two thermoseparating polymers. *Journal of Chromatography A*, *864*, 31–48.

Prabaharan, M., Grailer, J. J., Steeber, D. A., & Gong, S. (2008). Stimuli-responsive chitosan-graft-poly(*N*-vinylcaprolactam) as a promising material for controlled hydrophobic drug delivery. *Macromolecular Bioscience*, *8*, 843–851.

Prabaharan, M., Grailer, J. J., Steeber, D. A., & Gong, S. (2009). Thermosensitive micelles based on folate-conjugated poly(*N*-vinylcaprolactam)-*block*-poly (ethylene glycol) for tumor–targeted drug delivery. *Macromolecular Bioscience*, *9*, 744–753.

Price, C. (1983). Micelle formation by block co-polymers in organic-solvents. *Pure and Applied Chemistry*, *55*, 1563–1572.

Principi, T., Goh, C. C. E., Liu, R. C. W., & Winnik, F. M. (2000). Solution properties of hydrophobically modified copolymers of *N*-isopropylacrylamide and *N*-glycine acrylamide: a study by microcalorimetry and fluorescence spectroscopy. *Macromolecules*, *33*, 2958–2966.

Qin, S., Geng, Y., Discher, D. E., & Yang, S. (2006). Temperature-controlled assembly and release from polymer vesicles of poly(ethylene oxide)-*block*- poly(*N*-isopropylacrylamide). *Advanced Materials*, *18*, 2905–2909.

Qiu, Y., & Park, K. (2001). Environment-sensitive hydrogels for drug delivery. *Advanced Drug Delivery Reviews, 53*, 321–339.

Rangel-Yagui, C. O., Pessoa, A., Jr., & Tavares, L. C. (2005). Micellar solubilization of drugs. *Journal of Pharmacy and Pharmaceutical Sciences, 8*, 147–163.

Rassing, J., Mckenna, W. P., Bandyopadhyay, S., & Eyring, E. M. (1984). Ultrasonic and C-13-Nmr studies on gel formation in aqueous-solutions of the Aba block polymer pluronicr F-127. *Journal of Molecular Liquids, 27*, 165–178.

Reese, C. E., Mikhonin, A. V., Kamenjicki, M., Tikhonov, A., & Asher, S. A. (2004). Nanogel nanosecond photonic crystal optical switching. *Journal of the American Chemical Society, 126*, 1493–1496.

Reiersen, H., Clarke, A. R., & Rees, A. R. (1998). Short elastin-like peptides exhibit the same temperature-induced structural transitions as elastin polymers: implications for protein engineering. *Journal of Molecular Biology, 283*, 255–264.

Ricka, J., Meewes, M., Nyffenegger, R., & Binkert, T. (1990). Intermolecular and intramolecular solubilization–collapse and expansion of a polymer-chain in surfactant solutions. *Physical Review Letters, 65*, 657–660.

Rijcken, C. J., Veldhuis, T. F., Ramzi, A., Meeldijk, J. D., Van Nostrum, C. F., & Hennink, W. E. (2005). Novel fast degradable thermosensitive polymeric micelles based on PEG-*block*-poly(*N*-(2-hydroxyethyl) methacrylamide-oligolactates). *Biomacromolecules, 6*, 2343–2351.

Rito-Palomares, M. (2004). Practical application of aqueous two-phase partition to process development for the recovery of biological products. *Journal of Chromatography B, 807*, 3–11.

San Biagio, P. L., Bulone, D., Martorana, V., Dowd, S., & Newman, J. (2001). Agarose micro-gelation at very low concentrations in the presence and absence of salt. *Biophysical Journal, 80*, 399a.

Saunders, B. R., Laajam, N., Daly, E., Teow, S., Hu, X., & Stepto, R. (2009). Microgels: from responsive polymer colloids to biomaterials. *Advances in Colloid and Interface Science, 148*, 251–262.

Saunders, B. R., & Vincent, B. (1999). Microgel particles as model colloids: theory, properties and applications. *Advances in Colloid and Interface Science, 80*, 1–25.

Sawahata, K., Hara, M., Yasunaga, H., & Osada, Y. (1990). Electrically controlled drug delivery system using polyelectrolyte gels. *Journal of Controlled Release, 14*, 253–262.

Schild, H. G. (1992). Poly(*N*-Isopropylacrylamide) - experiment, theory and application. *Progress in Polymer Science, 17*, 163–249.

Schild, H. G., Muthukumar, M., & Tirrell, D. A. (1991). Cononsolvency in mixed aqueous-solutions of poly(*N*-isopropylacrylamide). *Macromolecules, 24*, 948–952.

Schild, H. G., & Tirrell, D. A. (1990). Microcalorimetric detection of lower critical solution temperatures in aqueous polymer-solutions. *Journal of Physical Chemistry, 94*, 4352–4356.

Schmidt, P., Dybal, J., Rodriguez-Cabello, J. C., & Reboto, V. (2005). Role of water in structural changes of poly(AVGVP) and poly(GVGVP) studied by FTIR and Raman spectroscopy and ab initio calculations. *Biomacromolecules, 6*, 697–706.

Shibayama, M., Suetoh, Y., & Nomura, S. (1996). Structure relaxation of hydrophobically aggregated poly(*N*-isopropylacrylamide) in water. *Macromolecules, 29*, 6966–6968.

Shimazu, M., Mulchandani, A., & Chen, W. (2003). Thermally triggered purification and immobilization of elastin–OPH fusions. *Biotechnology and Bioengineering, 81*, 74–79.

Shin, B. C., Jhon, M. S., Lee, H. B., & Yuk, S. H. (1998). pH/temperature dependent phase transition of an interpenetrating polymer network: anomalous swelling behavior above lower critical solution temperature. *European Polymer Journal, 34*, 1675–1681.

Soga, O., Van Nostrum, C. F., Fens, M., Rijcken, C. J., Schiffelers, R. M., Storm, G., et al. (2005). Thermosensitive and biodegradable polymeric micelles for paclitaxel delivery. *Journal of Controlled Release, 103*, 341–354.

Southall, N. T., Dill, K. A., & Haymet, A. D. J. (2002). A view of the hydrophobic effect. *Journal of Physical Chemistry B, 106*, 521–533.

Stiborova, H., Kostal, J., Mulchandani, A., & Chen, W. (2003). One–step metal–affinity purification of histidine–tagged proteins by temperature–triggered precipitation. *Biotechnology and Bioengineering, 82*, 605–611.

Sugawara, T., & Matsuda, T. (1994). Novel surface graft copolymerization method with micron order regional precision. *Macromolecules, 27*, 7809–7814.

Sun, S., & Wu, P. (2011). Infrared spectroscopic insight into hydration behavior of poly (*N*-vinylcaprolactam) in water. *The Journal of Physical Chemistry B, 115*, 11609–11618.

Suzuki, A., & Tanaka, T. (1990). Phase-transition in polymer gels induced by visible-light. *Nature, 346*, 345–347.

Suzuki, K., Yumura, T., Mizuguchi, M., Tanaka, Y., Chen, C.-W., & Akashi, M. (2000). Poly(*N*-isopropylacrylamide)-grafted silica as a support of platinum colloids: preparation method, characterization, and catalytic properties in hydrogenation. *Journal of Applied Polymer Science, 77*, 2678–2684.

Tamai, Y., Tanaka, H., & Nakanishi, K. (1996a). Molecular dynamics study of polymer–water interaction in hydrogels. 1. Hydrogen-bond structure. *Macromolecules, 29*, 6750–6760.

Tamai, Y., Tanaka, H., & Nakanishi, K. (1996b). Molecular dynamics study of polymer–water interaction in hydrogels. 2. Hydrogen-bond dynamics. *Macromolecules, 29*, 6761–6769.

Tanaka, T., & Fillmore, D. J. (1979). Kinetics of swelling of gels. *The Journal of Chemical Physics, 70*, 1214–1218.

Taylor, L. D., & Cerankowski, L. D. (1975). Preparation of films exhibiting a balanced temperature-dependence to permeation by aqueous-solutions - study of lower consolute behavior. *Journal of Polymer Science Part A-Polymer Chemistry, 13*, 2551–2570.

Topp, M. D. C., Dijkstra, P. J., Talsma, H., & Feijen, J. (1997). Thermosensitive micelle-forming block copolymers of poly(ethylene glycol) and poly(*N*-isopropylacrylamide). *Macromolecules, 30*, 8518–8520.

Torchilin, V. P. (2001). Structure and design of polymeric surfactant-based drug delivery systems. *Journal of Controlled Release: Official Journal of the Controlled Release Society, 73*, 137.

Urry, D. W. (1992). Free-Energy transduction in polypeptides and proteins based on inverse temperature transitions. *Progress in Biophysics & Molecular Biology, 57*, 23–57.

Urry, D. W. (1997). Physical chemistry of biological free energy transduction as demonstrated by elastic protein-based polymers. *Journal of Physical Chemistry B, 101*, 11007–11028.

Urry, D., Luan, C.-H., Harris, C. M., & Parker, T. M. (1997). *Protein-based materials with a profound range of properties and applications: the elastin ΔTt hydrophobic paradigm. Protein-based materials*. (pp. 133–177). Birkhäuser Boston: Springer.

Urry, D. W., Peng, S. Q., Parker, T. M., Gowda, D. C., & Harris, R. D. (1993). Relative significance of electrostatic-induced and hydrophobic-induced Pk(a) shifts in a model protein - the aspartic-acid residue. *Angewandte Chemie-International Edition in English, 32*, 1440–1442.

Urry, D. W., Trapane, T. L., Iqbal, M., Venkatachalam, C. M., & Prasad, K. U. (1985). C-13 Nmr relaxation studies demonstrate an inverse temperature transition in the elastin polypentapeptide. *Biochemistry, 24*, 5182–5189.

Urry, D. W., Trapane, T. L., & Prasad, K. U. (1985). Phase-structure transitions of the elastin polypentapeptide water-system within the framework of composition temperature studies. *Biopolymers, 24*, 2345–2356.

Vadnere, M., Amidon, G., Lindenbaum, S., & Haslam, J. L. (1984). Thermodynamic studies on the gel sol transition of some pluronic polyols. *International Journal of Pharmaceutics, 22*, 207–218.

Vihola, H., Laukkanen, A., Valtola, L., Tenhu, H., & Hirvonen, J. (2005). Cytotoxicity of thermosensitive polymers poly(*N*-isopropylacrylamide), poly(*N*-vinylcaprolactam) and amphiphilically modified poly(*N*-vinylcaprolactam). *Biomaterials, 26*, 3055–3064.

Volpert, E., Selb, J., & Candau, F. (1998). Associating behavior of polyacrylamides hydrophobically modified with dihexylacrylamide. *Polymer, 39*, 1025–1033.

Von Recum, H. A., Kim, S. W., Kikuchi, A., Okuhara, M., Sakurai, Y., & Okano, T. (1998). Novel thermally reversible hydrogel as detachable cell culture substrate. *Journal of Biomedical Materials Research, 40*, 631–639.

Wang, J., Chen, Z., Mauk, M., Hong, K.-S., Li, M., Yang, S., et al. (2005). Self-actuated, thermo-responsive hydrogel valves for lab on a chip. *Biomedical Microdevices, 7*, 313–322.

Wang, X. H., Qiu, X. P., & Wu, C. (1998). Comparison of the coil-to-globule and the globule-to-coil transitions of a single poly(*N*-isopropylacrylamide) homopolymer chain in water. *Macromolecules, 31*, 2972–2976.

Ward, M. A., & Georgiou, T. K. (2011). Thermoresponsive polymers for biomedical applications. *Polymers, 3*, 1215–1242.

Wei, H., Zhang, X. Z., Zhou, Y., Cheng, S. X., & Zhuo, R. X. (2006). Self-assembled thermoresponsive micelles of poly(*N*-isopropylacrylamide-*b*-methyl methacrylate). *Biomaterials, 27*, 2028–2034.

Wu, J. Y., Liu, S. Q., Heng, P. W. S., & Yang, Y. Y. (2005). Evaluating proteins release from, and their interactions with, thermosensitive poly(*N*-isopropylacrylamide) hydrogels. *Journal of Controlled Release, 102*, 361–372.

Wu, C., & Wang, X. H. (1998). Globule-to-coil transition of a single homopolymer chain in solution. *Physical Review Letters, 80*, 4092–4094.

Yakushiji, T., Sakai, K., Kikuchi, A., Aoyagi, T., Sakurai, Y., & Okano, T. (1999). Effects of cross-linked structure on temperature-responsive hydrophobic interaction of poly(*N*-isopropylacrylamide) hydrogel-modified surfaces with steroids. *Analytical Chemistry, 71*, 1125–1130.

Yamada, N., Okano, T., Sakai, H., Karikusa, F., Sawasaki, Y., & Sakurai, Y. (1990). Thermo-responsive polymeric surfaces; control of attachment and detachment of cultured cells. *Die Makromolekulare Chemie, Rapid Communications, 11*, 571–576.

Yan, L. F., Zhu, Q. S., & Kenkare, P. U. (2000). Lower critical solution temperature of linear PNIPA obtained from a yukawa potential of polymer chains. *Journal of Applied Polymer Science, 78*, 1971–1976.

Yoshida, R., Uchida, K., Kaneko, Y., Sakai, K., Kikuchi, A., Sakurai, Y., et al. (1995). Comb-type grafted hydrogels with rapid de-swelling response to temperature-changes. *Nature, 374*, 240–242.

Yuk, S. H., Cho, S. H., & Lee, S. H. (1997). pH/temperature-responsive polymer composed of poly((*N,N*-dimethylamino)ethyl methacrylate-*co*-ethylacrylamide). *Macromolecules, 30*, 6856–6859.

Yu, C., Mutlu, S., Selvaganapathy, P., Mastrangelo, C. H., Svec, F., & Frã©Chet, J. M. J. (2003). Flow control valves for analytical microfluidic chips without mechanical parts based on thermally responsive monolithic polymers. *Analytical Chemistry, 75*, 1958–1961.

Zeng, F., Tong, Z., & Feng, H. Q. (1997). Nmr investigation of phase separation in poly(*N*-isopropyl acrylamide)/water solutions. *Polymer, 38*, 5539–5544.

Environmentally responsive polyelectrolytes and zwitterionic polymers

2

M.T. Bernards
University of Missouri, Columbia, MO, USA

2.1 Introduction

Many biomedical polymers contain charged moieties, and the presence and location of these charged regions play an essential role in the biological response to devices that contain or are coated by these polymers. Interest in these types of polymeric systems has also been growing because of their similarity to natural biological macromolecules such as proteins (Alfrey, Morawetz, Fitzgerald, & Fuoss, 1950). This chapter provides an introduction to the responsive properties of three specific subclasses of charged biomedical polymers: polyelectrolytes, polyampholytes, and betaine polymers. Polyelectrolytes are polymers that are composed of a single type of charged monomer subunit, and therefore the overall polymer is either positively or negatively charged. Betaine polymers contain both a positively and a negatively charged region within the side chain of each individual monomer subunit, resulting in an overall neutral polymer chain. Finally, polyampholyte polymers are composed of mixtures of charged monomer subunits. These polymers can have a positive, negative, or neutral charge depending on their composition. Both polyampholyte and betaine polymer systems fall under the classification of zwitterionic polymers because they contain both positively and negatively charged regions. Polyampholyte polymers have also been further broken down into two subcategories. Annealed polyampholyte systems are those in which the monomer subunits are sensitive to pH, resulting in a change in the overall polymer charge as a function of pH. Quenched polyampholyte systems remain relatively unchanged with large changes in pH, maintaining their base charged state (Kudaibergenov, 2002).

Polyelectrolytes and zwitterionic polymers have many industrial applications, including cosmetics, advanced separations, and water treatment (Kudaibergenov, 2002; McCormick, 2000). These applications take advantage of the charge distribution throughout the underlying polymeric structure. However, the focus of this chapter is on biomedical applications of these polymeric systems, with a particular emphasis on polyelectrolytes and zwitterionic polymer systems that are responsive to environmental cues. First, it will provide an introduction to the typical monomers used to synthesize polyelectrolytes, polyampholytes, and betaine polymers, along with an overview of some of the polymerization and coating approaches. Next, it will outline the properties of these polymeric systems that make them attractive for biomedical

Switchable and Responsive Surfaces and Materials for Biomedical Applications. http://dx.doi.org/10.1016/B978-0-85709-713-2.00002-X

applications, with a focus on systems that have a desirable response to changes in pH, salt concentration, temperature, or other stimuli. Finally, the chapter will conclude with a summary of the future outlook for these polymers in biomedical applications.

2.2 Monomer subunits and polymerization approaches

It is well understood that the choice of the underlying monomers used to synthesize polyelectrolytes, polyampholytes, and betaine polymers have a significant impact on the final properties of the resulting material. This is especially important when identifying polymers for use in biomedical applications, as the underlying chemistry may not be tolerated well in the *in vivo* environment even if it has the most desirable physical characteristics. Because of this, only a subset of charged monomers has been adapted for biomedical applications.

2.2.1 *Betaine monomer subunits*

There are three primary betaine monomer subunits that have been widely investigated for biomedical applications; representative structures for all three are shown in Figure 2.1 (Jiang & Cao, 2010; Singh, Singh, & Singh, 2007). The phosphorylcholine (PC) functional group, shown in Figure 2.1(a), was originally synthesized by Ishihara et al. (1998), and its composition was inspired by the hydrophilic portion of naturally occurring phospholipids. Demonstrations of the great biocompatibility of PC-based polymers (Feng, Brash, & Zhu, 2004; Feng, Brash, & Zhu, 2006; Feng, Zhu, Ishihara,

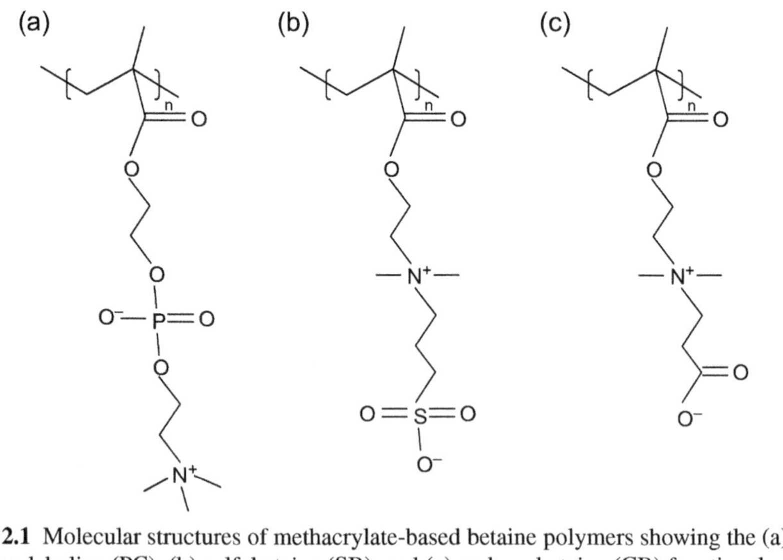

Figure 2.1 Molecular structures of methacrylate-based betaine polymers showing the (a) phosphorylcholine (PC), (b) sulfobetaine (SB), and (c) carboxybetaine (CB) functional groups.

& Brash, 2005; Hirota, Murakami, Nemoto, & Miyake, 2005; Ishihara et al., 1998) led to investigations of the biocompatibility of the sulfobetaine (SB) (Cheng, Zhang, Chen, Bryers, & Jiang, 2007; Ladd, Zhang, Chen, Hower, & Jiang, 2008; Zhang, Chen, Chang, & Jiang, 2006) and carboxybetaine (CB) (Ladd et al., 2008; Zhang, Chen, & Jiang, 2006; Zhang et al., 2008) functional groups, which are shown in Figure 2.1(b) and (c), respectively.

These three betaine functional groups can be found with multiple polymer backbone chemistries. The most common chemistries are acrylate and acrylamide based, either with or without pendant methyl groups. The structures shown in Figure 2.1 are the methacrylate-based polymer backbones for all three functional groups. It has been suggested that the acrylamide-based polymer backbones have better long-term stability (Cao, Li, Tang, & Cheng, 2013), yet the majority of the biomedical oriented work has been focused on acrylate-based systems.

In addition to the base structures provided for the betaine polymer systems shown in Figure 2.1, researchers have begun investigating variations to these functional groups. This work is centered around either varying the number of methyl groups between the charged subunits in order to better control the responsive properties of the resulting betaine polymer (Shao & Jiang, 2013; Zhang et al., 2008) or adding additional functional and/or cleavable groups to the end of the monomer side chain (Cao et al., 2012; Cao, Li et al., 2013; Cao, Tang et al., 2013; Cheng, Xue, Zhang, Chen, & Jiang, 2008). The CB-based functional group has the most sensitive pH responsive properties due to its terminal carboxylic acid group so much of the work has focused on modifying this monomer. This will be discussed in more detail below.

2.2.2 Polyelectrolyte and polyampholyte monomer subunits

Most of the biocompatible, charged monomer subunits have been investigated in both polyelectrolyte and polyampholyte polymers. Table 2.1 provides a summary of the terminal functional groups that are most commonly investigated. Similar to the betaine polymers, these terminal functional groups generally have acrylate- and acrylamide-based polymer chemistries; this portion of the molecules has been denoted as an R group in Table 2.1. The relative strengths of the functional groups for these

Table 2.1 Common monomer functional groups used to form polyelectrolytes and polyampholyte polymers

Strength of functional group	Terminal functional group
Weak cations	$R–NH_2$ $R–N(CH_3)_2$ $R–N(CH_2CH_3)_2$
Strong cations	$R–N(CH_3)_3$
Weak anions	R–COOH
Strong anions	$R–SO_3$

monomers are also divided into weak or strong categories. The weakly charged functional groups are more likely to have responsive properties, so they are the primary focus of the later sections of this chapter.

During the preparation of polyampholyte polymer systems it is possible to tailor the final material properties through both the monomer subunits used to form the resulting polymer and by controlling the polymer architecture. Polyampholyte polymers have historically been grouped into four subclasses based on the strength of the charge for the cationic and anionic functional groups: weak–weak, weak–strong, strong–weak, and strong–strong (Lowe & McCormick, 2002). However, Schroeder, Zurick, McGrath, and Bernards (2013) have recently proposed adding additional levels of control over material properties through the combination of three or more charged monomer subunits. This would rapidly expand the number of subclasses required to describe the possible combinations of charged functional groups.

The architecture of polyampholyte polymers also has a significant impact on the final material properties. Polyampholyte polymers have been prepared as copolymers with a statistical distribution of the charged monomer subunits or as block copolymers where the monomer subunits are localized together (Pafiti, Philippou, Loizou, Porcar, & Patrickios, 2011). In addition, polyampholyte polymers can be composed of unequal combinations of charged monomer subunits with either approach, resulting in a polymer with an overall positive or negative charge. These polyampholyte polymer systems typically behave similarly to polyelectrolytes. Examples of these approaches are discussed below in the areas where this level of control has led to beneficial responsive properties for biomedical applications.

2.2.3 General overview of polymerization and coating approaches

To create surfaces that are responsive to environmental cues, polyampholytes, polyelectrolytes, or betaine polymer coatings and/or materials must first be formed through controlled polymerization techniques. There are a number of excellent reviews on polymerization techniques (Barbey et al., 2009; Lowe & McCormick, 2002; Matyjaszewski & Xia, 2001), so only a brief overview of the most relevant approaches is provided here.

The first approach for developing controlled surfaces composed of polyelectrolytes or zwitterionic polymers begins with solution-based polymerization techniques including free radical polymerization, atom transfer radical polymerization, reversible addition-fragmentation chain transfer polymerization, and others (Lowe & McCormick, 2002; Matyjaszewski & Xia, 2001). These approaches allow control over the composition, molecular weight, and architecture of the polymer. The exact polymerization technique that is used often depends on the degree of control over the polymer characteristics that is required for the intended application. After forming and purifying the polymers, they can be applied to surfaces and devices through dip coating or grafting to approaches. Dip coating is a simple and rapid approach for applying polymer coatings to the surface of another material (Chen & Jiang, 2008), but it does not generally result in the formation of an irreversible coating because there are no specific interactions between the polymer and underlying substrate. Therefore, this approach often results in less robust coatings. "Grafting to" approaches require an additional modification to the end of the polymer chain, to attach a targeting molecule that will strongly bind the polymer to the underlying substrate (Waite & Tanzer, 1981). In one example of a grafting to approach, polyampholyte polymers were linked to a 3,4-dihydroxyphenyl-L-alanine (DOPA) molecule, which has strong binding interactions with a range of surface chemistries, resulting in the formation of a polyampholyte polymer brush coating (Li, Xue, Gao, Zhang, & Jiang, 2010).

The second approach for developing controlled surfaces composed of polyelectrolytes or zwitterionic polymers is based on a grafting from the surface approach. In this approach, polymerization initiation molecules are first attached to the surface of interest and then a polymer brush coating is grown directly from the surface. This approach is commonly used to form polymer brush coatings on silicon using a silane-based chemistry or noble metals using a thiol-based chemistry (Barbey et al., 2009; Dyer, 2006; Orski, Fries, Sontag, & Locklin, 2011; Xu, Neoh, & Kang, 2009). The "grafting from" approach allows careful control over the polymer brush density and thickness, both of which come into play in the final properties of these coatings (Azzaroni, Brown, & Huck, 2006; Huang, Li, Krause, Brault, & Jiang, 2012; Hucknall, Rangarajan, & Chilkoti, 2009; Tah & Bernards, 2012; Yang et al., 2008; Yang, Xue, Li, Zhang, & Jiang, 2009). In addition, there are several micro- and nano-patterning techniques that can be used to control the localization of polymer brushes of differing compositions, as reviewed elsewhere (Hook, Voelcker, & Thissen, 2009; Orski et al., 2011; Xu et al., 2009).

The final approach for providing a coating composed of polyelectrolytes or zwitterionic polymers is to form a hydrogel capsule around a substrate of interest. The hydrogels can be formed through chemical initiators or photoinitiators using standard hydrogel polymerization procedures (Drury & Mooney, 2003; Lee & Mooney, 2001; Peppas, Hilt, Khademhosseini, & Langer, 2006). However, in this approach it is important to tailor the cross-linking density for the intended application. For example, in an implantable glucose sensor, a lightly cross-linked hydrogel is required to allow sufficient diffusion through the hydrogel to the sensor (Yang et al., 2012). At the same time, in other applications, a more robust hydrogel may be required (Carr, Cheng, Xue, & Jiang, 2010; Dobbins, McGrath, & Bernards, 2012).

2.3 General solution properties

Polyelectrolytes and zwitterionic polymers have interesting solution-based properties that have been well documented in the literature (Kudaibergenov, 2002; McCormick, 2000). These solution-based properties are briefly covered below because there is significantly more literature discussing these effects, and they carry over to surfaces and coatings composed of polyelectrolytes and zwitterionic polymers.

2.3.1 The polyelectrolyte effect

When a polyelectrolyte is dissolved in water, it can be found in an extended conformation. This occurs because of electrostatic repulsions that occur between the like-charged polymer side chains. However, when small amounts of salt or other electrolytes are added, the conformation of the dissolved polyelectrolyte begins to contract. This reduction in size occurs because oppositely charged ions from the dissolved salt or electrolyte associate with the charged regions of the polyelectrolyte, reducing the electrostatic repulsions that are observed in pure water. This is known as the polyelectrolyte effect, and it is shown schematically in Figure 2.2. The polyelectrolyte effect has also been shown to influence the viscosity of solutions containing polyelectrolytes as well as the solution phase diffusion of these polymers (McCormick, 2000). As the amount of salt is increased the viscosity of an aqueous solution containing a polyelectrolyte will continue to decrease until it reaches the solution minimum. This is directly associated with the chain conformation and the interactions between neighboring chains. If the polyelectrolyte is synthesized from weak cationic or anionic species, a similar phenomenon has also been observed for changes in the solution pH with a fixed concentration of salt. Again, this is due to the impact that the solution has on the charge state of the polymer itself.

2.3.2 The antipolyelectrolyte effect

In biomedical applications, zwitterionic polymers are generally synthesized with an overall neutral charge. This is a natural result of the structure of the betaine monomers, and it is a design choice that results in desirable characteristics for polyampholyte polymers, which are discussed below. As a result, there is an opposite set of solution phase behaviors as compared to those of the charged polyelectrolytes. In what is known as the antipolyelectrolyte effect, zwitterionic polymers adopt a more linear and extended conformation with increasing levels of salt and a more compact conformation at lower levels of salt (An et al., 2011; Lowe & McCormick, 2002; Sanjuan & Tran, 2008; Xiao, Tao, Cui, Zhang, & Li, 2013). As shown in Figure 2.3, the presence of salt ions disrupts intra- and interchain interactions that naturally occur between the oppositely charged regions within neighboring polymer side chains, leading to the changes in conformation. The impacts of the polymer conformation on the solution phase viscosity and the polymer diffusivity are identical to those described above for the polyelectrolytes, the only difference being the concentration of salt required to reach each conformational state. Furthermore, it has been suggested by Xiao et al. (2013)

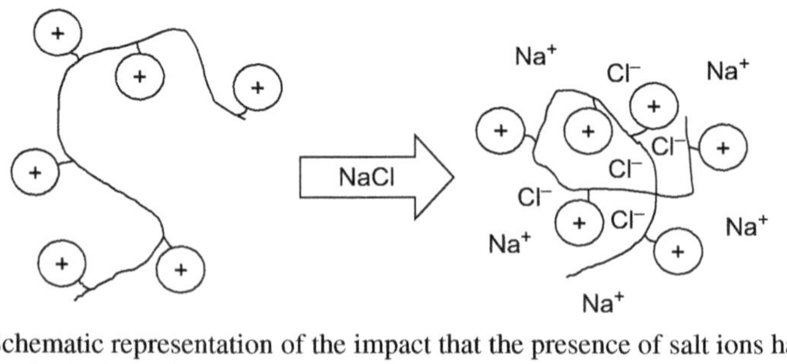

Figure 2.2 Schematic representation of the impact that the presence of salt ions have on the conformation of polyelectrolytes.

that the viscosity changes as a function of salt concentration are more sensitive to the specific anion as compared to the cation species.

A related but more complex behavior is also seen when a solution containing zwitterionic polymers undergoes a change in pH. This response is more prevalent when the zwitterionic polymer contains weakly charged cation and anion species. At the isoelectric point (IEP) of the polymer, there is an overall charge balance in the polymer and the number of intra- and interchain interactions is at a maximum. This results in the most compact conformation of the polymer. As the pH is changed either above or below the IEP, one of the charged groups begins to be neutralized and the polymer begins to develop an overall charged state. This results in an increase in electrostatic repulsions which causes the polymer to swell. This is also shown schematically in Figure 2.3. This property is clearly demonstrated in work by Pafiti et al. (2011), who compared the solution-based swelling behavior of four different polyampholyte polymer architectures with identical overall compositions; these results are shown in Figure 2.4. Polymers were prepared from combinations of a weak cationic monomer, 2-(dimethylamino) ethyl methacrylate (D), and a weak anionic monomer, methacrylic acid (A), which were cross-linked with ethylene glycol dimethacrylate (E). The four polymer architectures that were compared are block copolymers with the sequences of E_3-grad-A_{25}-grad-D_{50}-grad-A_{25}-grad-E_3 and E_3-grad-D_{25}-grad-A_{50}-grad-D_{25}-grad-E_3, and statistical copolymers with sequences of E_3-grad-(A_{50}-*co*-D_{50})-grad-E_3 and E_6-*co*-A_{50}-*co*-D_{50}. This work is highlighted to emphasize both the pH-dependent swelling properties of zwitterionic polymers and their composition-dependent behaviors.

2.3.3 General characteristics relevant to biomedical applications

As stated in the introduction, polyelectrolytes and zwitterionic polymers have been pursued for biomedical applications because of their similarity to natural biological molecules. This similarity has allowed researchers to tailor the polymer properties to promote

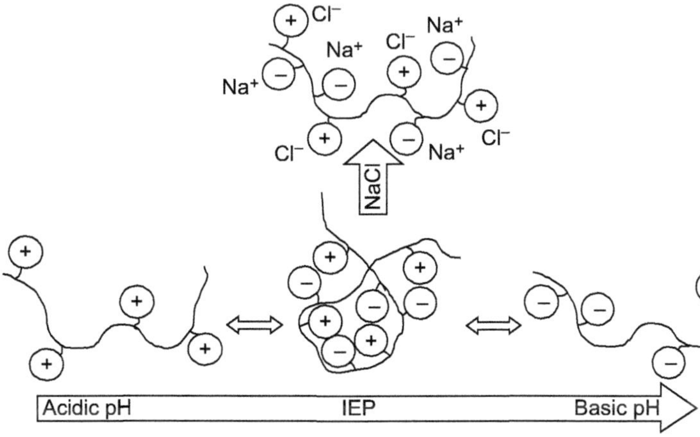

Figure 2.3 Schematic representation of the impact that the presence of salt ions or pH changes have on the conformation of zwitterionic polymers.

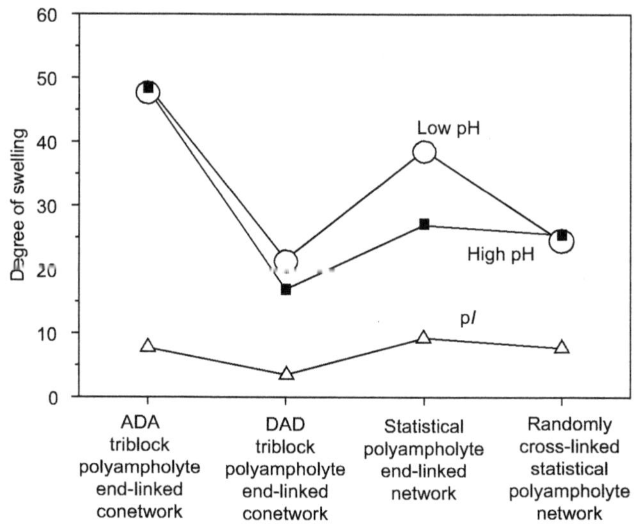

Figure 2.4 Swelling behaviors of polyampholyte copolymers as a function of the solution pH and their polymer architecture.
This figure is reprinted from Pafiti et al. (2011) with permission. Copyright 2011 American Chemical Society.

or to prevent specific interactions with molecules such as proteins and cells. In the case of polyelectrolytes, their charged state allows them to play roles as protein separation and segregation agents, protein solubility enhancers, and protein mimics (Kayitmazer, Seeman, Minsky, Dubin, & Xu, 2013). A similar role has also been seen for polyampholyte polymers that do not have charge balance. Alternatively, there has been an exponential growth of studies that have demonstrated that zwitterionic polymers with overall charge balance prevent nonspecific interactions with proteins, cells, and bacteria (Jiang & Cao, 2010; Zurick & Bernards, 2014). This has been demonstrated for all three of the primary betaine functional groups and for multiple polyampholyte copolymer compositions. Although the protein-resistant characteristics are not the primary focus of this chapter, many of the investigations into this phenomenon led to the discovery of some of the responsive properties of polyelectrolytes and zwitterionic polymer systems.

2.4 Stimuli-induced changes and their applications

In the biomedical community, materials that are responsive to external stimuli play an important role because of the varying conditions that are found throughout the *in vivo* environment. In this section, many of the responsive properties of polyelectrolytes and zwitterionic polymers will be discussed, and examples of applications for each responsive property will be provided if they have been demonstrated. If not, proposed applications are given. The subsections are ordered based on the relative level of activity in the scientific community to develop and utilize each stimulus for biomedical purposes.

2.4.1 pH-induced responsiveness

The most widely investigated stimulus for inducing changes in polyelectrolytes and zwitterionic polymer systems is the solution pH. The solution pH affects the protonation and deprotonation state of the charged groups for both polyelectrolytes and zwitterionic polymers, thereby affecting the overall charge state of the polymers. This effect is more prominent with weak cation and anion species, but it is also seen for the strong cation and anion groups under more extreme basic and acidic conditions, respectively.

A change in the charge state of the polymer affects the electrostatic interactions that occur between these polymeric systems and biomolecules (e.g., protein, bacteria, cells). For example, many zwitterionic polymer brush coatings have been demonstrated to have nonfouling properties when they have an overall neutral charge balance (Bernards, Cheng, Zhang, & Jiang, 2008; Ishihara et al., 1998; Ladd et al., 2008), but they lose these nonfouling properties when they become charged through changes in pH or structure (Bernards et al., 2008; Chang et al., 2009). These nonfouling properties have been attributed to the presence of a tightly bound hydration layer (He et al., 2008; Hoffman, 1999; Hower, He, Bernards, & Jiang, 2006; Hower et al., 2008; Zheng et al., 2005), which was recently shown by Yu, Yun-Fei, Huan-Lin, and Hui-Min (2010) to be effected by pH. As a result, it is possible to tailor the capture and release of biomolecules through pH adjustments (Leal Denis et al., 2008). This is a common approach for completing protein separations with polyelectrolyte coatings by taking advantage of the different charged states of individual proteins (Jia, Wildes, & Titmuss, 2011; Kayitmazer et al., 2013) and it has also recently been extended to a betaine polymer-based platform (Sundaram, Ella-Menye, Brault, Shao, & Jiang, 2014). A similar concept has been demonstrated for the capture and release of charged nanoparticles as well (Das, Mordoukhovski, & Kumacheva, 2008). More interestingly, this concept was applied to more complex biological entities by Mi, Bernards, Cheng, Yu, and Jiang (2010), who demonstrated a polyampholyte polymer brush coating with pH-dependent bacteria "catch and release" capabilities. The polyampholyte polymer was formed from an equimolar mixture of positively charged [2-(acryloyloxy) ethyl] trimethylammonium chloride (TMA) monomers and negatively charged 2-carboxyethyl acrylate (CAA) monomers. As shown in Figure 2.5(a), at neutral pH, these polyampholyte polymers were neutral and resistant to *Staphylococcus epidermidis* (*S. epidermidis*). However, at pH 4.5, these polyampholyte polymers were positively charged and they promoted the attachment of the negatively charged *S. epidermidis*. The most unique result was the demonstration of the release of adherent bacteria through a basic buffer detachment rinse. These results are shown in Figure 2.5(b). In a related extension of this application, Cao, Li, et al. (2013) and Cao, Tang, et al. (2013) recently synthesized a betaine-based monomer that switches from a neutral and nonfouling state to a cationic and antimicrobial state under acidic conditions. This system was shown to be capable of killing more than 99.99996% of the adherent *Escherichia coli* K12 and then releasing the dead bacteria through neutral or basic rinsing.

As discussed above, changes in the charge state of the polymer also result in changes in the conformation of the polymer. This responsive swelling behavior has been demonstrated in both hydrogel (Dogu, Kilic, & Okay, 2009; Xu et al., 2008) and

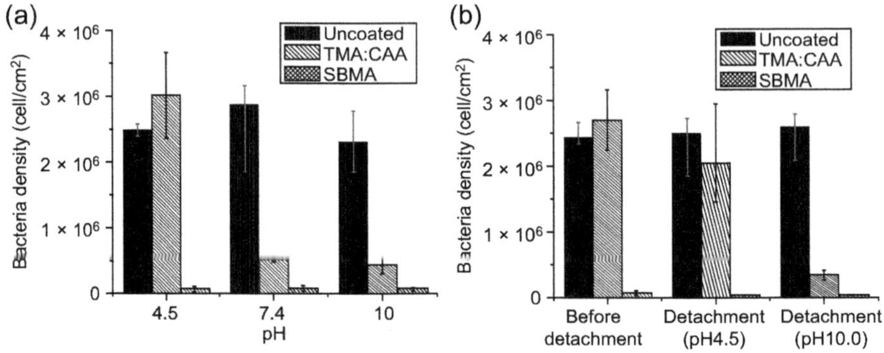

Figure 2.5 *Staphylococcus epidermidis* (a) attachment and (b) release from TMA:CAA polyampholyte copolymer brushes as a function of pH.
This figure is reprinted from Mi et al. (2010) with permission. Copyright 2010 Elsevier.

polymer brush systems (Jhon, Arifuzzaman, Ozcam, Kiserow, & Genzer, 2012; Sanjuan & Tran, 2008; Yu & Han, 2009), and it has been proposed to be beneficial for drug delivery applications. The basic concept is that polymers can be swelled under appropriate conditions to assist in drug loading by diffusion, and then the pH can be changed to shrink the polymers to control the release of the encapsulated drug molecules. A permutation of this approach is to shrink the polymer enough to prevent the escape of the drug until the delivery system has reached certain cellular microenvironments where the natural change in pH will result in swelling and a burst release of drug molecules. This approach lies heavily on polyampholyte polymer–based systems because the pH responsive properties are more easily tuned when a mixture of positively and negatively charged monomers are included in the final polymer formulation. To demonstrate this concept, Ng and Ng (2008) investigated the pH dependent release of three model drugs from a polymeric delivery system based on a poly(2-hydroxyethyl methacrylate) (pHEMA) core coated with a polyampholyte copolymer composed of negatively charged acrylic acid (AA) and positively charged 2-(diethylamino) ethyl methacrylate (DEAEMA) with a range of compositions. The model drugs were methylene blue, metanil yellow, and caffeine, which are cationic, anionic, and neutral, respectively, at neutral pH. Based on the drug release profiles as a function of pH and polyampholyte copolymer composition, the authors concluded that the drug release rate was controlled by electrostatic interactions between the drug and the polyampholyte copolymer coating. Therefore, it should be possible to tailor the composition of a polyampholyte copolymer drug delivery vehicle for specific environmental pH values. Others have probed the pH-dependent release of indomethacin (Mishra, Ramasamy, Ban, & Majeed, 2013), doxorubicin (Iatridi, Mattheolabakis, Avgoustakis, & Tsitsilianis, 2011), chitosan (Deng et al., 2009), and paclitaxel (Lee & Geckeler, 2012; Lee, Graeser, Kratz, & Geckeler, 2011), among others. One additional approach for mediating drug delivery from polyelectrolytes and zwitterionic polymer systems has been pursued, again using pH-responsive properties. However, in this approach, pH changes are used to mediate the self-assembly of micelles or micro-gel structures with encapsulated drugs (Agut, Brulet, Schatz, Taton, & Lecommandoux, 2010; Ekblad, Andersson, Tai, Ederth, & Liedberg, 2009; Iatridi & Tsitsilianis, 2011; Rao, Zhang, Zhang, & Liu, 2008; Xiong et al., 2011). In one such example, Popescu, Mourtas, Pampalakis, Antimisiaris, and Tsitsilianis (2011) demonstrated the pH-dependent formation of micelles composed of poly(2-vinylpyridine)-block-poly(acrylic acid)-block-poly(*n*-butyl methacrylate) copolymers and the release of calcein from these systems.

The final potential application for the pH-responsive properties of zwitterionic molecules is focused on a subset of the polymers that contain terminal carboxylic acid groups. This subset includes the CB-based family of betaines and polyampholyte copolymers containing CAA monomers or similar. It has been demonstrated that the pH-controlled protonation state of the carboxylic acid group can be used to covalently conjugate proteins within a polymer matrix (Zhang, Chen, & Jiang, 2006) for both chemistry platforms, using *N*-(3-dimethylaminopropyl)-*N'*-ethycarbodiimide hydrochloride/*N*-hydroxysuccinimide (EDC/NHS) bioconjugation chemistry (Hermanson, 2008). This allows the specific attachment of proteins or other signaling molecules within a nonfouling or protein-resistant background in CB polymer systems and polyampholyte copolymer systems with overall charge neutrality. This has been demonstrated in biosensor-based platforms with both chemistries, as a platform for highly sensitive and selective diagnostics (Tah & Bernards, 2012; Vaisocherová et al., 2008, 2009). It has also been demonstrated *in vitro* with hydrogel-based platforms (Schroeder et al., 2013; Zhang, Chen, & Jiang, 2006). These systems have been proposed as a smart biomaterial platform for inducing an improved wound-healing response in regenerative medicine applications, but they have not been widely investigated in the *in vivo* environment with attached signaling molecules (Zhang et al., 2013). In the final proposed application, the pH-mediated conjugation capacity has been used for encapsulating drug molecules, to improve their circulation time in the *in vivo* environment (Zhang et al., 2012). The initial studies in this application have been very promising and they represent an exciting new direction for these polymers. The key benefit in all of these applications is the unique dual functionality (nonfouling and protein conjugation) of the underlying polymer, which is not seen in most of the zwitterionic polymer systems.

2.4.2 Salt concentration induced responsiveness

The second most widely investigated environmental stimulus is salt. This stimulus includes the impacts of both salt concentration and composition and their impact on polyelectrolytes and zwitterionic polymers. In fact, many of the recent studies investigating the ability to control the properties of polyelectrolytes and zwitterionic polymers through pH have also included a component based on salt. However, it is much more challenging to take advantage of the responsiveness to salt for many biomedical applications, so most of the work has focused on one primary application.

In general, much of the biomedical related work is focused on using changes in the charged state of polyelectrolytes and zwitterionic polymer species for ion exchange chromatography applications. In this technique, biological molecules are passed through a column coated with either cationic or anionic species. The molecules are eluted from the column by passing through a gradient of salt concentrations. The salt species neutralize the electrostatic interactions between the molecules and the coating, at concentrations specific to each molecule. A schematic representation outlining this technique is shown in Figure 2.6. This technique has been used to separate DNA, proteins, and peptides, among other species. There is a wide array of literature reviewing this technique, so the reader is referred to additional sources for more information (Di Palma, Hennrich, Heck, & Mohammed, 2012; Ghanem, Healey, & Adly, 2013; Uppsala, 1986).

Additional applications utilizing the impact of salt concentration on polyelectrolytes and zwitterionic polymers for biomedical applications are generally based on taking advantage of the polyelectrolyte effect or the antipolyelectrolyte effect, which were discussed earlier. The ability to tailor the swelling behavior with salt concentration was proposed as a mechanism for controlling the self-assembly of a polyampholyte polymer–based drug delivery system, but actual drug delivery was not demonstrated (Yin, Li, Yin, Miao, & Jiang, 2009). Wang, Zhang, Wang, Sun, and Shen (2010) did demonstrate the impact of salt on the loading of two probe drug molecules in a multi-polymer system capped with a polyampholyte copolymer. In this investigation, the amount of methyl orange (negatively charged) and rhodamine 6G (positively charged) loaded into the drug carrier system was increased as the concentration of NaCl was increased in the loading buffer. This is likely due to the neutralization of the charged groups in the polyampholyte copolymer cap, which allowed better diffusion of the drug into the delivery system. A similar study utilizing the antipolyelectrolyte effect was completed with chitosan as a probe drug molecule, with similar results (Deng et al., 2009). However, additional studies of this type have not been heavily pursued because an identical response is also seen with pH, which has more relevance to the *in vivo* environment.

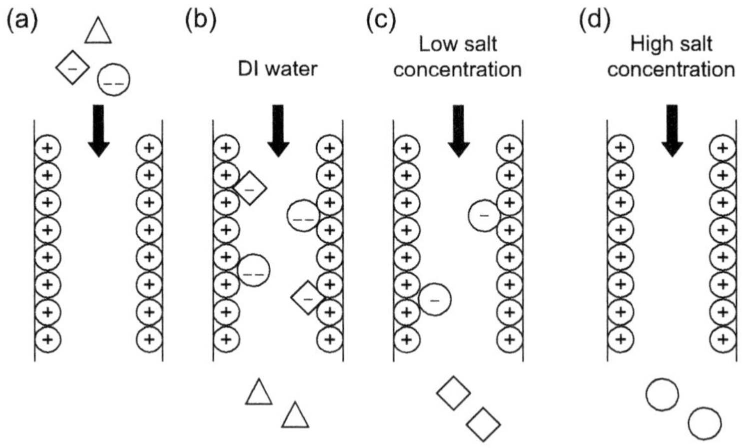

Figure 2.6 Schematic outlining the basic steps of ion exchange chromatography. (a) A cationic column is exposed to a mixture of charged and uncharged biomolecules, leading to (b) the purification of the uncharged molecules. (c) A change in the salt concentration of the buffer neutralizes some of the charged species, leading to their purification, followed by (d) additional changes in the salt concentration, resulting in the complete recovery of the biomolecules.

2.4.3 Temperature-induced responsiveness

The third environmental stimulus that has been used with polyelectrolytes and zwitterionic polymer systems is temperature. As with pH and salt, the primary temperature-induced response is a change in the conformational state of the polymer structure. The conformational state change is caused by one or more of the monomers passing through its lower critical solution temperature (LCST). This has been demonstrated in polyelectrolytes and zwitterionic polymers, where the charged monomers are responsible for the change, and in systems in which an additional temperature-sensitive monomer is included as a copolymer.

The most widely investigated temperature-responsive biomedical polymer is poly(*N*-isopropyl acrylamide) (pNIPAM). This polymer is the focus of Chapter 1 in this book, and therefore it will not be discussed in depth here. However, pNIPAM has been paired with polyampholyte copolymers and applied to nanoparticle separations (Das et al., 2008), drug delivery (Bradley, Liu, Keddie, Vincent, & Burnett, 2009; Bradley, Vincent, & Burnett, 2009), and tissue engineering applications (Xu et al., 2008). In a related system, Iatridi et al. (2011) also used the LCST-responsive properties of polyethylene glycol methacrylate (PEGMA) copolymerized with methacrylic acid and 2-(diethylamino) ethyl methacrylate in a temperature- and pH-sensitive doxorubicin drug delivery system. However, the primary focus of this study was to demonstrate the pH-dependent release properties as discussed earlier.

Temperature responsive polyampholyte copolymer systems without any additional uncharged monomers have also been demonstrated, although to a lesser extent. Studies of this type have generally included either methacrylic acid (MAA) monomers (Ho, Tan, Tan, & Tam, 2008; Mishra et al., 2013) or 2-(dimethylamino) ethyl methacrylate (DMAEMA) monomers (Agut et al., 2010; Xiong et al., 2011), which both have temperature-stimulated responsiveness. The MAA and DMAEMA monomers have been copolymerized into systems that have temperature-controlled swelling behaviors or temperature-induced self-assembly or aggregation. In one example, Mishra et al. (2013) demonstrated the temperature-induced swelling behavior of a copolymer composed of MAA and 3-(methacryloylamino) propyl trimethyl ammonomium chloride, and proposed it as a drug delivery vehicle. Although the temperature-induced responsive properties were demonstrated for this copolymer, all of the drug release results that were provided again relied on pH as the stimulus. A similar limitation has been seen in much of the literature regarding temperature-induced aggregation. There have been several proposed drug delivery systems, but no real demonstrated applications of the temperature-induced response (Agut et al., 2010; Xia, Gao, Chen, Jia, & Liang, 2011; Xiong et al., 2011).

2.4.4 Other stimuli-induced responsiveness

Other stimuli have also been used to control polyelectrolytes and zwitterionic polymer systems, but they are much less common in biomedical applications. These stimuli include solvent and electric current cues. Many of the responses that are induced by these stimuli are similar to those discussed above, but they have the drawback that they are primarily limited to potential *in vitro* applications such as biosensing, micelle formation, and separations. Therefore, a detailed description of the response to these environmental cues is not provided here, and the reader is referred to other sources for more information on these stimuli (Kudaibergenov, 2002).

2.5 Future trends

As outlined throughout this chapter, there is great potential for polyelectrolytes and zwitterionic polymers to be applied to needs in the biomedical community. Many of the advantages to these polymers lie in the ability to tune the polymer composition and architecture to take control of the responsive properties to meet the intended application. From this perspective, it is essential that scientists continue to develop controlled polymerization procedures to gain even further control over the architecture and exact composition of the polymer components, especially for polyampholyte copolymer systems. As the chemistry of these systems is improved, it will be equally important to develop cost-effective methods for applying these polymers to biomedical applications. This will allow further testing of these responsive properties in the *in vivo* environment as a feedback loop to improve the chemistry. In fact, the greatest opportunity for these polymers lies not in their general biomedical uses, which have been widely demonstrated, but rather in demonstrating the effective use of the responsive properties discussed in this chapter in the *in vivo* environment. Herein lies the biggest challenge and opportunity for the widespread adaptation of polyelectrolytes and zwitterionic polymers as responsive biomedical systems.

2.6 Sources of further information

There is a significant body of literature demonstrating the biomedical applications of polyelectrolytes and zwitterionic polymers. These texts provide excellent insight into the properties of these systems that make them desirable for biomedical applications and they discuss the performance of these polymers from broader perspective. For example, Kudaibergenov (2002) has an excellent book discussing polyampholyte copolymers and their applications across many fields of interest. McCormick (2000) has done extensive work with these polymeric systems, and he has edited a book and prepared a review discussing many aspects (Lowe & McCormick, 2002). Similarly, there have been a number of extensive review articles prepared that are related to polyelectrolytes, polyampholytes, and betaine polymers (Jiang & Cao, 2010; Kayitmazer et al., 2013; Kudaibergenov, Jaeger, & Laschewsky, 2006; Singh et al., 2007; Zurick & Bernards, 2014). Finally, when one is envisioning potential applications for these responsive properties, it will be important to plan for the integration of the polymers into the final biomedical device. Albeit not specific to the polymers that are the focus of this chapter, there are many excellent reviews in the general areas of hydrogels (Drury & Mooney, 2003; Lee & Mooney, 2001; Peppas et al., 2006) and polymer brushes (Barbey et al., 2009; Dyer, 2006; Hook et al., 2009; Orski et al., 2011; Xu et al., 2009) for biomedical applications, along with reviews of polymerization techniques (Barbey et al., 2009; Dyer, 2006; Kudaibergenov et al., 2006; Lowe & McCormick, 2002; Matyjaszewski & Xia, 2001).

References

Agut, W., Brulet, A., Schatz, C., Taton, D., & Lecommandoux, S. (2010). pH and temperature responsive polymeric micelles and polymersomes by self-assembly of poly(2-(dimethylamino)ethyl methacrylate)-b-poly(glutamic acid) double hydrophilic block copolymers. *Langmuir, 26*, 10546–10554.

Alfrey, T., Morawetz, H., Fitzgerald, E. B., & Fuoss, R. M. (1950). Synthetic electrical analog of proteins. *Journal of the American Chemical Society, 72*, 1864.

An, H., Lu, C., Wand, P., Li, W., Tan, Y., Xu, K., et al. (2011). A novel hydrophobically associating polyampholytes of poly(AM/AA/AMQC12): preparation, characterization, and solution properties. *Polymer Bulletin, 67*, 141–158.

Azzaroni, O., Brown, A. A., & Huck, W. T. S. (2006). UCST wetting transitions of polyzwitterionic brushes driven by self-association. *Angewandte Chemie International Edition, 45*, 1770–1774.

Barbey, R. L., Lavanant, L., Paripovic, D., Schüwer, N., Sugnaux, C., Tugulu, S., et al. (2009). Polymer brushes via surface-initiated controlled radical polymerization: synthesis, characterization, properties, and applications. *Chemical Reviews, 109*, 5437–5527.

Bernards, M. T., Cheng, G., Zhang, Z., & Jiang, S. (2008). Mixed charge, overall neutral nonfouling polymer brushes formed by surface-initiated atom transfer radical polymerization. *Macromolecules, 41*, 4216–4219.

Bradley, M., Liu, D., Keddie, J. L., Vincent, B., & Burnett, G. (2009). The uptake and release of cationic surfactant from polyampholyte microgel particles in dispersion and as an adsorbed monolayer. *Langmuir, 25*, 9677–9683.

Bradley, M., Vincent, B., & Burnett, G. (2009). Uptake and release of surfactants from polyampholyte microgel particles. *Colloid and Polymer Science, 287*, 345–350.

Cao, B., Li, L., Tang, Q., & Cheng, G. (2013). The impact of structure on elasticity, switchability, stability and functionality of an all-in-one carboxybetaine elastomer. *Biomaterials, 34*, 7592–7600.

Cao, Z., Mi, L., Mendiola, J., Ella-Menye, J.-R., Zhang, L., Xue, H., et al. (2012). Reversibly switching the function of a surface between attacking and defending against bacteria. *Angewandte Chemie International Edition, 51*, 2602–2605.

Cao, B., Tang, Q., Li, L., Humble, J., Wu, H., Liu, L., et al. (2013). Switchable antimicrobial and antifouling hydrogels with enhanced mechanical properties. *Advanced Healthcare Materials, 2*, 1096–1102.

Carr, L., Cheng, G., Xue, H., & Jiang, S. (2010). Engineering the polymer backbone to strengthen nonfouling sulfobetaine hydrogels. *Langmuir, 26*, 14793–14798.

Chang, Y., Shu, S.-H., Shih, Y.-J., Chu, C.-W., Ruaan, R.-C., & Chen, W.-Y. (2009). Hemocompatible mixed-charge copolymer brushes of pseudozwitterionic surfaces resistant to nonspecific plasma protein fouling. *Langmuir, 26*, 3522–3530.

Cheng, G., Xue, H., Zhang, Z., Chen, S., & Jiang, S. (2008). A switchable biocompatible polymer surface with self-sterilizing and nonfouling capabilities. *Angewandte Chemie, 120*, 8963–8966.

Cheng, G., Zhang, Z., Chen, S., Bryers, J. D., & Jiang, S. (2007). Inhibition of bacterial adhesion and biofilm formation on zwitterionic surfaces. *Biomaterials, 28*, 4192–4199.

Chen, S., & Jiang, S. (2008). A new avenue to nonfouling materials. *Advanced Materials, 20*, 335–338.

Das, M., Mordoukhovski, L., & Kumacheva, E. (2008). Sequestering gold nanorods by polymer microgels. *Advanced Materials, 20*, 2371–2375.

Deng, L., Zhai, Y., Guo, S., Jin, F., Xie, Z., He, X., et al. (2009). Investigation on properties of P((MAA-*co*-DMAEMA)-g-EG) polyampholyte nanogels. *Journal of Nanoparticle Research, 11*, 365–374.

Di Palma, S., Hennrich, M. L., Heck, A. J. R., & Mohammed, S. (2012). Recent advances in peptide separation by multidimensional liquid chromatography for proteome analysis. *Journal of Proteomics, 75*, 3791–3813.

Dobbins, S., McGrath, D. E., & Bernards, M. T. (2012). Nonfouling hydrogels formed from charged monomer subunits. *Journal of Physical Chemistry B, 49*, 14346–14352.

Dogu, S., Kilic, M., & Okay, O. (2009). Collapse of acrylamide-based polyampholyte hydrogels in water. *Journal of Applied Polymer Science, 113*, 1375–1382.

Drury, J. L., & Mooney, D. J. (2003). Hydrogels for tissue engineering: scaffold design variables and applications. *Biomaterials, 24*, 4337–4351.

Dyer, D. J. (2006). Photoinitiated synthesis of grafted polymers. *Advances in Polymer Science, 197*, 47–65.

Ekblad, T., Andersson, O., Tai, F.-I., Ederth, T., & Liedberg, B. (2009). Lateral control of protein adsorption on charged polymer gradients. *Langmuir, 25*, 3755–3762.

Feng, W., Brash, J., & Zhu, S. (2004). Atom-transfer radical grafting polymerization of 2-methacryloyloxyethyl phosphorylcholine from silicon wafer surfaces. *Journal of Polymer Science Part A: Polymer Chemistry, 42*, 2931–2942.

Feng, W., Brash, J. L., & Zhu, S. (2006). Non-biofouling materials prepared by atom transfer radical polymerization grafting of 2-methacryloloxyethyl phosphorylcholine: separate effects of graft density and chain length on protein repulsion. *Biomaterials, 27*, 847–855.

Feng, W., Zhu, S., Ishihara, K., & Brash, J. L. (2005). Adsorption of fibrinogen and lysozyme on silicon grafted with poly(2-methacryloyloxyethyl phosphorylcholine) via surface-initiated atom transfer radical polymerization. *Langmuir, 21*, 5980–5987.

Ghanem, A., Healey, R., & Adly, F. G. (2013). Current trends in separation of plasmid DNA vaccines: a review. *Analytica Chimica Acta, 760*, 1–15.

He, Y., Hower, J., Chen, S., Bernards, M. T., Chang, Y., & Jiang, S. (2008). Molecular simulation studies of protein interactions with zwitterionic phosphorylcholine self-assembled monolayers in the presence of water. *Langmuir, 24*, 10358–10364.

Hermanson, G. T. (Ed.). (2008). *Bioconjugate techniques*. Amsterdam: Academic Press, Inc.

Hirota, K., Murakami, K., Nemoto, K., & Miyake, Y. (2005). Coating of a surface with 2-methacryloyloxyethyl phosphorylcholine (MPC) co-polymer significantly reduces retention of human pathogenic microorganisms. *FEMS Microbiology Letters, 248*, 37–45.

Ho, B. S., Tan, B. H., Tan, J. P. K., & Tam, K. C. (2008). Inverse microemulsion polymerization of sterically stabilized polyampholyte microgels. *Langmuir, 24*, 7698–7703.

Hoffman, A. S. (1999). Non-fouling surface technologies. *Journal of Biomaterials Science–Polymer Edition, 10*, 1011.

Hook, A. L., Voelcker, N. H., & Thissen, H. (2009). Patterned and switchable surfaces for biomolecular manipulation. *Acta Biomaterialia, 5*, 2350–2370.

Hower, J. C., Bernards, M. T., Chen, S., Tsao, H.-K., Sheng, Y.-J., & Jiang, S. (2008). Hydration of "Nonfouling" functional groups. *The Journal of Physical Chemistry B, 113*, 197–201.

Hower, J. C., He, Y., Bernards, M. T., & Jiang, S. (2006). Understanding the nonfouling mechanism of surfaces through molecular simulations of sugar-based self-assembled monolayers. *Journal of Chemical Physics*, *125*, 214704.

Huang, C.-J., Li, Y., Krause, J. B., Brault, N. D., & Jiang, S. (2012). Internal architecture of zwitterionic polymer brushes regulates nonfouling properties. *Macromolecular Rapid Communications*, *33*, 1003–1007.

Hucknall, A., Rangarajan, S., & Chilkoti, A. (2009). Pursuit of zero: polymer brushes that resist the adsorption of proteins. *Advanced Materials*, *21*, 2441–2446.

Iatridi, Z., Mattheolabakis, G., Avgoustakis, K., & Tsitsilianis, C. (2011). Self-assembly and drug delivery studies of pH/thermo-sensitive polyampholytic (A-*co*-B)-b-C-b-(A-*co*-B) segmented terpolymers. *Soft Matter*, *7*, 11160–11168.

Iatridi, Z., & Tsitsilianis, C. (2011). pH responsive self assemblies from an An-core-(B-b-C)n heteroarm star block terpolymer bearing oppositely charged segments. *Chemical Communications*, *47*, 5560–5562.

Ishihara, K., Nomura, H., Mihara, T., Kurita, K., Iwasaki, Y., & Nakabayashi, N. (1998). Why do phospholipid polymers reduce protein adsorption? *Journal of Biomedical Materials Research*, *39*, 323–330.

Jhon, Y. K., Arifuzzaman, S., Ozcam, A. E., Kiserow, D. J., & Genzer, J. (2012). Formation of polyampholyte brushes via controlled radical polymerization and their assembly in solution. *Langmuir*, *28*, 872–882.

Jia, H., Wildes, A., & Titmuss, S. (2011). Structure of pH-responsive polymer brushes grown at the gold-water interface: dependence on grafting density and temperature. *Macromolecules*, *45*, 305–312.

Jiang, S., & Cao, Z. (2010). Ultralow-fouling, functionalizable, and hydrolyzable zwitterionic materials and their derivatives for biological applications. *Advanced Materials*, *22*, 920–932.

Kayitmazer, A. B., Seeman, D., Minsky, B. B., Dubin, P. L., & Xu, Y. (2013). Protein-polyelectrolyte interactions. *Soft Matter*, *9*, 2553–2583.

Kudaibergenov, S. E. (2002). *Polyampholytes: synthesis, characterization, and application*. New York: Kluwer Academic/Plenum Publishers.

Kudaibergenov, S. E., Jaeger, W., & Laschewsky, A. (2006). Polymeric betaines: synthesis, characterization, and application. *Advances in Polymer Science*, *201*, 157–224.

Ladd, J., Zhang, Z., Chen, S., Hower, J. C., & Jiang, S. (2008). Zwitterionic polymers exhibiting high resistance to nonspecific protein adsorption from human serum and plasma. *Biomacromolecules*, *9*, 1357–1361.

Leal Denis, M. F., Carballo, R. R., Spiaggi, A. J., Dabas, P. C., Campo Dall' Orto, V., Martínez, J. M. L., et al. (2008). Synthesis and sorption properties of a polyampholyte. *Reactive and Functional Polymers*, *68*, 169–181.

Lee, Y., & Geckeler, K. E. (2012). Cellular interactions of a water-soluble supramolecular polymer complex of carbon nanotubes with human epithelial colorectal adenocarcinoma cells. *Macromolecular Bioscience*, *12*, 1060–1067.

Lee, Y., Graeser, R., Kratz, F., & Geckeler, K. E. (2011). Paclitaxel-loaded polymer nanoparticles for the reversal of multidrug resistance in breast Cancer cells. *Advanced Functional Materials*, *21*, 4211–4218.

Lee, K. Y., & Mooney, D. J. (2001). Hydrogels for tissue engineering. *Chemical Reviews*, *101*, 1869–1879.

Li, G., Xue, H., Gao, C., Zhang, F., & Jiang, S. (2010). Nonfouling polyampholytes from an ion-pair comonomer with biomimetic adhesive groups. *Macromolecules*, *43*, 14–16.

Lowe, A. B., & McCormick, C. L. (2002). Synthesis and solution properties of zwitterionic polymers. *Chemical Reviews*, *102*, 4177–4190.

Matyjaszewski, K., & Xia, J. (2001). Atom transfer radical polymerization. *Chemical Reviews*, *101*, 2921–2990.

McCormick, C. L. (Ed.). (2000). *Stimuli-responisive water soluble and amphilphilic polymers.* Washington, DC: American Chemical Society.

Mi, L., Bernards, M. T., Cheng, G., Yu, Q., & Jiang, S. (2010). pH responsive properties of non-fouling mixed-charge polymer brushes based on quaternary amine and carboxylic acid monomers. *Biomaterials*, *31*, 2919–2925.

Mishra, R. K., Ramasamy, K., Ban, N. N., & Majeed, A. B. A. (2013). Synthesis of poly(3-(methacryloylamino) propyl trimethylammonium chloride-*co*-methacrylic acid) copolymer hydrogels for controlled indomethacin delivery. *Journal of Applied Polymer Science*, *128*, 3365–3374.

Ng, L.-T., & Ng, K.-S. (2008). Photo-cured pH-responsive polyampholyte-coated membranes for controlled release of drugs with different molecular weights and charges. *Radiation Physics and Chemistry*, *77*, 192–199.

Orski, S. V., Fries, K. H., Sontag, S. K., & Locklin, J. (2011). Fabrication of nanostructures using polymer brushes. *Journal of Materials Chemistry*, *21*, 14135–14149.

Pafiti, K. S., Philippou, Z., Loizou, E., Porcar, L., & Patrickios, C. S. (2011). End-linked poly(2-(dimethylamino)ethyl methacrylate)-poly(methacrylic acid) polyampholyte conetworks: synthesis by sequential RAFT polymerization and swelling and SANS characterization. *Macromolecules*, *44*, 5352–5362.

Peppas, N. A., Hilt, J. Z., Khademhosseini, A., & Langer, R. (2006). Hydrogels in biology and medicine: from molecular principles to bionanotechnology. *Advanced Materials*, *18*, 1345–1360.

Popescu, M.-T., Mourtas, S., Pampalakis, G., Antimisiaris, S. G., & Tsitsilianis, C. (2011). Ph-responsive hydrogel/liposome soft nanocomposites for tuning drug release. *Biomacromolecules*, *12*, 3023–3030.

Rao, J., Zhang, Y., Zhang, J., & Liu, S. (2008). Facile preparation of well-defined AB2 Y-shaped Miktoarm star polypeptide copolymer via the combination of ring-opening polymerization and click chemistry. *Biomacromolecules*, *9*, 2586–2593.

Sanjuan, S., & Tran, Y. (2008). Stimuli-responsive interfaces using random polyampholyte brushes. *Macromolecules*, *41*, 8721–8728.

Schroeder, M. E., Zurick, K. M., McGrath, D. E., & Bernards, M. T. (2013). Multifunctional polyampholyte hydrogels with fouling resistance and protein conjugation capacity. *Biomacromolecules*, *14*, 3112–3122.

Shao, Q., & Jiang, S. (2013). Effect of carbon spacer length on zwitterionic carboxybetaines. *The Journal of Physical Chemistry B*, *117*, 1357–1366.

Singh, P. K., Singh, V. K., & Singh, M. (2007). Zwitterionic polyelectrolytes: a review. *e-Polymers*, *7*, 30.

Sundaram, H. S., Ella-Menye, J.-R., Brault, N. D., Shao, Q., & Jiang, S. (2014). Reversibly switchable polymer with cationic/zwitterionic/anionic behavior through synergistic protonation and deprotonation. *Chemical Science*, *5*, 200–205.

Tah, T., & Bernards, M. T. (2012). Nonfouling polyampholyte polymer brushes with protein conjugation capacity. *Colloids and Surfaces B: Biointerfaces*, *93*, 195–201.

Uppsala. (1986). *Ion exchange chromatography: principles and methods.* Uppsala: Pharmacia Fine Chemicals.

Vaisocherová, H., Yang, W., Zhang, Z., Cao, Z., Cheng, G., Piliarik, M., et al. (2008). Ultralow fouling and functionalizable surface chemistry based on a zwitterionic polymer enabling sensitive and specific protein detection in undiluted blood plasma. *Analytical Chemistry*, *80*, 7894–7901.

Vaisocherová, H., Zhang, Z., Yang, W., Cao, Z., Cheng, G., Taylor, A. D., et al. (2009). Functionalizable surface platform with reduced nonspecific protein adsorption from full blood plasma–material selection and protein immobilization optimization. *Biosensors and Bioelectronics*, *24*, 1924–1930.

Waite, J. H., & Tanzer, M. L. (1981). Polyphenolic substance of *Mytilus edulis*: novel adhesive containing L-Dopa and hydroxyproline. *Science*, *212*, 1038–1040.

Wang, X., Zhang, L., Wang, L., Sun, J., & Shen, J. (2010). Layer-by-layer assembled polyampholyte microgel films for simultaneous release of anionic and cationic molecules. *Langmuir*, *26*, 8187–8194.

Xia, Y., Gao, M., Chen, Y., Jia, X., & Liang, D. (2011). Mimic of protein: a highly pH-sensitive and thermoresponsive polyampholyte. *Macromolecular Chemistry and Physics*, *212*, 2268–2274.

Xiao, H., Tao, R., Cui, W., Zhang, S., & Li, R. H. (2013). Solution properties of a novel ampholytic polyphenylene sulfide. *Journal of Applied Polymer Science*, *127*, 4052–4060.

Xiong, Z., Peng, B., Han, X., Peng, C., Liu, H., & Hu, Y. (2011). Dual-stimuli responsive behaviors of diblock polyampholyte PDMAEMA-b-PAA in aqueous solution. *Journal of Colloid and Interface Science*, *356*, 557–565.

Xu, F. J., Neoh, K. G., & Kang, E. T. (2009). Bioactive surfaces and biomaterials via atom transfer radical polymerization. *Progress in Polymer Science*, *34*, 719–761.

Xu, K., Wang, J., Chen, Q., Yue, Y., Zhang, W., & Wang, P. (2008). Spontaneous volume transition of polyampholyte nanocomposite hydrogels based on pure electrostatic interaction. *Journal of Colloid and Interface Science*, *321*, 272–278.

Yang, W., Bai, T., Carr, L. R., Keefe, A. J., Xu, J., Xue, H., et al. (2012). The effect of lightly crosslinked poly(carboxybetaine) hydrogel coating on the performance of sensors in whole blood. *Biomaterials*, *33*, 7945–7951.

Yang, W., Chen, S., Cheng, G., Vaisocherová, H., Xue, H., Li, W., et al. (2008). Film thickness dependence of protein adsorption from blood serum and plasma onto poly(sulfobetaine)-grafted surfaces. *Langmuir*, *24*, 9211–9214.

Yang, W., Xue, H., LI, W., Zhang, J., & Jiang, S. (2009). Pursuing "Zero" protein adsorption of poly(carboxybetaine) from undiluted blood serum and plasma. *Langmuir*, *25*, 11911–11916.

Yin, Q., Li, Y., Yin, Q.-J., Miao, X., & Jiang, B. (2009). Synthesis and rheological behavior of a novel *N*-sulfonate ampholyte chitosan. *Journal of Applied Polymer Science*, *113*, 3382–3387.

Yu, K., & Han, Y. (2009). Effect of block sequence and block length on the stimuli-responsive behavior of polyampholyte brushes: hydrogen bonding and electrostatic interaction as the driving force for surface rearrangement. *Soft Matter*, *5*, 759–768.

Yu, C., Yun-Fei, L., Huan-Lin, T., & Hui-Min, T. (2010). Study of carboxymethyl chitosan based polyampholyte superabsorbent polymer I: optimization of synthesis conditions and pH sensitive property study of carboxymethyl chitosan-g-poly(acrylic acid-*co*-dimethyldiallylammonium chloride) superabsorbent polymer. *Carbohydrate Polymers*, *81*, 365–371.

Zhang, L., Cao, Z., Bai, T., Carr, L., Ell-Menye, J.-R., Irvin, C., et al. (2013). Zwitterionic hydrogels implanted in mice resist the foreign-body reaction. *Nature Biotechnology*, *31*, 553–556.

Zhang, L., Cao, Z., Li, Y., Ella-Menye, J.-R., Bai, T., & Jiang, S. (2012). Softer zwitterionic nanogels for longer circulation and lower splenic accumulation. *ACS Nano*, *6*, 6681–6686.

Zhang, Z., Chen, S., Chang, Y., & Jiang, S. (2006). Surface grafted sulfobetaine polymers via atom transfer radical polymerization as superlow fouling coatings. *Journal of Physical Chemistry B*, *110*, 10799–10804.

Zhang, Z., Chen, S., & Jiang, S. (2006). Dual-functional biomimetic materials: nonfouling poly(carboxybetaine) with active functional groups for protein immobilization. *Biomacromolecules*, *7*, 3311–3315.

Zhang, Z., Vaisocherová, H., Cheng, G., Yang, W., Xue, H., & Jiang, S. (2008). Nonfouling behavior of polycarboxybetaine-grafted surfaces: structural and environmental effects. *Biomacromolecules*, *9*, 2686–2692.

Zheng, J., Li, L., Tsao, H. K., Sheng, Y. J., Chen, S., & Jiang, S. (2005). Strong repulsive forces between protein and oligo (ethylene glycol) self-assembled monolayers: a molecular simulation study. *Biophysical Journal*, *89*, 158–167.

Zurick, K. M., & Bernards, M. T. (2014). Recent biomedical advances with polyampholyte polymers. *Journal of Applied Polymer Science*, *131*, 40069.

Peptide-based switchable and responsive surfaces

3

M. Zelzer[1,2]
[1]University of Nottingham, Nottingham, UK; [2]National Physical Laboratory, Middlesex, UK

3.1 Introduction

The biological role of stimulus-responsive materials in general and peptide materials in particular are presented first, before discussing the importance of responsive interfaces and surfaces in biology.

3.1.1 The importance of stimulus responsiveness in biology

Responsiveness to changes in the immediate environment is an integral part of biological systems. The effect of these responses can, in some instances, be observed on a macroscopic level, such as the ability of cuttlefish to adjust the color and patterns of their body surface to their surroundings via a neurally controlled, visually driven chromatophore apparatus (Yoshida & Lahann, 2008). On a molecular level, stimuli responsiveness of biological materials in living systems is essential for the regulation of a large number of processes. Two prominent examples are the regulation of enzyme activity and biological self-assembly processes. The activity of enzymes, one of the core mechanisms by which biological processes are regulated, can be adjusted by changes in the biological environment. Many enzymes therefore display responsiveness to changes in environmental conditions such as pH, temperature, or the presence or absence of other regulatory biomolecules (Ulijn, 2006; Zelzer, Todd, Hirst, McDonald, & Ulijn, 2013). The self-assembly of small (e.g., peptides, lipids) and large (e.g., proteins) molecules is another important mechanism that is subject to tight control through the biological environment (Bromley, Channon, Moutevelis, & Woolfson, 2008; Murtas, 2013; Zelzer & Ulijn, 2010). Factors such as salt concentration, pH, and temperature can play a significant role in the properties of the final self-assembled material or in the ability of the material to self-assemble.

The propensity of biological systems to compartmentalise functionalities creates boundaries through which communication and interaction has to take place to allow the system to function as a single unit. This can, for example, be illustrated by the confinement of cellular functions by the cell membrane, which not only acts as a confinement barrier but also represents a mechanism through which the cell interacts with the extracellular environment (Escriba et al., 2008). Interfaces and their response to external stimuli therefore play a crucial role in biological systems, including aspects such as cell signaling (Sun, Qing, Su, & Jiang, 2011). Interfacial interactions of cell membranes are not restricted to liquid–liquid interfaces; they may also take place at

Switchable and Responsive Surfaces and Materials for Biomedical Applications. http://dx.doi.org/10.1016/B978-0-85709-713-2.00003-1

solid–liquid interfaces. Certain cells are adhesion dependent, that is, the development of their cellular functions depends on the formation of cellular adhesions with a solid surface. These interactions are mediated by integrins, self-assembled protein complexes that are formed between the solid substrate surfaces and the cell membrane (Giancotti & Ruoslahti, 1999; Stevens & George, 2005). Integrin formation is also a process that is induced by the availability of external stimuli, and thus the ability of a cell to generate cell adhesions and the nature of these adhesions depend on the stimuli (i.e., the chemical functionalities and other properties) presented by the surface (Geiger, Spatz, & Bershadsky, 2009).

3.1.2 Peptides as natural stimulus-responsive materials

Proteins are widely used in biology as components of stimulus-responsive systems (Lowik, Leunissen, van den Heuvel, Hansen, & van Hest, 2010). Amino acids, the building blocks of proteins, are structurally and functionally highly versatile molecules (Zelzer & Ulijn, 2010). They have previously been termed "nature's expression language" because of their ability to form proteins that are responsive to stimuli with high specificity, such as ion channels in cell membranes or light-harvesting proteins. Moreover, proteins (i.e., enzymes) can perform highly specific functions themselves that are often controlled by the presence of cofactors or activators and inhibitors. The action of the stimulus on a protein typically results in conformational changes of the protein structure, thus altering the delicate balance of intramolecular interactions within the protein and hence affecting the protein's functionality (Lowik et al., 2010).

The range of stimuli exploited by nature to affect material changes is extremely broad and encompasses many of the stimuli found in man-made systems such as light, pH, ion concentrations, temperature, electrical potential, and the presence of specific biomolecules (Lowik et al., 2010; Mart, Osborne, Stevens, & Ulijn, 2006). Table 3.1 provides a selection of stimuli and corresponding natural responsive materials.

3.1.3 Responsive peptides on surfaces

The desire to improve the interaction between man-made biomaterials and living biological systems, in terms of both understanding and controlling these interactions, has caused biomaterials to evolve from a passive to a responsive component that actively engages with its biological surrounding (Bryers, Giachelli, & Ratner, 2012). It is generally accepted that the surface of a material is crucial in determining the response and interaction between the man-made material and the living system, as it is the first point of contact between these two components (Castner & Ratner, 2002; Roach, Eglin, Rohde, & Perry, 2007). Consequently, tremendous effort has been placed on designing and investigating solid surfaces for the material–biology interface.

Initial efforts in the design of biomaterial surfaces focused on the decoration of surfaces with functionalities that promote a predetermined biological response, such as modulating the adhesion of bacteria or cells, controlling the functionality of cells, or binding specific biomolecules for quantitative and/or qualitative analysis

Table 3.1 Natural peptide (protein)–based stimuli-responsive materials

Natural peptide-based material	Stimulus	Response	Function
Ion channels and receptors	Electrochemical potential Small biomolecules (e.g., acetylcholine, glutamate) Temperature pH	Conformational change	Control of ion transport across the cell membrane
Light-harvesting complex	Light	Generation of excited electrons	Part of the photosynthetic machinery
Enzymes	Cofactors (e.g., flavine adenine dinucleotide (FAD), adenosine triphosphate (ATP), ions) Enzymes	Activation/deactivation of the enzyme	Biocatalysis

(Bryers et al., 2012). The range of functionalities used to decorate these surfaces is indeed extremely large and includes peptides, proteins, DNA, and carbohydrates as well as artificial molecules and polymers. A considerable number of reviews are available on this topic (e.g., Stevens & George, 2005; von der Mark, Park, Bauer, & Schmuki, 2010; Bacakova, Filova, Parizek, Ruml, & Svorcik, 2011), and biomolecules other than peptides and proteins will therefore not be discussed in this chapter.

As a consequence of advancements in the area of stimuli-responsive materials and surfaces, increasingly efforts focus on designing stimuli-responsive biomaterial surfaces, that is, man-made surfaces that actively interact with their biological environment to perform a specific function (Cole, Voelcker, Thissen, & Griesser, 2009; Mendes, 2008; Sun et al., 2011; Wischerhoff, Badi, Lutz, & Laschewsky, 2010). The term "bioactive surfaces" is increasingly used in this context. A recent review defines bioactive surfaces as "intelligent materials capable of performing sophisticated interactions (i.e. reversible, selective or specific) with biological compounds" (Wischerhoff et al., 2010).

The advent of stimuli-responsive surfaces in the biomaterials community has also attracted attention to the question as to whether responsive surfaces that are based on natural stimuli-responsive materials such as peptides could be designed. Compared to artificial stimuli-responsive materials such as polymers, peptide-based stimuli-responsive surfaces are expected to have better biocompatibility, to integrate more seamlessly with existing biological interactions and processes, to have improved interaction characteristics such as specificity and selectivity, and to display high versatility (Table 3.2). Although these advantages are significant, biomolecule-based surfaces also pose challenges such

Table 3.2 Advantages and disadvantages of peptide-based stimuli-responsive materials

Advantages	Disadvantages
Biocompatible	More costly fabrication than some synthetic materials
Based on the same materials used for naturally occurring response and control mechanisms	Complex interaction mechanisms with other biological components
Versatile building blocks	Often require more stringent operating conditions than other artificial materials
Interactions can be very specific and/or selective	

as the cost associated with the use of complex proteins, the complexity of the biological interactions, and the requirement for specific operating conditions to prevent denaturation or degradation of the biomolecules. Because of this, it is perhaps not surprising that many stimuli-responsive surfaces are not based on proteins directly—one of nature's original stimuli-responsive materials—but on shorter peptide sequences that can be prepared in a well-controlled manner at reasonable cost (Lowik et al., 2010).

Because of the advantages laid out above, an increasing amount of work has focused on the development of peptide-based stimuli-responsive materials. Most of these efforts are based on soluble or bulk materials (Lowik et al., 2010; Mart et al., 2006); the translation of responsive peptides to a surface has only recently begun to be explored in more detail. As with other materials, the attachment or fabrication of the stimuli-responsive peptide on a surface poses several additional challenges in terms of preparation, optimisation, and characterisation. In the following sections, the different types of responsive peptide surfaces prepared thus far will be introduced. The responsive peptide surfaces will be discussed in two main groups. The first group are surfaces in which the peptide itself reacts to a stimulus and changes its property in response to that stimulus. These are considered here to be genuine stimuli-responsive peptide surfaces. A large body of literature is also available on surfaces in which a non–peptide-based responsive material governs the availability of a peptide sequence on the surface. Here, responsiveness does not stem from the peptide itself but is built in the surface through the second component. Finally, possible applications of these materials and current challenges will be discussed.

3.2 Preparation of peptide surfaces

When preparing responsive peptide surfaces, a number of points need to be considered to ensure functionality and compatibility with the response mechanism and the application. These considerations are laid out first before introducing the various fabrication strategies that have been used to prepare responsive peptide surfaces.

3.2.1 General considerations

Preparation strategies for peptide surfaces have to meet several general criteria. First, the immobilisation of the peptide-based material to the surface has to be stable under the conditions used for the intended application. Second, the responsive material has to retain its function when immobilised, and the immobilisation procedure must be nondestructive to the peptides. It should be noted that the responsiveness of the peptide material may also be directly related to the surface attachment, for example, if the response is provided by a change in orientation of the peptide in relation to the substrate surface (Yeung et al., 2010), or by electrically interfacing with the surface (Yasutomi, Morita, Imanishi, & Kimura, 2004). In these instances, the peptide surface constitutes a whole new functional material that would not exist if the peptide were merely used in solution. Finally, it is desirable and sometimes necessary to control the orientation and surface density of the immobilised peptide to obtain functional and reproducible responsive materials (Yeung et al., 2010).

Most immobilisation strategies used to date are based on self-assembled monolayers (SAMs). These are predominantly prepared from thiols on gold or via silanisation of glass surfaces. If the peptide is attached to a polymer, film deposition methods may also be used to generate the peptide surface. The general strategies used thus far for the preparation of peptide-based responsive materials are explained in the following sections and illustrated in Figures 3.1–3.3.

3.2.2 Gold-based SAMs

Gold surfaces are convenient for the preparation of peptide-based surfaces (Figure 3.1) for two reasons: (1) they can act as electrodes for electrochemically responsive peptide materials and (2) the availability of an essential amino acid that carries a thiol side group (cysteine) allows peptide attachment without the need to incorporate other synthetic moieties. Incorporation of cysteine units in artificially designed responsive peptides is indeed a common way to covalently immobilise the peptide to the surface (Wang, Shen, Cao, & Li, 2010; Yeung et al., 2010) even if the material is not designed to be electrochemically responsive (Figure 3.1(a)) (Minelli, Liew, Muthu, & Andresen, 2013).

Thiols have also been artificially introduced at the end of peptide sequences, with the advantage that the orientation of the peptide on the surface is not restricted by the geometry of cysteine. Furthermore, the use of disulfide linkers provides stronger adhesion forces per molecule, as they result in the formation of two gold-thiol links per molecule (Yasutomi et al., 2004; Yasutomi, Morita, & Kimura, 2005).

If direct modification of the peptide-based material is not desired, synthetic thiols can also be used to modify the gold surface with other chemical functionalities. For example, amine-terminated alkanethiols are used to decorate gold surfaces with R—NH_2 groups that act as initiators for the graft polymerisation of *N*-carboxy anhydride (NCA) amino acid derivatives (Figure 3.1(b)). In this manner, polypeptides are grown directly from the modified gold surfaces (Koga, Nagaoka, & Higashi, 2006).

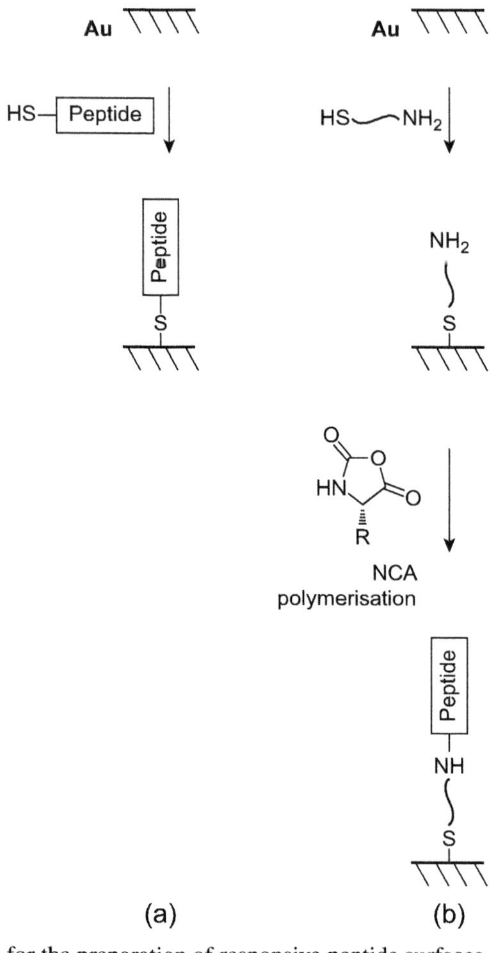

Figure 3.1 Strategies for the preparation of responsive peptide surfaces—gold-based self-assembled monolayers (SAMs). (a) Direct attachment and (b) pre-functionalisation.

3.2.3 Silanisation of glass surfaces

The modification of glass surfaces can be readily achieved through silanisation chemistry (Figure 3.2). In contrast to gold-based SAMs, the silanisation reactions typically require slightly harsher conditions. They are therefore generally not carried out in the presence of the peptides but are used to modify the glass surface with other functionalities before attachment of the peptide material.

In peptide surface fabrication, the most commonly performed modification of glass surfaces is the attachment of molecules that display reactive functionalities such as amines, carboxylic acids, or aldehydes. Amine-terminated silanes such as (3-aminopropyl)triethoxysilane are typically used as the first step. Homobifunctional linkers, that is, linkers that have the same amine reactive group on both

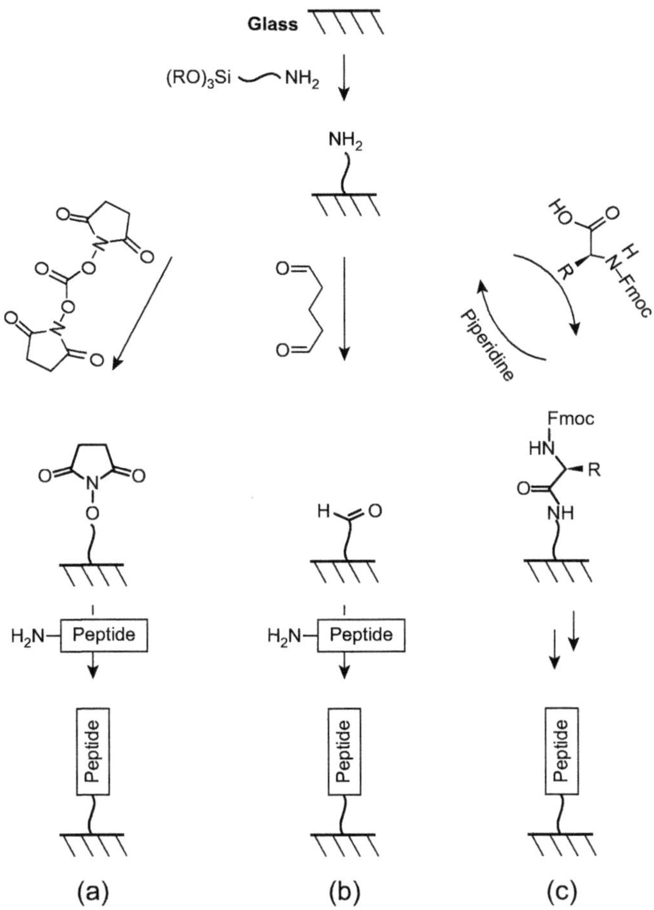

Figure 3.2 Strategies for the preparation of responsive peptide surfaces—silanisation. (a) Activation, (b) functionalisation, and (c) direct build-up.

ends, are subsequently used to decorate the surface with a reactive functionality. *N*,*N*′-disuccinimidyl carbonate is used to introduce *N*-hydroxysuccinimide (NHS) groups to which a polypeptide is attached via its *N*-terminus (Figure 3.2(a)) (Nath & Chilkoti, 2003). As an alternative reagent, glutaraldehyde can be used to accomplish immobilisation of a polypeptide (Figure 3.2(b)) (Teeuwen, Zuilhof, de Wolf, & van Hest, 2009). Direct immobilisation of a (non–peptide-based) responsive unit to glass can be carried out by synthesizing a silane derivative of the functional molecule (Gong et al., 2011).

The above strategies are designed to immobilise peptide materials onto the glass substrate after the peptide has been synthesised (an example of "grafting to"). As an alternative, a "grafting from" approach in which the peptide sequence is built up stepwise from the surface using standard solid-phase peptide synthesis chemistry

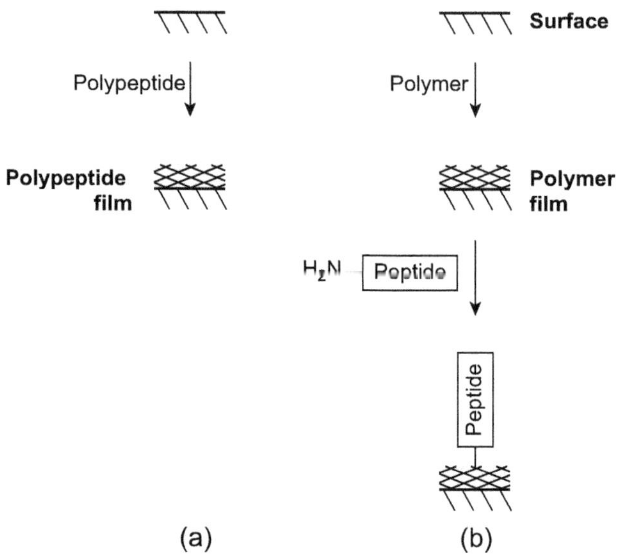

Figure 3.3 Strategies for the preparation of responsive peptide surfaces—adsorption. (a) Pre-functionalised polymer and (b) post-functionalised polymer.

(i.e., amide bond formation in the presence of *N,N'*-diisopropylcarbodiimide) can be used (Figure 3.2(c)) (Todd, Scurr, Gough, Alexander, & Ulijn, 2009; Zelzer, Scurr, Alexander, & Ulijn, 2012).

3.2.4 Adsorption or deposition on a surface

Noncovalent attachment of the peptide-based material can be used for both a responsive polypeptide and a responsive non–peptide-based polymer that is decorated with peptide sequences (Figure 3.3). The polypeptide is electrostatically adsorbed to a gold surface that is modified with carboxylic acid–terminated alkanethiols (Figure 3.3(a)) (Nath & Chilkoti, 2001). The non–peptide-based polymer is first prepared as a film on a substrate by solvent casting before using free carboxylic acid functionalities to covalently attach the peptide sequences via standard peptide chemistry, using 1-ethyl-3-(3-dimethylaminopropyl)-carbodiimide hydrochloride as the coupling reagent (Figure 3.3(b)) (Ebara et al., 2004).

3.3 Responsive peptide surfaces

Responsive peptide surfaces may originate from biologically inspired peptide sequences or they may be artificially designed based on our knowledge of the interactions of amino acids within a peptide chain. Both classes will be discussed here. Naturally inspired responsive peptide surfaces are based either on a protein mimic (a long peptide sequence that mimics the properties of a responsive protein) or on short sequences of minimal biologically active parts of a protein. Artificial peptides are either *de novo* designed sequences based on theoretical design rules or polymers whose building blocks are amino acids.

3.3.1 Naturally inspired peptides

Responsive peptides that are derived from naturally inspired motifs fall into two categories. First, responsive polypeptides can be prepared based on the structure of elastin, an extracellular matrix protein that is found in connective tissue such as skin and arteries (MacEwan & Chilkoti, 2010). These materials typically respond to a stimulus through a change in conformation of the polymer. The second category of naturally inspired peptides consists of short peptide sequences that represent substrates for enzymes. In these materials, the change induced by the stimulus (the enzyme) is an alteration of the chemical functionality displayed on the material surface. Both response mechanisms are shown in Figure 3.4.

Elastin-like polypeptides (ELPs) have been extensively studied due to the fact that they combine similar stimulus response properties to other artificial polymers such as poly(*N*-isopropylacrylamide) (pNIPAM) with the advantages of a biologically derived material, that is, it is biocompatible, modular in its composition, and can be obtained by biological processes. ELPs are polypeptides that contain a short, repetitive peptide sequence, most commonly $(VPGXG)_n$ that is derived from tropoelastin, the precursor of elastin. In this sequence, X represents any amino acid sequence except proline. Polypeptides composed of the pentapeptide repeat unit VPGXG possess a reversible lower critical solution temperature (LCST). Below the LCST, the peptide is soluble

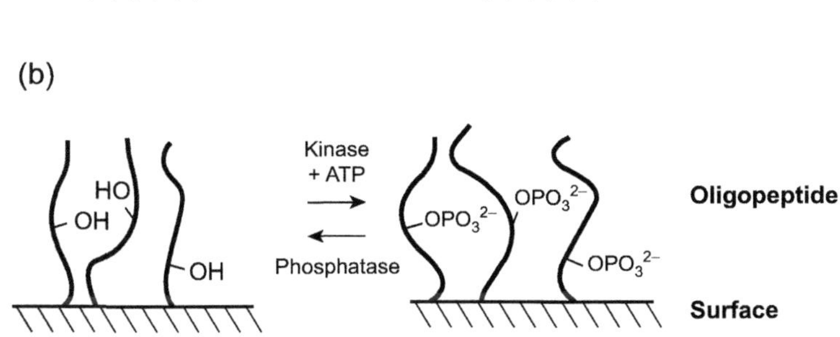

Figure 3.4 Response mechanisms for naturally inspired peptide surfaces. (a) Phase transition and (b) surface chemistry.

in an aqueous solution. When increasing the temperature above the LCST, demixing of the polypeptide from the aqueous solvent causes the polymer to aggregate (Figure 3.4(a)) (MacEwan & Chilkoti, 2010). It has been shown that the LCST depends on the nature of the amino acid in the fourth position (denoted as X) (Urry et al., 1991). Similar to pNIPAM, ELPs are not solely thermo-responsive; under isothermal conditions, some ELP variants respond to changes in pH, light, salt concentrations, or redox triggers (MacEwan & Chilkoti, 2010).

ELP surfaces can be prepared both by adsorption of the polypeptide to a modified gold surface (Hyun, Lee, Nath, Chilkoti, & Zauscher, 2004; Nath & Chilkoti, 2001) and by covalent attachment at the *N*-terminus of the peptide via NHS coupling chemistry (Nath & Chilkoti, 2003). Furthermore, a layer-by-layer deposition technique can be used to fabricate mixed layers of ELP and chitosan that maintain the thermo-responsiveness of the polypeptide (Barbosa et al., 2009). Chilkoti and co-workers used ELP-modified surfaces to induce temperature-triggered aggregation of gold particles. Subsequently, building on their advances in preparing ELP fusion proteins (Meyer & Chilkoti, 1999), the same group demonstrated that ELP-modified glass surfaces can be used to thermally trigger the capture of these ELP fusion proteins on the surface, based on the hydrophobic interaction between the surface-bound ELPs and the ELPs attached to the proteins (Nath & Chilkoti, 2003). It was shown that the process is reversible, allowing release of the ELP fusion protein upon reversal of the stimulus. Furthermore, the process can also be triggered by varying the salt (NaCl) concentration while keeping the temperature constant (Hyun et al., 2004; Nath & Chilkoti, 2003). In a similar manner, Hyun and co-workers patterned glass surfaces with ELP and an antimicrobial polymer, and demonstrated that bacteria whose surface had been decorated with ELP can be reversibly attached to the ELP-patterned domains on the surface (Lee, Jung, Na, Heo, & Hyun, 2009).

In an extension of the ELP fusion protein work carried out in Chilkoti's group, van Hest and co-workers demonstrated that two protein populations, each fused to an ELP with a different LCST, could be immobilised on a uniform ELP-modified glass surface by locally controlling the temperature (Teeuwen et al., 2009). Using a temperature gradient on the surface, one fusion protein was immobilised, first on one half of the surface. After washing the surface and inversing the temperature gradient, the second fusion protein was successfully attached to the second half of the surface, generating a heterogeneous protein surface from a homogenous substrate.

Responsive surfaces based on short peptides are inspired by peptide sequences that act as natural enzymatic substrates in proteins. Consequently, these surfaces are designed to be responsive to enzymatic stimuli (Figure 3.4(b)). Li and co-workers immobilised the peptide CALRRASLGW on a gold surface via the cysteine-thiol (C) (Wang et al., 2010). The heptapeptide LRRASLG (also known as Kemptide) is a synthetic analogue to the peptide sequence in pig liver pyruvate kinase that serves as a natural substrate for protein kinase A (PKA) (Kemp, Graves, Benjamini, & Krebs, 1977). Li et al. showed that the two positive charges of arginine (R) in Kemptide effectively act as a barrier for a solubilised redox agent, $[Ru(NH_3)_5Cl]^{2+}$, preventing its reaction with the gold electrode surface (Wang et al., 2010). In the presence of PKA, adenosine triphosphate (ATP), and cyclic adenosine monophosphate, the serine in

Kemptide becomes phosphorylated. The phosphorylation introduced negative charges in the peptide surface that counteract the charge barrier previously presented by the positive charges of arginine, allowing the $[Ru(NH_3)_5Cl]^{2+}$ redox agent to approach the electrode surface and an electrical signal to be generated. Hence, an enzyme responsive peptide surface was designed that enabled the generation of an electrical signal in response to a change in its chemical functionality, that is, its phosphorylation state. Li and co-workers further demonstrated that the system could be regenerated by exposure to alkaline phosphatase (AP), an enzyme that cleaves the phosphate groups from the phosphorylated serine residue.

Another approach towards enzyme responsive peptide surfaces has been reported by Zelzer and co-workers. Phosphorylated peptide surfaces with the sequence RGDS and YRGDS were prepared on glass via a grafting-from method wherein the peptide sequences were built up directly on the surface (Zelzer, McNamara, et al., 2012; Zelzer, Scurr, et al., 2012). The chemical functionality of the surface was changed by exposure to both exogenous (i.e., artificially added) and endogenous (i.e., cell secreted) AP. Importantly, this demonstrates that the concept of enzyme responsive peptide surfaces is in principle suitable for the design of interfaces between artificial materials and living biological systems (Zelzer, McNamara, et al., 2012).

3.3.2 Designed peptides

The design of artificial peptide sequences for the preparation of responsive surfaces takes advantage of the wealth of literature and information available on how the sequence of amino acids in a peptide chain affect the overall properties of the peptide (Boyle & Woolfson, 2011; Bromley et al., 2008). The two general strategies employed so far are the exploitation of charge effects on the secondary conformation of the peptide and the ability of dipole moments of helical peptides to direct an electric current (Figure 3.5).

Charge dependent conformational changes are incorporated in a peptide surface by introducing ionisable amino acids such as lysine and glutamic acid into the peptide chain. The amine functionality in the side group of lysine has a pK of 10.5, and is therefore positively charged at a $pH < 10.5$. Conversely, a pK of 4.1 renders the carboxylic acid side group functionality in glutamic acid negatively charged at $pH > 4.1$ (Minelli et al., 2013). The exact pK and hence the isoelectric points (IEPs) of these functional groups in a peptide material are a function of their chemical environment. In a neutral state, both lysine and glutamic acid tend to form α-helices, whereas ionisation triggers a conformational transition to a random coil (Appel & Yang, 1965).

Peptide surfaces that change their conformation in response to pH (Figure 3.5(a)) can be prepared by two different approaches. Higashi and co-workers prepared block polymers of lysine and glutamic acid via NCA graft polymerisation from an amine terminated gold surfaces (Koga et al., 2006). NCA polymerisation enables the preparation of polypeptides from the NCA derivative of the corresponding amino acid and is initiated by amine functionalities, enabling polymerisation directly from amine functionalised surface. Recently, several strategies including reduced pressure, catalyst design,

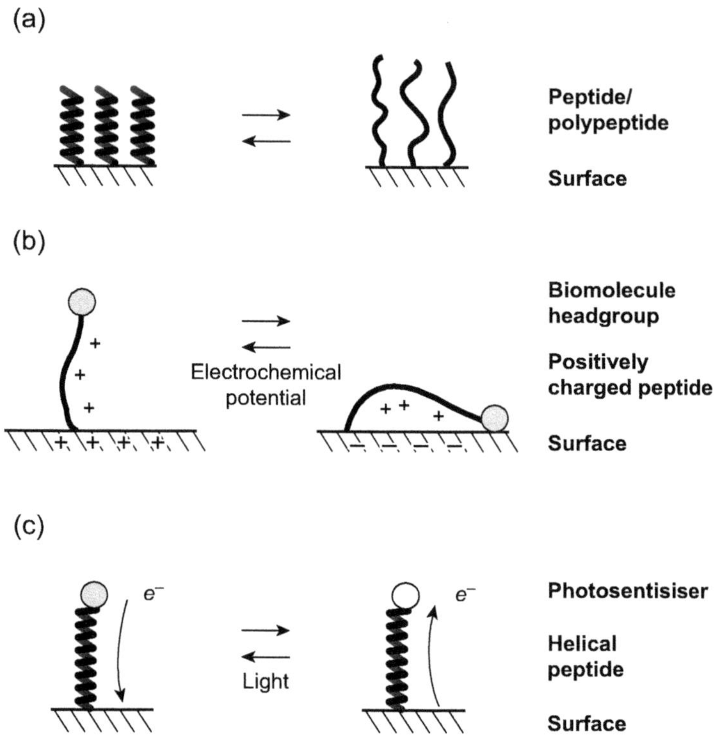

Figure 3.5 Response mechanisms for designed peptide surfaces. (a) Change in conformation, (b) change in orientation, and (c) change in current direction.

and low temperature have been proposed to improve the control over the molecular weight and polydispersity of the resulting polypeptides, making NCA polymerisation an attractive tool for the preparation of responsive peptide–based polymers (Deming, 2007; Hadjichristidis, Iatrou, Pitsikalis, & Sakellariou, 2009; Huang & Heise, 2013). The artificial polypeptide surfaces prepared by Higashi et al. are block polymers, poly(-glutamic acid)-*b*-poly(lysine) and poly(lysine)-*b*-poly(glutamic acid) that are covalently bound on the gold substrate at a 70° angle. The poly(glutamic acid) block is negatively charged and in a coiled coil conformation at a pH of 6; it adapts an α-helical conformation when neutral at pH 3. Conversely, for poly(lysine), a pH of 6 results in a positively charged block that forms a coiled coil; this is converted to an α-helical conformation at pH 9, at which the polymer block is neutral (Koga et al., 2006).

An alternative approach to prepare pH-responsive peptide surfaces is the rational design of the whole peptide sequence as demonstrated by Minelli and co-workers. Taking inspiration from the leucine zipper design of the yeast transcriptional activator GCN4, a *de novo* peptide was designed that contained lysine and glutamic acid in strategic positions in the peptides helix. The positioning of the amino acids was intended to support α-helical conformation at low and intermediate pH due to hydrophobic (no charge) and electrostatic (glutamic acid and lysine display opposite charges and attract each other) interactions, respectively. At high pH, lysine is neutral, and thus the negative charge of glutamic acid is expected to disrupt the helical conformation. This was indeed confirmed by circular dichroism (CD) in solution and implied on the surface (where the peptide was attached through a cysteine end group) using quartz crystal microbalance (QCM) measurements (Minelli et al., 2013).

The ability of peptides to carry charge has also inspired the use of electrical stimuli to change the properties of a peptide surface (Figure 3.5(b)). Mendes and co-workers used an electrical potential applied to a gold surface to control the orientation of immobilised oligolysine molecules. The peptide-conjugate used, Biotin-KKKKC displays a thiol group (cysteine, C) for surface attachment, a positively charged lysine sequence and a biotin endgroup. When a negative potential is applied to the surface, the biotin is unavailable for binding events with neutravidin due to the affinity of the positively charged peptide with the negatively charged surface. Application of a positive potential changed the orientation of the peptide and made the biotin available for interactions with labelled neutravidin, as shown by fluorescence microscopy and surface plasmon resonance (SPR) (Yeung et al., 2010).

Peptides can not only be designed to be responsive to electrical potentials, but they are also able to control the direction of electrical current and to alter it in response to an external stimulus (Figure 3.5(c)). Kimura and co-workers demonstrated that the direction of the dipole moment in helical peptides that are attached to a gold electrode (via a disulfide linker at a 60° angle from the surface) determines the direction of a photoinduced current along the length of the peptide helix. The current was induced by photoexcitation of a photosensitiser that is attached to the end of the peptide helix (Morita, Kimura, Kobayashi, & Imanishi, 2000). This approach can be used to prepare a surface with a mixed population of α helices, both containing eight repeat units of a dipeptide made of leucine and α-aminoisobutyric acid but using either the D or the L form of leucine to change the chirality and hence the direction of the dipole moment of the α-helix. By attaching two different photosensitisers to the two peptide helices, selective irradiation of the mixed peptide surface with light of a wavelength of either 351 nm or 459 nm, anodic (current to the surface) or cathodic (current away from the surface) currents can be induced on demand (Yasutomi et al., 2004). In subsequent work, Kimura and co-workers also demonstrated that by introducing a carboxylic acid endgroup to the peptide next to the photosensitiser, the same peptide can be triggered to change the direction of the current in response to the pH of the solution (Yasutomi et al., 2005).

3.4 Peptides attached to responsive surfaces

Biomaterial surfaces may be peptide based without relying on the peptide to be stimulus responsive. This may be advantageous when current technology does not furnish us with responsive peptides that react to the stimulus dictated by a selected application. Other considerations such as response time, robustness, and cost may also contribute to the selection of non–peptide-based response mechanisms for peptide surfaces. In these instances, the peptide sequences will typically be chosen for their biological importance, the prevailing sequences used being RGD (a sequence found in the cell

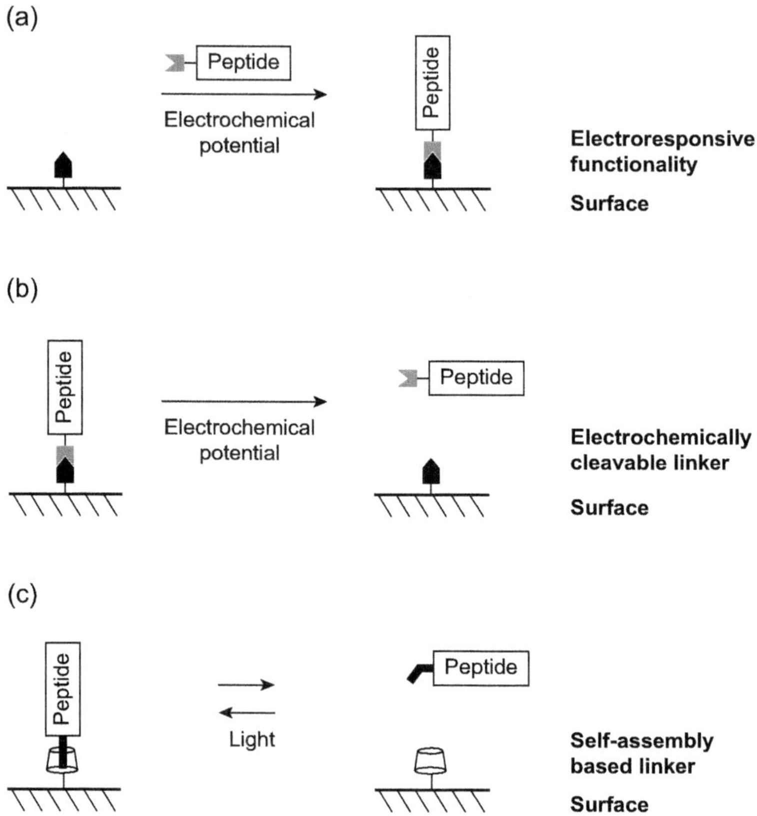

Figure 3.6 Strategies to attach or release peptides to or from surfaces. (a) Covalent attachment, (b) detachment via bond scission, and (c) supramolecular interactions.

binding domain of fibronectin that is associated with cell adhesion) or derivatives thereof to promote the adhesion of cells to surfaces in response to a stimulus. Two types of responsive surfaces are distinguished below. First, surfaces may immobilise or release the peptide sequence in response to a trigger event; this type of response is generally irreversible and can be carried out only once (Figure 3.6). Second, surfaces may hide or reveal peptide sequences; these are mostly reversible events that allow repeated switching of surface properties (Figure 3.7).

3.4.1 Release/immobilisation of peptides

The electrochemically induced release and immobilisation of peptides was extensively researched by Mrksich and co-workers. They designed alkanethiols that are terminated with hydroquinone moieties, which can be oxidised to the quinone form by applying an electrochemical potential to the gold substrate. The quinone-terminated SAMs then readily undergo Diels-Alder reactions with cyclopentadienes that are attached to a peptide sequence (RGD), thus binding the peptide covalently to the surface (Figure 3.6(a)) (Yousaf, Houseman, & Mrksich, 2001). This approach was extended such that

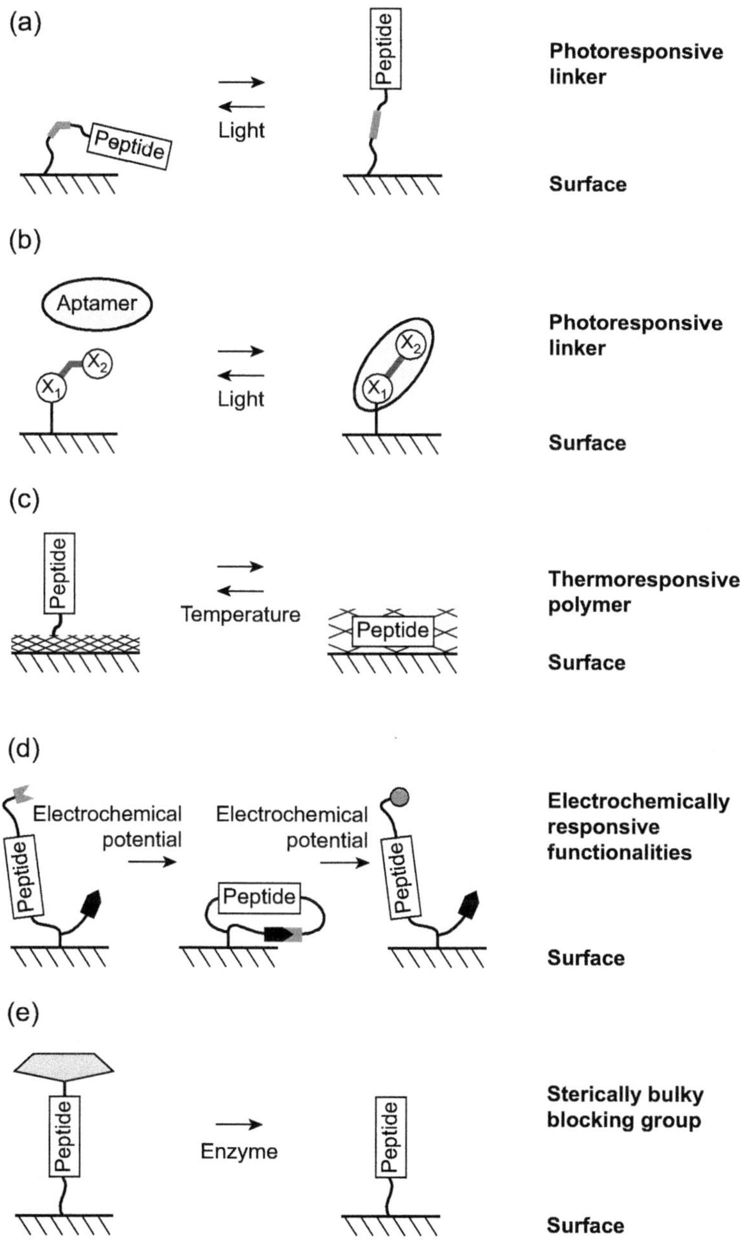

Figure 3.7 Strategies to hide or reveal peptides on a surface. (a) Orientation, (b) conformation, (c) segregation, (d) cyclisation, and (e) unblocking.

a peptide that is attached to the hydroquinone via an O-silyl hydroquinone ether bond can be released in response to an oxidizing potential (Figure 3.6(b)). The resulting quinone terminated surface can subsequently be used to immobilise another peptide sequence as described above (Yeo, Yousaf, & Mrksich, 2003). In addition, Mrksich and co-workers also described a variation of their quinone system that allows release of the peptide sequence in response to a reducing potential, thus enabling the orthogonal release of peptides in response to either a negative or a positive potential from the same surface (Yeo & Mrksich, 2006). The systems discussed so far are all irreversible. Yousaf and co-workers designed an electrochemically responsive peptide immobilisation/release system that can, in principle, be reused multiple times. While using the same hydroquinone surfaces, the peptide sequence that is to be attached is now modified with an oxyamine (R—ONH_2) functionality. This group also readily reacts with the electrochemically produced quinone on the surface and covalently binds the peptide via an oxime (O—N═C) bond. Although the oxime bond is chemically stable at pH 0, under physiological conditions it undergoes electrochemical reduction to reform the hydroquinone and to release the peptide from the surface (Chan, Park, & Yousaf, 2008).

A different approach to trigger the release of peptides from a surface was presented by Zhang and co-workers. These researchers made use of the non-covalent host–guest interaction between α-cyclodextrin (α-CD) and azobenzene to design a photoresponsive surface. The α-CD is immobilised on a glass surface through an alkanesilane, while the azobenzene is attached to the peptide sequence (an RGDS derivative). In its thermodynamically stable trans configuration, azobenzene is able to form a supramolecular host–guest complex with α-CD, thus immobilizing the attached peptide on the surface noncovalently. Upon irradiation with UV light (λ=365 nm), azobenzene isomerises to its cis form. This conformational change causes a mismatch in the geometry of the azobenzene and the cavity of the α-CD, resulting in the expulsion of the azobenzene from the supramolecular complex and the removal of the peptide sequence from the surface (Figure 3.6(c)). As the underlying process relies on non-covalent interactions, it is, in principle, reversible, and peptides can be immobilised and released repeatedly using the same surface (Gong et al., 2011).

3.4.2 Triggered display of peptides

Strategies to reveal and/or to hide peptide sequences on a surface in response to a trigger event mostly rely on conformational changes of the functional material on the surface. Similar to the light-responsive supramolecular complexes discussed above, azobenzenes have been used by Kessler and co-workers as covalent linkers to immobilise cell adhesion–modulating peptide sequences (a cyclic RGD peptide, RGDfK) on poly(methylmethacrylate) surfaces (Figure 3.7(a)) (Auernheimer, Dahmen, Hersel, Bausch, & Kessler, 2005). It has been shown that the distance between the RGD peptide and the cell affects cell adhesion (Kantlehner et al., 2000). Thus, cell adhesion can be reduced by the photoinduced change of the azobenzene linker from the trans to the cis form due to the increase in the distance between the cell and the adhesion peptide (Auernheimer et al., 2005). The effects of photoinduced geometrical changes can also be used to trigger the ability of a peptide to bind to its RNA aptamer. Nakatina and co-workers incorporated the azobenzene unit in between two arginine moieties and immobilised the modified peptide on a surface via a lysine residue at the *N*-terminus (Hayashi, Hagihara, Dohno, & Nakatani, 2007). The azobenzene in the middle of the two arginine functionalities allows photoinduced changes in the conformation of the peptide on the surface. As the geometry of the RNA aptamer is only able to accommodate the peptide in its trans isomer, the binding and release of the aptamer at the surface can be triggered by irradiation with light (Figure 3.7(b)) as shown by SPR.

Thermoresponsive control over the display of RGD peptides has been demonstrated by Okano and co-workers (Ebara et al., 2004). A temperature-sensitive copolymer, poly(*N*-isopropylacrylamide-*co*-2-carboxyisopropylacrylamide), is grafted on a surface and then decorated with RGD by attachment of the peptide to the carboxylic acid functionalities of the polymer. Cell adhesion is promoted at temperatures above the LCST but reduced below the LCST. This is explained with a modulation of the availability of the peptide at the material surface; above the LCST, the polymer is collapsed and the peptide is available at the surface, whereas below the LCST, the polymer becomes soluble and the expanding polymer chains prevent access to the peptide (Figure 3.7(c)).

Although the surfaces discussed thus far in this section all display reversible material changes, two examples of nonreversible modulation of peptide display have been reported. In a continuation of the electrochemically responsive surfaces discussed in Section 3.4.1, Yousaf and co-workers designed a sophisticated system (Figure 3.7(d)) in which a series of chemical reactions allow first the attachment of an RGD-based peptide to the surface via a Huisgen Click reaction (Lamb & Yousaf, 2011). Upon application of the electrochemical trigger, the free end of the peptide is reacted with the surface via an oxime link to form a cyclic peptide that alters the bioavailability of the immobilised RGD sequence. The cyclic form of the peptide could subsequently be reverted to the linear form by a second electrochemical trigger event. This strategy allows switching back and forth between the availability of the peptide a single time by forming a cyclic structure but it is not reversible. A different approach for a single trigger event to reveal RGD peptides for cell adhesion was taken by Ulijn and co-workers (Figure 3.7(e)) (Todd et al., 2009). They grafted the peptide sequence Fmoc-AARGD on a glass surface; Fmoc (fluorenylmethyloxycarbonyl) is a bulky endgroup that prevents cells from interacting with the RGD sequence, whereas the AA peptide introduced sensitivity to an enzymatic stimulus. Upon exposure of the surface to elastase, which cleaves between the two alanine (A) moieties, the Fmoc group is removed, causing the RGD sequence to be revealed and cell adhesion to be promoted.

3.5 Protein surfaces

A small number of reports have shown that natural proteins can also be harnessed to generate stimulus responsive surfaces. Although it can be expected that protein surfaces are less robust than synthetic peptides due to their more complex structure, the advantages of harnessing existing biological machinery may outweigh these reservations. Two types of responsive protein surfaces have been reported so far; surfaces in which the protein itself displays responsiveness to a trigger, and surfaces in which the protein is attached to another responsive unit (Figure 3.8).

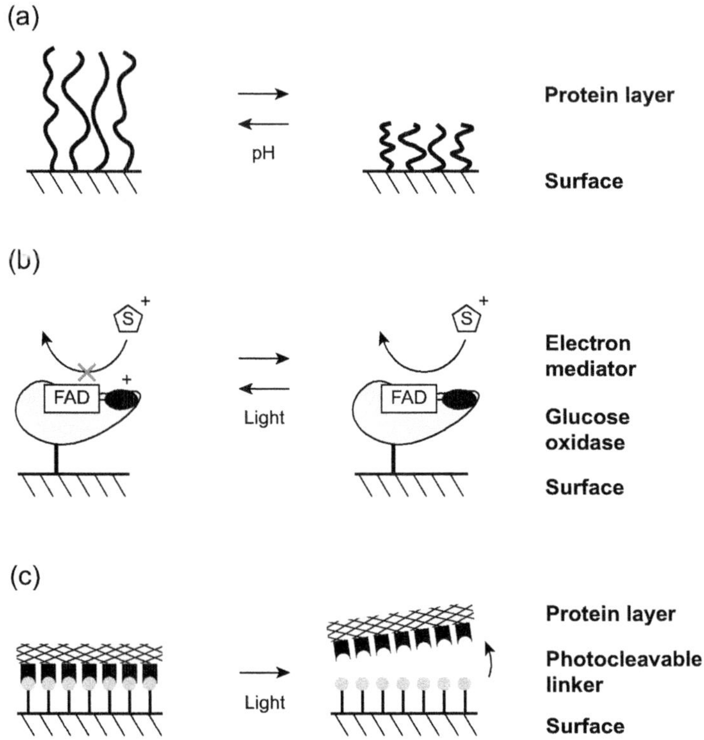

Figure 3.8 Approaches to designing responsive protein-based surfaces. (a) Conformation, (b) enzyme substrate accessibility, (c) protein layer detachment.

3.5.1 Stimulus-responsive proteins

Research by Dutta, Choudhury, and co-workers demonstrated that the resilin-mimetic protein *rec1-resilin* can be used to prepare a pH-responsive surface (Figure 3.8(a)) (Truong et al., 2010). *Rec1-resilin* possesses both an LCST and an upper critical solution temperature, the latter being pH dependent. It was initially observed that the adsorption of *rec1-resilin* is pH dependent. Different conformations and packing densities of the protein on a gold substrate can be obtained at different pH; the adsorption properties are governed by the IEP of the protein. Importantly, these researchers showed that, once adsorbed, the conformation of *rec1-resilin* can be changed by altering the pH of the solution. Thus, the protein conformation was reversibly changed from a globular (pH 4.9) to a coil (pH 2) and an extended coil (pH 12).

The second example of a stimulus-responsive protein is based on an immobilised enzyme. Willner and co-workers designed a light-responsive enzyme surface by immobilising modified glucose oxidase (GOx) onto a glutaraldehyde functionalised gold surface (Blonder, Katz, Willner, Wray, & Buckmann, 1997). GOx is an oxidoreductase that requires the presence of a cofactor, FAD, to perform its catalytic activity. Willner and co-workers modified the FAD cofactor by attaching a nitrospiropyrane to its adenine moiety. This enabled them to induce a positive charge into the cofactor in response to irradiation with light (360 nm < λ < 380 nm). The redox activity of the enzyme on the surface also relies on the close proximity of a soluble electron mediator, a charged ferrocene derivative. The charge on the nitrospiropyrane-modified FAD determines whether the electron mediator is able to approach close enough for the enzymatic redox reaction to take place. As the charge of the modified FAD can be changed photochemically, a system was obtained in which the enzymatic activity is directly controlled by light (Figure 3.8(b)).

3.5.2 Protein release

The photochemically induced release of a protein layer (bovine serum albumin (BSA)) from a surface was demonstrated in the groups of Takarada and Maeda (Nakanishi et al., 2004). Through silanisation, a glass surface was decorated with a photocleavable 2-nitrobenzyl ester. After adsorption of BSA, the protein film is removed upon irradiation at 365 nm (Figure 3.8(c)), allowing repopulation of the cleaved surface with a different protein.

3.6 Physical properties and characterisation

Here, the various stimuli used to trigger responses from peptide surfaces will be discussed. Emphasis is placed on addressing possible issues that are particular to the peptide material or the biological environment in which the stimulus may be used. Subsequently, various techniques and strategies to analyse both the surface and the material response will be presented.

3.6.1 Stimuli and responses of peptide surfaces

From the previous sections of this chapter, it becomes evident that, despite the early stages of research on peptide-based responsive surfaces, the diversity of stimuli that are being used match those of other responsive materials. Here, these stimuli will be shortly discussed with respect to their benefits and limitations when used with peptide materials.

3.6.1.1 Temperature

For stimulus-responsive peptides, temperature has thus far exclusively been used to induce phase transitions of ELP (Hyun et al., 2004; Nath & Chilkoti, 2003). As with other temperature-responsive biomaterials such as peptide-modified pNIPAM-based polymers (Ebara et al., 2004), the transition temperature has to be close to the physiological temperature if the material is to be used in conjunction with a living system in which the denaturing of naturally occurring proteins becomes an issue.

3.6.1.2 pH

Changes in the solution pH are typically used to elicit a change in the conformation of a peptide. As the formation of secondary structures in peptides strongly depends on the charge of ionisable amino acid residues (e.g., lysine and glutamic acid), a change in pH is a straightforward way to affect the conformation of the peptide surface (Minelli et al., 2013). An alternative aspect is the ability of pH changes to control the direction of a photoinduced current through a layer of an immobilised peptide helix by changing the dipole moment of the peptide helix (Yasutomi et al., 2005). In some instances, pH is also used to prevent the occurrence of a surface response that would take place under physiological conditions (Lamb & Yousaf, 2011). It should be noted that some pH values (Chan et al., 2008; Truong et al., 2010) can be rather extreme from a biological point of view, and their application as biomaterials has to be carefully considered.

3.6.1.3 Light

No inherently photoresponsive peptides have yet been used to design photoresponsive surfaces, and all photoresponsive peptide surfaces to date require the introduction of a synthetic photosensitive moiety. The larger part of light-responsive peptide-based materials are designed around azobenzene and are therefore based on a conformational change of the material due to photoisomerisation. The effect on the material is diverse and includes a change in the availability of a peptide ligand (Auernheimer et al., 2005) and the inability to bind to an aptamer (Hayashi et al., 2007) or to form a supramolecular complex (Gong et al., 2011). Other light-responsive peptide–based surfaces use nitrobenzene as a photocleavable linker (Nakanishi et al., 2004), nitrospyropyrane to reversibly introduce a charge (Blonder et al., 1997), and photosensitisers to generate a photocurrent on an electrode surface (Yasutomi et al., 2004). Although the use of light is advantageous because it is noninvasive and can be spatially and temporally well controlled, some photoresponsive systems require high-energetic wavelengths that may raise concern with respect to photoinduced damage of the biological components in a living system.

3.6.1.4 Electrochemical potential

Electrochemical potentials are used extensively to induce attachment or detachment of peptide sequences on a surface (Yeo et al., 2003; Yousaf et al., 2001). In these materials, the peptide itself is not responsive to the electrochemical potential. The responsiveness stems from redoxactive functionalities that have been immobilised on an electrode surface. In addition, with one exception (Chan et al., 2008), the reactions used thus far are not reversible. Electrochemical potentials are also used to change the orientation of a peptide on a surface (Yeung et al., 2010). This is accomplished by attracting the inherently positively charged oligopeptide sequence to the surface, and thus represents an example in which the peptide itself displays responsiveness to the electrochemical potential. Electrochemical stimuli are highly attractive for sensing applications, as they are noninvasive, and as signals are easily transduced and read out. For other biological applications, the need for an electrode and electrical wires may present a considerable obstacle.

3.6.1.5 Enzymes

Enzymes are explored as stimuli for peptide surfaces to alter the chemical properties of the surfaces. This includes the introduction and removal of charges in response to

PKA and AP, respectively (Wang et al., 2010; Zelzer, McNamara, et al., 2012), as well as the triggered availability of a peptide sequence on the surface (Todd et al., 2009). As peptides are natural substrates for enzymes, enzymatic response of a peptide surface appears to be an ideal way to interface a biomaterial surface with a living system. Enzymatic stimulation of peptide surfaces has the advantage that the environmental conditions required (physiological conditions) are ideal for both the peptide and the stimulus. For applications in living systems, emphasis must be placed on a well-designed peptide surface to prevent unwanted interaction with other enzymes.

3.6.1.6 Response types

The material response from surfaces in which the peptide itself reacts to the stimulus can be classed into five types: (1) change of conformation (i.e., helix to coil transition); (2) change of orientation with respect to the surface; (3) phase change (aggregated to soluble); (4) change in the chemical functionality; and (5) the directionality of an induced photocurrent. By making use of non–peptide-based responsive materials, other response types can be accessed. These include (1) the attachment and removal of peptide sequences from the surface and (2) the hiding/revealing of peptide sequence.

3.6.2 Characterisation of responsive peptide surfaces

The characterisation of a stimulus responsive surface in general includes two aspects: verification of the surface composition and evaluation of the materials response due to the presence of the stimulus. Although a variety of techniques are available to characterise peptides and their stimulus-responsive properties in solution and in bulk, many of these are not compatible with surface-immobilised peptides. Hence, a common approach is to characterise the peptide material in solution before attachment to the surface. UV-based turbidity measurements (Lee et al., 2009; Nath & Chilkoti, 2003; Teeuwen et al., 2009) and calorimetry (Barbosa et al., 2009) are used to determine the LCST of ELPs. The isomerisation of azobenzene can be studied with UV absorption, nuclear magnetic resonance spectroscopy, and high-performance liquid chromatography (Auernheimer et al., 2005; Hayashi et al., 2007), and CD is used to determine the presence of helices in a peptide (Minelli et al., 2013; Yasutomi et al., 2005). Non–solution-based methods that can be used to characterise responsive peptide surfaces will be discussed in more detail below.

3.6.2.1 Surface characterisation

The different stages of the preparation of peptide surfaces can be confirmed with surface-sensitive physical and chemical analysis techniques. For gold-based SAMs, Mrksich and co-workers have introduced a matrix-assisted laser desorption ionisation time-of-flight (MALDI-TOF) mass spectrometry–based analysis procedure with which they are able to identify the presence of various surface functional groups via their mass (Yeo & Mrksich, 2006; Yeo et al., 2003). Although this method is applicable to SAMs, it is not strictly a surface sensitive technique, as the desorption process in MALDI is not confined to the uppermost layer of a material.

A rapid and inexpensive method to analyse the top layer of a surface is the determination of the surface tension through water contact angle measurements (Gong et al., 2011; Koga et al., 2006; Teeuwen et al., 2009; Zelzer, Scurr, et al., 2012). Although the contact angle depends on the chemical composition/state of the surface, it is also affected by the topography of the surface. Surface sensitive chemical information can be obtained by X-ray photoelectron spectroscopy (XPS) and time-of-flight secondary ion mass spectrometry (ToF-SIMS). Both methods gather information from the top layer of a surface. XPS provides quantitative and qualitative information about the elements and their chemical environment present on a surface. Its surface sensitivity is typically quoted to be around 10 nm, but this can be reduced by changing the incident angle of the exciting X-rays on the surface (Watts & Wolstenholme, 2003). ToF-SIMS provides mass spectra that originate from molecules present in the top layer (about 1 nm) of the sample (Bubert & Jenett, 2003). The two methods (XPS and ToF-SIMS) provide complementary information and are powerful tools to fully characterise the chemical composition of a surface. They have been used in several instances to confirm the composition of the final responsive peptide surfaces, and are equally useful to monitor the individual surface modification steps (Gong et al., 2011; Koga et al., 2006; Teeuwen et al., 2009; Todd et al., 2009; Yeung et al., 2010; Zelzer, McNamara, et al., 2012).

When using XPS and ToF-SIMS for the analysis of peptide surfaces, a number of factors have to be considered. Because of its larger analysis depth, standard XPS measurements are more readily used on thicker peptide layers such as ELP. For thin layers, e.g., short oligopeptides sequences whose molecular dimensions are considerably smaller than the analysis depth, XPS analysis becomes challenging because of the low signal intensities generated by the peptides. Although it has been shown that XPS measurements on such peptide surfaces are possible (Stevens et al., 2013; Todd et al., 2009), they are far from trivial and require well-established surface preparation and analysis procedures. Nonetheless, the ability to perform quantitative surface-sensitive measurements makes XPS an attractive tool to study peptide surfaces, and even allows the estimation of the thickness of peptide layers on a substrate surface (Koga et al., 2006; Teeuwen et al., 2009). Although ToF-SIMS is currently not considered a strictly quantitative method, it is powerful because of its high surface sensitivity. A landmark study by Wagner and Castner has presented a database of ToF-SIMS mass fragments for the essential amino acids (Wagner & Castner, 2001). Furthermore, ToF-SIMS has also been exploited to gain information about the conformation and orientation of proteins on surfaces (Castner & Ratner, 2002), an aspect that has not yet been used for the analysis of a stimulus-responsive peptide that undergoes such changes on a surface.

3.6.2.2 Response characterisation

In addition to being surface sensitive, the analysis method used to characterise the response of the peptide surface to a stimulus has to be able to measure a physical parameter related to that response. It should be noted that, in some instances, the response of the peptide-based materials is analysed only in solution; the response of the immobilised material to the stimulus is then inferred by observation of a biological response such as the adhesion/detachment of cells, bacteria, or other biomolecules. Although this is indirect evidence supporting the responsive nature of the surface, here only techniques that provide a direct measurement of the surface response will be discussed.

Two widely used techniques to analyse peptide surface responses are QCM and SPR. Both are able to detect changes in the amount of adsorbed material on a surface. Hence both techniques can be used to demonstrate the stimulus-induced attachment/detachment of peptides on a surface (Minelli et al., 2013; Truong et al., 2010). In addition, the triggered availability/concealment of surface bound peptides for binding events with other biomolecules can be measured (Chan et al., 2008; Hayashi et al., 2007; Minelli et al., 2013; Nath & Chilkoti, 2003; Yeung et al., 2010). QCM is also able to detect changes in the viscoelastic properties of the adsorbed material, thus providing access to information about conformational changes (Truong et al., 2010). Both SPR and QCM can provide time-dependent information on the surface response and, in principle, response times can be obtained from these measurements if they are within the measured timeframe. Furthermore, the techniques lend themselves to studying the reversibility of an adsorption event.

Cyclic voltammetry (CV) is the method of choice for electrochemically responsive peptide surfaces. It provides evidence of the presence of redoxactive compounds at the surface (Lamb & Yousaf, 2011; Yeo & Mrksich, 2006; Yeo et al., 2003; Yeung et al., 2010) and can be used to demonstrate the reversibility of the responsive reaction (Blonder et al., 1997; Wang et al., 2010). If CV data are collected at specific time points, kinetic information about the response time becomes available (Chan et al., 2008; Wang et al., 2010).

The conformation of peptides on surfaces has been studied by reflection absorption Fourier transform infrared spectroscopy. The two amide bond signals present in the IR spectra provide information about the secondary structure of the peptide as well as the orientation of a helical peptide with respect to the surface (Koga et al., 2006; Yasutomi et al., 2004, 2005).

Topographical changes caused by temperature-induced phase changes (Barbosa et al., 2009; Hyun et al., 2004) or by the adsorption of compounds on the surface in response to a stimulus (Koga et al., 2006) can be monitored by atomic force microscopy (AFM). Altered adhesion properties of a responsive peptide surface can also be probed by AFM force measurements (Hyun et al., 2004).

The methods mentioned thus far to study the response of a peptide surface all have the potential (and are in some cases required) to be conducted in a solvated environment. Surface chemical tools such as XPS and ToF-SIMS that typically require high-vacuum conditions for their operation can be used to analyse the chemistry of the peptide surfaces before and after the triggering event took place, and are thus complementing the *in situ* methods (Gong et al., 2011; Todd et al., 2009; Zelzer, McNamara, et al., 2012). Finally, the response time for a trigger event can also be reported from a biological point of view by measuring the progressive detachment of cells from the surface (Ebara et al., 2004).

3.7 Emerging applications

By their nature, peptide-based responsive surfaces have been predominantly considered for biological applications. Three applications areas have been explored thus far: (1) biosensors; (2) control of the adhesion of proteins, cells, and other biomolecules; and (3) bioseparation. The current state of the art for these three application areas will be outlined below.

3.7.1 *Biosensor fabrication*

The ability to prepare responsive peptide surfaces on conductive electrodes such as gold enables the use of responsive peptide surfaces as components for electronic biosensors. Even though electrochemically responsive peptide surfaces have been prepared, they have yet to be translated into an actual sensing device. One peptide-based biosensor reported to date is based on an enzyme-responsive mechanism and was designed to detect and measure the activity of PKA (Wang et al., 2010). Although the trigger is enzymatic, the output ultimately generated by the enzymatic action is an electrical signal. This sensor has a detection limit for PKA of 1×10^{-4} units/mL (approximately 0.2 ng/mL), which is claimed to be three orders of magnitude lower than previously reported PKA sensors. The main advantages of this sensor are that it requires no additional chemicals or surface treatments and the surface can be regenerated and reused.

A second sensor based on ELP was reported by Chen and co-workers (Gao et al., 2005). The ELP is fused to antibodies for a tumor marker (cancer antigen 19-9, CA 19-9) and subsequently used to immobilise the antibodies to a surface. In this case, the temperature responsiveness of the ELP was used for the fabrication of the sensor; detection occurred via the addition of a second, fluorescently labelled antibody to the tumor marker after the marker had been immobilised. This biosensor had a detection limit of 21 units/mL for CA 19-9.

3.7.2 *Control of adhesion events*

The availability of peptides at the material surface can be used to control biological adhesion events. Protein adhesion, and consequently cell adhesion, was modulated by Takarada and Maeda via a light-responsive mechanism that allows spatiotemporal control over the cell adhesion process (Nakanishi et al., 2004). This provides a route towards the fabrication of patterns in a cell culture. Similarly, the electrochemically responsive surfaces prepared by Mrksich and co-workers enable control of the cell adhesiveness of the cell culture substrate *in situ*, allowing attachment or detachment of cells on demand (Yeo & Mrksich, 2006; Yeo et al., 2003; Yousaf et al., 2001). Cell detachment in response to temperature by hiding the cell-adhesive RGD sequence has also been exploited by Okano and co-workers to harvest cells for tissue engineering applications without the need for enzymatic digestion (Ebara et al., 2004). Finally, the reversible binding/detachment of bacteria has also been demonstrated using ELP-based surfaces (Lee et al., 2009).

3.7.3 *Bioseparation*

It has been suggested that ELP-based materials are suited for the separation of other proteins, as they can reversibly entrap other biomolecules (Meyer & Chilkoti, 1999). Although such an application has not yet been designed to use an ELP-based surface, Chilkoti and co-workers demonstrated that immobilisation of an ELP fusion protein can be accomplished directly from cell lysate without previous purification steps (Nath & Chilkoti, 2003). Van Hest and co-workers have elegantly demonstrated that ELPs with different LCSTs can be used to spatially control the immobilisation of fusion proteins in a channel, thus preparing spatially separated areas with two different proteins (Teeuwen et al., 2009).

3.8 Future trends

During the last decade, various clever concepts for the design and application of peptide-based responsive materials have emerged that promise exciting developments of the area in the near future. Initial efforts to design these surfaces, such as ELP- and RGD-based materials, were inspired by naturally existing functionalities (Nath & Chilkoti, 2001; Yousaf et al., 2001). More recent work has shown that we are now in a position where we can make constructive use of the design rules for peptidic materials that have been developed thus far to create novel and unprecedented responsive peptide surfaces such as the pH-controlled photocurrent direction (Yasutomi et al., 2005) and conformational changes with a specific, predesigned response range (Minelli et al., 2013). This trend can be expected to continue, with new insight into peptide design (Boyle & Woolfson, 2011) feeding into the development of novel responsive peptide materials, allowing for more diverse and customisable trigger events and response mechanisms.

The characterisation of these materials is likely to move closer into the focus of our attention, as a fundamental understanding of the material characteristics and the interactions of these novel materials with a biological environment are vital but poorly understood. Although a number of powerful tools are available to characterise peptide surfaces and their responses, direct observation of the material–protein or material–cell interaction remains challenging. The peptide surface community will certainly stand to benefit enormously by the current drive in the surface analysis community to push the limits of their instruments to enable surface analysis under ambient or physiological conditions. Furthermore, reliable and quantitative surface measurements and standard procedures will be required to facilitate uptake of these techniques in the peptide surface community. Finally, the development of 3D depth profiling of organic materials, in particular with TOF-SIMS, is likely to have an impact on our ability to analyse and understand complex peptide surfaces in more detail. These analytical developments and insights are essential to answer open questions regarding the exact nature of the responsive interaction in a biological environment, and, importantly, how to control it.

In the longer term, it can be expected that responsive peptides will find their way onto more complex, 3D dimensional surfaces such as microfluidic devices or porous scaffolds that are frequently used in biomedical engineering. This transition will pose a new set of complex challenges in terms of surface modification and analysis, but is likely to be accompanied by significant benefits for biomaterials devices in tissue engineering, bioseparation, sensing, and microfluidic reactors. Possibly the ultimate vision of a responsive peptide-based surface is one that truly mimics a natural process such as the environmental sensing and signal transduction in a cell membrane. If such functionality can be accomplished even in parts, it may prove to be useful as a component for constructs in synthetic biology and the design of artificial cells.

References

Appel, P., & Yang, J. T. (1965). *Biochemistry*, *4*, 1244–1249. http://dx.doi.org/10.1021/bi00883a005.

Auernheimer, J., Dahmen, C., Hersel, U., Bausch, A., & Kessler, H. (2005). *Journal of the American Chemical Society*, *127*, 16107–16110. http://dx.doi.org/10.1021/ja053648q.

Bacakova, L., Filova, E., Parizek, M., Ruml, T., & Svorcik, V. (2011). *Biotechnology Advances*, *29*, 739–767. http://dx.doi.org/10.1016/j.biotechadv.2011.06.004.

Barbosa, J. S., Costa, R. R., Testera, A. M., Alonso, M., Rodriguez-Cabello, J. C., & Mano, J. F. (2009). *Nanoscale Research Letters*, *4*, 1247–1253. http://dx.doi.org/10.1007/s11671-009-9388-5.

Blonder, R., Katz, E., Willner, I., Wray, V., & Buckmann, A. F. (1997). *Journal of the American Chemical Society*, *119*, 11747–11757. http://dx.doi.org/10.1021/ja972663v.

Boyle, A. L., & Woolfson, D. N. (2011). *Chemical Society Reviews*, *40*, 4295–4306. http://dx.doi.org/10.1039/c0cs00152j.

Bromley, E. H. C., Channon, K., Moutevelis, E., & Woolfson, D. N. (2008). *ACS Chemical Biology*, *3*, 38–50. http://dx.doi.org/10.1021/cb700249v.

Bryers, J. D., Giachelli, C. M., & Ratner, B. D. (2012). *Biotechnology and Bioengineering*, *109*, 1898–1911. http://dx.doi.org/10.1002/bit.24559.

Bubert, H., & Jenett, H. (2003). *Surface and thin film analysis*. Wiley-VCH.

Castner, D. G., & Ratner, B. D. (2002). *Surface Science*, *500*, 28–60. http://dx.doi.org/10.1016/s0039-6028(01)01587-4.

Chan, E. W. L., Park, S., & Yousaf, M. N. (2008). *Angewandte Chemie-International Edition*, *47*, 6267–6271. http://dx.doi.org/10.1002/anie.200800166.

Cole, M. A., Voelcker, N. H., Thissen, H., & Griesser, H. J. (2009). *Biomaterials*, *30*, 1827–1850. http://dx.doi.org/10.1016/j.biomaterials.2008.12.026.

Deming, T. J. (2007). *Progress in Polymer Science*, *32*, 858–875. http://dx.doi.org/10.1016/j.progpolymsci.2007.05.010.

Ebara, M., Yamato, M., Aoyagi, T., Kikuchi, A., Sakai, K., & Okano, T. (2004). *Biomacromolecules*, *5*, 505–510. http://dx.doi.org/10.1021/bm0343601.

Escriba, P. V., Gonzalez-Ros, J. M., Goni, F. M., Kinnunen, P. K. J., Vigh, L., Sanchez-Magraner, L., et al. (2008). *Journal of Cellular and Molecular Medicine*, *12*, 829–875. http://dx.doi.org/10.1111/j.1582-4934.2008.00281.x.

Gao, D., McBean, N., Schultz, J. S., Yan, Y., Mulchandani, A., & Chen, W. (2005). *Journal of the American Chemical Society*, *128*, 676–677. http://dx.doi.org/10.1021/ja056364e.

Geiger, B., Spatz, J. P., & Bershadsky, A. D. (2009). *Nature Reviews Molecular Cell Biology*, *10*, 21–33. http://dx.doi.org/10.1038/nrm2593.

Giancotti, F. G., & Ruoslahti, E. (1999). *Science*, *285*, 1028–1032. http://dx.doi.org/10.1126/science.285.5430.1028.

Gong, Y. H., Li, C., Yang, J., Wang, H. Y., Zhuo, R. X., & Zhang, X. Z. (2011). *Macromolecules, 44*, 7499–7502. http://dx.doi.org/10.1021/ma201676w.
Hadjichristidis, N., Iatrou, H., Pitsikalis, M., & Sakellariou, G. (2009). *Chemical Reviews, 109*, 5528–5578. http://dx.doi.org/10.1021/cr900049t.
Hayashi, G., Hagihara, M., Dohno, C., & Nakatani, K. (2007). *Journal of the American Chemical Society, 129*, 8678–8679. http://dx.doi.org/10.1021/ja071298x.
Huang, J., & Heise, A. (2013). *Chemical Society Reviews, 42*, 7373–7390. http://dx.doi.org/10.1039/c3cs60063g.
Hyun, J., Lee, W. K., Nath, N., Chilkoti, A., & Zauscher, S. (2004). *Journal of the American Chemical Society, 126*, 7330–7335. http://dx.doi.org/10.1021/ja049721e.
Kantlehner, M., Schaffner, P., Finsinger, D., Meyer, J., Jonczyk, A., Diefenbach, B., et al. (2000). *Chembiochem, 1*, 107–114. http://dx.doi.org/10.1002/1439-7633.
Kemp, B. E., Graves, D. J., Benjamini, E., & Krebs, E. G. (1977). *Journal of Biological Chemistry, 252*, 4888–4894.
Koga, T., Nagaoka, A., & Higashi, N. (2006). *Colloids and Surfaces a-Physicochemical and Engineering Aspects, 284*, 521–527. http://dx.doi.org/10.1016/j.colsurfa.2005.10.013.
Lamb, B. M., & Yousaf, M. N. (2011). *Journal of the American Chemical Society, 133*, 8870–8873. http://dx.doi.org/10.1021/ja203198y.
Lee, J., Jung, J., Na, K., Heo, P., & Hyun, J. (2009). *ACS Applied Materials & Interfaces, 1*, 1359–1363. http://dx.doi.org/10.1021/am9002364.
Lowik, D. W. P. M., Leunissen, E. H. P., van den Heuvel, M., Hansen, M. B., & van Hest, J. C. M. (2010). *Chemical Society Reviews, 39*, 3394–3412. http://dx.doi.org/10.1039/b914342b.
MacEwan, S. R., & Chilkoti, A. (2010). *Biopolymers, 94*, 60–77. http://dx.doi.org/10.1002/bip.21327.
von der Mark, K., Park, J., Bauer, S., & Schmuki, P. (2010). *Cell and Tissue Research, 339*, 131–153. http://dx.doi.org/10.1007/s00441-009-0896-5.
Mart, R. J., Osborne, R. D., Stevens, M. M., & Ulijn, R. V. (2006). *Soft Matter, 2*, 822–835. http://dx.doi.org/10.1039/b607706d.
Mendes, P. M. (2008). *Chemical Society Reviews, 37*, 2512–2529. http://dx.doi.org/10.1039/b714635n.
Meyer, D. E., & Chilkoti, A. (1999). *Nature Biotechnology, 17*, 1112–1115.
Minelli, C., Liew, J. X., Muthu, M., & Andresen, H. (2013). *Soft Matter, 9*, 5119–5124. http://dx.doi.org/10.1039/c3sm50379h.
Morita, T., Kimura, S., Kobayashi, S., & Imanishi, Y. (2000). *Journal of the American Chemical Society, 122*, 2850–2859. http://dx.doi.org/10.1021/ja992769l.
Murtas, G. (2013). *Molecular Biosystems, 9*, 195–204. http://dx.doi.org/10.1039/c2mb25375e.
Nakanishi, J., Kikuchi, Y., Takarada, T., Nakayama, H., Yamaguchi, K., & Maeda, M. (2004). *Journal of the American Chemical Society, 126*, 16314–16315. http://dx.doi.org/10.1021/ja044684c.
Nath, N., & Chilkoti, A. (2001). *Journal of the American Chemical Society, 123*, 8197–8202. http://dx.doi.org/10.1021/ja015585r.
Nath, N., & Chilkoti, A. (2003). *Analytical Chemistry, 75*, 709–715. http://dx.doi.org/10.1021/ac0261855.
Roach, P., Eglin, D., Rohde, K., & Perry, C. C. (2007). *Journal of Materials Science-Materials in Medicine, 18*, 1263–1277. http://dx.doi.org/10.1007/s10856-006-0064-3.
Stevens, J. S., de Luca, A. C., Pelendritis, M., Terenghi, G., Downes, S., & Schroeder, S. L. M. (2013). *Surface and Interface Analysis, 45*, 1238–1246. http://dx.doi.org/10.1002/sia.5261.
Stevens, M. M., & George, J. H. (2005). *Science, 310*, 1135–1138. http://dx.doi.org/10.1126/science.1106587.

Sun, T., Qing, G., Su, B., & Jiang, L. (2011). *Chemical Society Reviews*, *40*, 2909–2921. http://dx.doi.org/10.1039/c0cs00124d.
Teeuwen, R. L. M., Zuilhof, H., de Wolf, F. A., & van Hest, J. C. M. (2009). *Soft Matter*, *5*, 2261–2268. http://dx.doi.org/10.1039/b903087e.
Todd, S. J., Scurr, D. J., Gough, J. E., Alexander, M. R., & Ulijn, R. V. (2009). *Langmuir*, *25*, 7533–7539. http://dx.doi.org/10.1021/la900376h.
Truong, M. Y., Dutta, N. K., Choudhury, N. R., Kim, M., Elvin, C. M., Hill, A. J., et al. (2010). *Biomaterials*, *31*, 4434–4446. http://dx.doi.org/10.1016/j.biomaterials.2010.02.019.
Ulijn, R. V. (2006). *Journal of Materials Chemistry*, *16*, 2217–2225. http://dx.doi.org/10.1039/b601776m.
Urry, D. W., Luan, C. H., Parker, T. M., Gowda, D. C., Prasad, K. U., Reid, M. C., et al. (1991). *Journal of the American Chemical Society*, *113*, 4346–4348. http://dx.doi.org/10.1021/ja00011a057.
Wagner, M. S., & Castner, D. G. (2001). *Langmuir*, *17*, 4649–4660. http://dx.doi.org/10.1021/la001209t.
Wang, J., Shen, M., Cao, Y., & Li, G. (2010). *Biosensors & Bioelectronics*, *26*, 638–642. http://dx.doi.org/10.1016/j.bios.2010.07.006.
Watts, J. F., & Wolstenholme, J. (2003). *An introduction to surface analysis by XPS and AES*. Wiley.
Wischerhoff, E., Badi, N., Lutz, J.-F., & Laschewsky, A. (2010). *Soft Matter*, *6*, 705–713. http://dx.doi.org/10.1039/b913594d.
Yasutomi, S., Morita, T., Imanishi, Y., & Kimura, S. (2004). *Science*, *304*, 1944–1947. http://dx.doi.org/10.1126/science.1098489.
Yasutomi, S., Morita, T., & Kimura, S. (2005). *Journal of the American Chemical Society*, *127*, 14564–14565. http://dx.doi.org/10.1021/ja055624p.
Yeo, W. S., & Mrksich, M. (2006). *Langmuir*, *22*, 10816–10820. http://dx.doi.org/10.1021/la061212y.
Yeo, W. S., Yousaf, M. N., & Mrksich, M. (2003). *Journal of the American Chemical Society*, *125*, 14994–14995. http://dx.doi.org/10.1021/ja038265b.
Yeung, C. L., Iqbal, P., Allan, M., Lashkor, M., Preece, J. A., & Mendes, P. M. (2010). *Advanced Functional Materials*, *20*, 2657–2663. http://dx.doi.org/10.1002/adfm.201000411.
Yoshida, M., & Lahann, J. (2008). *ACS Nano*, *2*, 1101–1107. http://dx.doi.org/10.1021/nn800332g.
Yousaf, M. N., Houseman, B. T., & Mrksich, M. (2001). *Angewandte Chemie-International Edition*, *40*, 1093–1096. http://dx.doi.org/10.1002/1521-3773.
Zelzer, M., McNamara, L. E., Scurr, D. J., Alexander, M. R., Dalby, M. J., & Ulijn, R. V. (2012). *Journal of Materials Chemistry*, *22*, 12229–12237. http://dx.doi.org/10.1039/c2jm31666h.
Zelzer, M., Scurr, D. J., Alexander, M. R., & Ulijn, R. V. (2012). *ACS Applied Materials & Interfaces*, *4*, 53–58. http://dx.doi.org/10.1021/am2015266.
Zelzer, M., Todd, S. J., Hirst, A. R., McDonald, T. O., & Ulijn, R. V. (2013). *Biomaterials Science*, *1*, 11–39. http://dx.doi.org/10.1039/c2bm00041e.
Zelzer, M., & Ulijn, R. V. (2010). *Chemical Society Reviews*, *39*, 3351–3357.

Photonic sensitive switchable materials

4

L. Jiang, L. Wen
Chinese Academy of Sciences, Beijing, P.R. China

4.1 Introduction

Many biological processes such as photosynthesis (Blankenship, 2002; Leonova, Fufina, Vasilieva, & Shuvalov, 2011) and vision (Chechile, 2000; Stryer, 1988) on the earth are triggered by light. Light has been widely used as a predominant stimulus to control chemical reactions, special physical properties, and biological evolution. Under light illumination, photonic sensitive materials can often reversibly change their molecular geometric and electronic structures (Kim, Kim, & Kim, 2006). In the past decade, a series of photonic sensitive materials have been developed, including everything from inorganic nanomaterials to small organic molecules and photoactive polymers to photonic sensitive biological materials and others. For inorganic complexes, some oxides such as zinc oxide (Aghamalyan, Hovsepyan, Poghosyan, & Lazaryan, 2004), titanium dioxide (Wang et al., 1997), tin oxide (Choi & Park, 2011; Li, Zhao, Zhang, & Chen, 2006), nickel oxide (Mehdaoui, Kröner, Pykavy, Freund, & Klüner, 2005), and tungsten oxide (Wang, Feng, Yao, & Jiang, 2006) exhibit photo induced hydrophilicity, which is attributed to the mechanism that involves photogenerated electrons and holes (Chen, Shen, Guo, & Mao, 2010; Palmisano, Augugliaro, Pagliaro, & Palmisano, 2007). For organic complexes, various classes of reversibly photoisomerizable substances are available, including diarylalkenes, spiropyrans, and azobenzenes (Bouas-Laurent, 2001; Heller, 1983). The two isomers in organic complexes differ from one another not only in their absorption spectra but also in several other properties such as conductivity, redox potentials, and dielectric constants (Bouas-Laurent, 2001; Feringa, Jager, & Delange, 1993). For biomolecular-based photonic sensitive materials, chlorophyll molecules and bacteriorhodopsin (bR) are the most fundamental photoreceptors, and photosensitive units have been used to design protein and enzyme optobiomaterials (Wang, Liu, & Willner, 2012). The chemical and physical properties of these materials have been extensively explored, including photoswitching of electrical conductivity (Kawai, Nakashima, & Irie, 2005) and magnetism (Matsuda & Irie, 2000a,b), photocontrollable micro patterning (Itoga, Yamato, Kobayashi, Kikuchi, & Okano, 2004), photoresponsive biocatalysis (Roy, Sharma, & Gupta, 2004), and cell adhesion (Auernheimer, Dahmen, Hersel, Bausch, & Kessler, 2005). Their applications cover photoswitchable wettability of surfaces, optobioelectronic devices, and biomedical fields. In this chapter, we review and highlight parts of noticeable achievements over the past years in the field of photonic sensitive materials.

Switchable and Responsive Surfaces and Materials for Biomedical Applications. http://dx.doi.org/10.1016/B978-0-85709-713-2.00004-3

4.2 Photonic sensitive switchable materials

4.2.1 Inorganic compound-based photonic sensitive materials

Many families of organic and inorganic photoresponsive compounds have been discovered (He & Yao, 2006; Irie, Sakemura, Okinaka, & Uchida, 1995; Kumar & Neckers, 1989). In comparison with photonic sensitive organic materials, inorganic photoresponsive compounds are comparatively few. Typical examples include the metal oxides (Mazurkevich & Kobasa, 1997), alkali metal azides (Deb, 1963), sodalite (Williams, Hodgson, & Brinen, 2006), mercury salts (Khalil, Rophael, & Mourad, 2002), zinc sulfide (Becker & Bard, 1983), Ni–Al-layered double hydroxides (Wei et al., 2006), and valence-tautomeric Prussian blue analogs (Liu et al., 2003), among others. For example, Fujishima and co-workers reported the photogeneration of a highly amphiphilic (both hydrophilic and oleophilic) titanium dioxide surface in 1997 (Wang et al., 1997). They used friction force microscopy images and Fourier transform infrared spectroscopy to study the formation mechanism of TiO_2 single-crystal, highly amphiphilic surfaces (see Figure 4.1).

4.2.2 Organic compound-based photonic sensitive materials

Many photosensitive molecules are known to be transformed under photoirradiation into other isomers, which return to the initial state either thermally or photochemically. The isomerizations are always accompanied by certain physical and chemical property changes. The well-known organic photoresponsive compounds include, among others, diarylethenes (Feringa et al., 1993; Uchida, Saito, Murakami, Nakamura, & Irie, 2003), spiropyrans (Feringa et al., 1993), azobenzene compounds (Kumar & Neckers, 1989; Renner, Kusebauch, Löweneck, Milbradt, & Moroder, 2005), and triphenylmethane leuco derivatives (Irie & Kunwatchakun, 1986), as discussed below. These photonic sensitive molecules can receive and convert the light signal appropriately and can be used as photoreceptors. The light-induced isomerisation and photochemical reactions are illustrated in Figure 4.2.

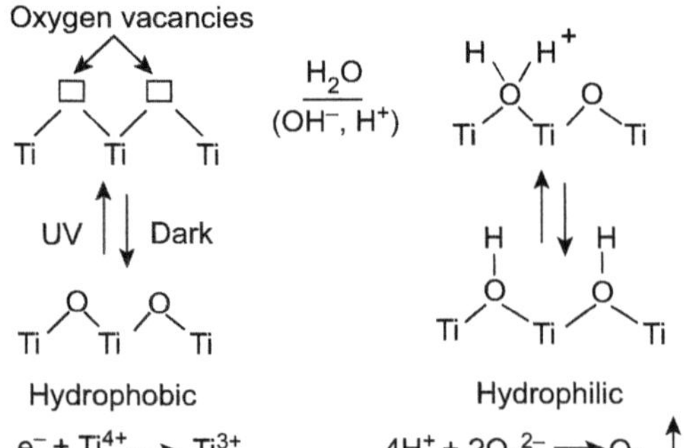

Figure 4.1 Formation mechanism of TiO_2 single-crystal highly amphiphilic surfaces.

1. Photochromic diarylethenes underwent reversible photocyclization between two isomers with different absorption spectra upon irradiation with light of appropriate wavelengths. Photochromic diarylethenes have been extensively studied for optoelectronic applications because of their notable thermally irreversible photochromic behavior and fatigue resistance.
2. The photochromic behavior of spirobenzopyran derivatives has been investigated by many researchers. The spirobenzopyran molecule underwent reversible photochemical cleavage of the C–O bond. The compound is stable in its closed form, and its solution is colorless. Another interesting feature of the photoinduced reaction is that UV irradiation induces electric charges in the open form of the molecule.
3. Azobenzene and its derivatives can undergo cis–trans isomerization, and they generally assume the trans-configuration under dark condition. The cis–trans isomerization of azobenzene derivatives is characterized by monitoring the intense UV absorption that is attributed to the π–π^* transition of the trans-isomer and the weak band at longer wavelengths arising from the n–π^* transition of the cis isomer.
4. Triphenylmethane leuco derivatives are well-known photochromic molecules that dissociate into ion pairs under UV irradiation, producing intensely colored triphenylmethyl cations. The cations thermally recombine with counter ions as above. Triphenylmethane leuco derivatives have been widely used as photoreceptor molecules.

Photoresponsive polymers are quite special polymers, able to respond to light and dark conditions and thus give rise to reversible variations in their structure and conformation. They can be obtained by introducing photochromic units, such as azobenzene or spiropyan group, into the macromolecules of polymeric compounds (Irie, 1990). Several papers have described light-induced changes in the shape of polymers,

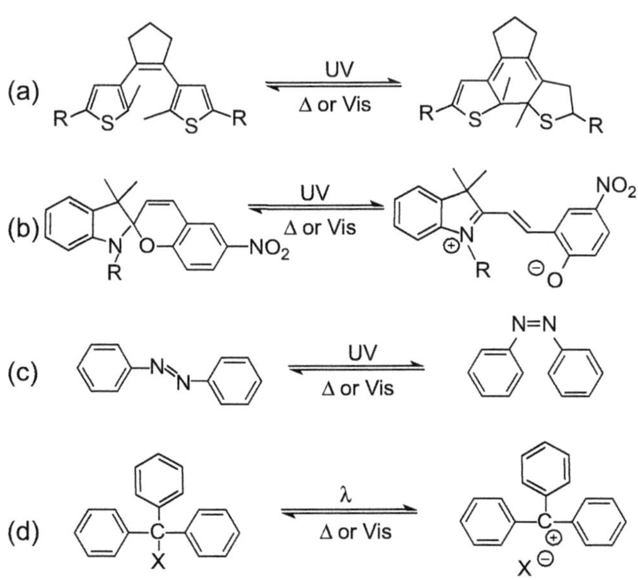

Figure 4.2 Schematic illustration of light-induced isomerisation and photochemical reactions. (a) Photocyclization; (b) zwitter ion formation; (c) trans–cis isomerization; and (d) ionic dissociation.

such as contraction (Finkelmann, Nishikawa, Pereira, & Warner, 2001; Li, Keller, Li, Wang, & Brunet, 2003), bending (Camacho-Lopez, Finkelmann, Palffy-Muhoray, & Shelley, 2004; Ikeda, Nakano, Yu, Tsutsumi, & Kanazawa, 2003), or volume changes (Juodkazis et al., 2000; Suzuki & Tanaka, 1990). Yu et al. showed that a single film of a liquid-crystal network containing an azobenzene chromophore can be repeatedly and precisely bent along any chosen direction by using linearly polarized light (see Figure 4.3) (Yu, Nakano, & Ikeda, 2003). This striking photomechanical effect results from a photoselective volume contraction and may be useful in the development of high-speed actuators for microscale or nanoscale applications, for example, in micro-robots in medicine or optical microtweezers.

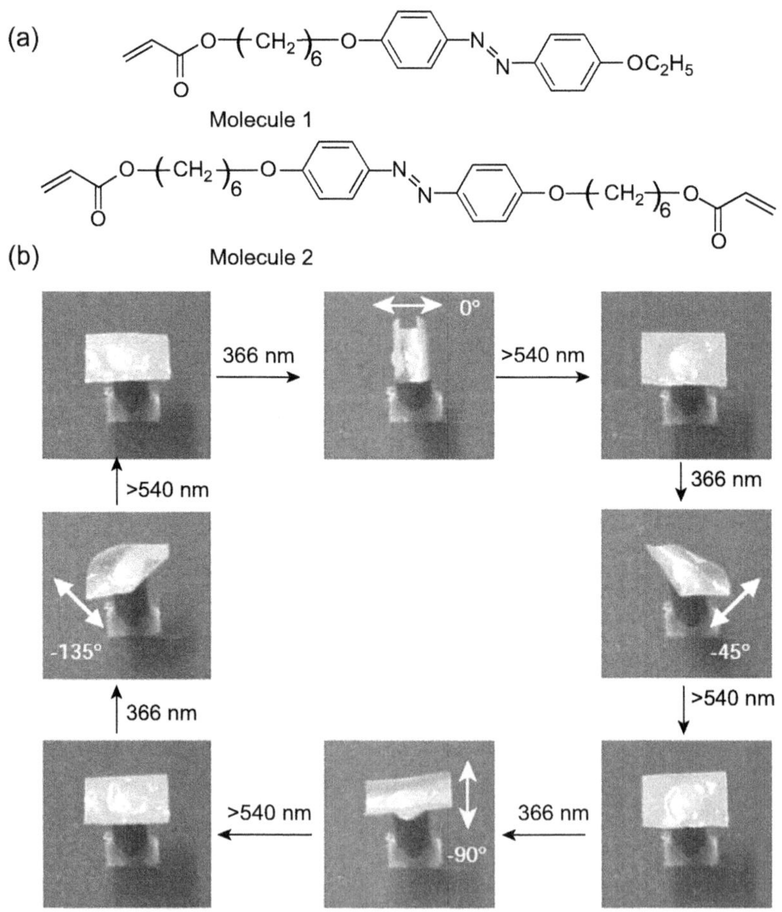

Figure 4.3 (a) Chemical structures of the liquid-crystal monomer (molecule 1) and cross-linker (molecule 2), which possess azobenzene moieties used for preparation of the photoresponsive polymer film. (b) Photographic frames of the film bending in different directions in response to irradiation by linearly polarized light.
Adapted with permission from Yu et al. (2003).

4.2.3 Biomolecular-based photonic sensitive materials

1. Photosynthetic pigment: Chlorophyll molecules acting as the main producers of photosynthetic pigment are bound to proteins of the photosynthetic membranes and harvest the sunlight (Von Wettstein, Gough, & Kannangara, 1995). In oxygenic plants, photon captures and conversion of light energy into chemical energy takes place in pigment–protein complexes known as Photosystem I (PSI) and Photosystem II (PSII) reaction centers (Munekage et al., 2004). The two reaction centers use a special pair of chlorophyll molecules as the primary electron donor and chlorophyll or pheophytin as the primary electron acceptor. The integration of natural photosynthetic reaction centers with electrodes and the construction of photobioelectrochemical cells and photobiofuel cells has been well developed (Wang, Liu, et al., 2012). Bacteriorhodopsin is another important photosynthetic pigment that serves as a light-driven proton pump in the purple membrane of *Halobacterium salinarum* (Oesterhelt & Stoeckenius, 1973). It has been shown that the protein is composed of seven transmembrane helices with a retinal chromophore covalently bound in the central region via a protonated Schiff base to a lysine residue (Figure 4.4(a)) (Jin et al., 2008). After light absorption, bR converts to a light-adapted state, with the retinal chromophore in the all-trans configuration (Balashov, Imasheva, Govindjee, & Ebrey, 1996). Light absorption by this form triggers isomerisation of the all-trans configuration to the 13-cis isomer (see Figure 4.4(b)).
2. Semiconductor-biomolecule quantum dots (QDs) hybrids: Semiconductor nanocrystals, also known as QDs, exhibiting unique photonic, electronic, and catalytic properties, provide invaluable building blocks for the construction of functional structures (Li & Zhang, 2009). Both group II–VI (e.g., CdSe, CdTe, CdS, and ZnSe) and group III–V (e.g., InP and InAs) nanocrystals have been synthesized and studied extensively in the past (Leutwyler, Bürgi, & Burgl, 1996; Wang et al., 2012). QDs in the 2- to 6-nm size range are of considerable current interest, not only because of their unique size-dependent properties but also because of their dimensional similarities with biological macromolecules (e.g., nucleic acids and proteins) (Nirmal & Brus, 1999). These similarities could allow an integration of nanotechnology and biology, leading to major advances in medical diagnostics, targeted therapeutics, molecular biology, and cell biology (Niemeyer, 2001). Some research reported by several groups has linked colloidal nanoparticles to biomolecules such as peptides (Whaley, English, Hu,

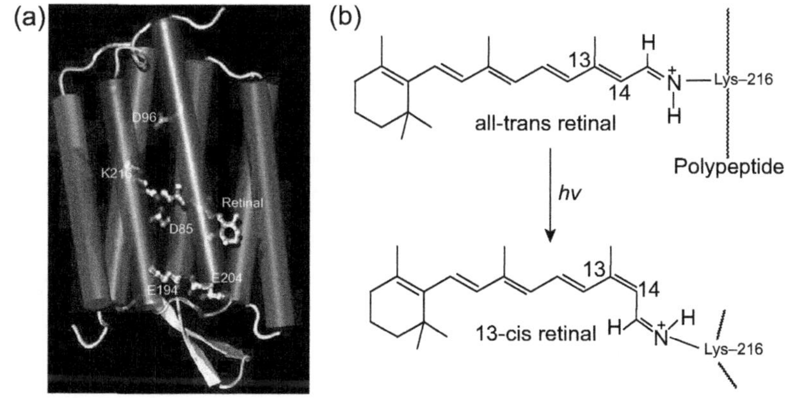

Figure 4.4 (a) Schematic representation of the three-dimensional structure of bacteriorhodopsin (bR). (b) Photoisomerization of all-trans to 13-cis retinal in bR.
Adapted with permission from Jin et al. (2008).

Barbara, & Belcher, 2000), proteins (Bruchez, Moronne, Gin, Weiss, & Alivisatos, 1998; Chan & Nie, 1998; Mattoussi et al., 2000), and DNA (Mitchell, Mirkin, & Letsinger, 1999; Pathak, Choi, Arnheim, & Thompson, 2001). Current bioconjugation methods are schematically illustrated in Figure 4.5 (Chan et al., 2002). For example, Lee et al. constructed a hybrid bR/(QDs) bionanosystem of thin films, and a stationary current is generated from the modified photocycle (Li, Li, Bao, Bao, & Lee, 2007). They proposed a model to explain that QDs could act as nanoscaled light sources embedded in bR to assist its generation of a stationary photocurrent.

3. Photoswitchable biomaterials: Nature has evolved many sophisticated photoresponsive systems, such as vision, photosynthesis, and photomorphogenesis (Kubasek et al., 1992). These natural photoresponsive systems are generally composed of a photosensitive element, which captures optical signals and converts them to physicochemical signals; and a second functional element (e.g., a protein domain), which senses the physicochemical signals and exhibits new output functions. A common feature is a photochromic molecule (chromophore) embedded in a biomolecular matrix. The absorbed light activates a chemical transformation in the chromophore (e.g., photoisomerization), which subsequently controls the conformation and/or assembly of the surrounding biomolecule or biomembrane. In general, there are two fundamental classes of photoswitchable biomolecules that have been developed: single-cycle and multicycle photoswitches, respectively. Single-cycle photoswitches are a class of biological photoswitches in which the biomaterial is deactivated by the attachment of photosensitive chemical–protecting groups. Figure 4.6 shows a biomaterial whose

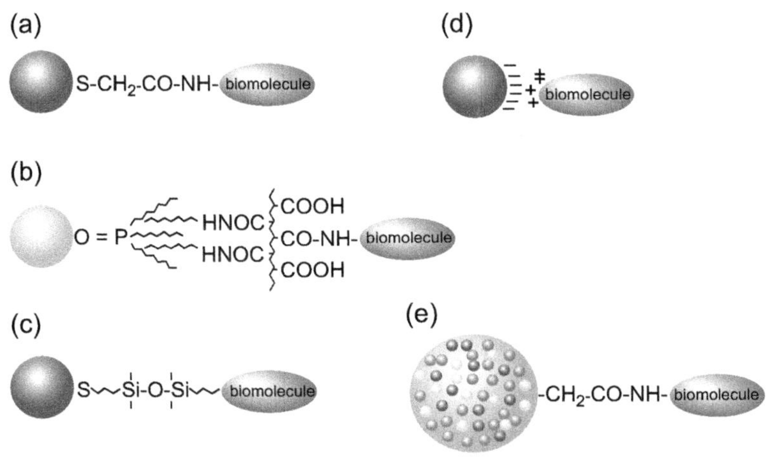

Figure 4.5 Schematic illustrations of bioconjugation methods: (a) bifunctional linkage, (b) electrostatic attraction, (c) hydrophobic attraction, (d) silanization, (e) nanobeads.
Adapted with permission from Chan et al. (2002).

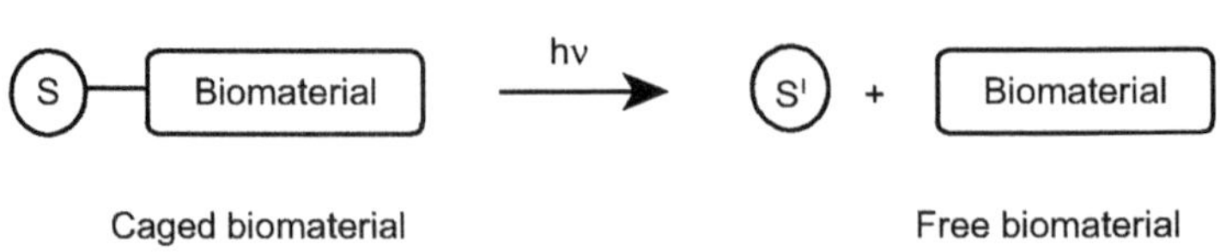

Figure 4.6 Photochemical activation of a biomaterial by light-induced cleavage of a photoprotective group.
Adapted with permission from Feringa and Browne (2001).

function has been blocked by chemical modification ("caged" biomaterial) (Feringa & Browne, 2001). The biomaterial is activated by light-stimulated removal of the protective group to restore the biologically active structure (Bochet, 2001). Activation of enzymes (Turner, Pizzo, Rozakis, & Porter, 1988), photoinduced formation of specific ion-chelators (Adams, Kao, & Tsien, 1989), and light-triggered activation of important biological components such as CAMP (Nargeot, Nerbonne, Engels, & Lester, 1983), CGMP (Nerbonne, Richard, Nargeot, & Lester, 1984), ATP (Kaplan & Hollis, 1980), and InP_3 have been used in this approach. Several review articles summarize the subject of single-cycle photoswitches (Morrison, 1993; Shimoboji et al., 2002). However, this kind of photobiological switch can only move from "closed" to "on" or from "open" to "closed," as an irreversible cycle, limiting its use range, so there has been more interest in multicycle photobiological switches.

Multicycle photobiological switches are the results of efforts to transform biomaterials or biological environments into light-activated operating reversibly between a mute bioactive state (switch "off") and an activated biological function (switch "on") (Feringa & Browne, 2001). Three different methodologies to reversibly photostimulate biomaterials are depicted in Figure 4.7. Willner and Rubin summarized the development of photobiological switches based on covalent attachment of photoisomerizable units to biological materials and on immobilization of biomaterials in photoisomerizable environments (Willner & Rubin, 1996).

4.3 Potential applications

4.3.1 Switchable wettability on photoresponsive surfaces

Inorganic oxides always occupy a paramount position in many industrial applications because of their intrinsic advantages, such as a good chemical and mechanical stability, low cost, and outstanding photo/electro properties. Recently, the photoinduced changes of surface chemical states of inorganic oxide materials have made it possible to create an intelligent surface with controllable wettability. The photoinduced superhydrophilicity of titanium dioxide was first reported by the Fujishima's group (Wang et al., 1997). On titanium dioxide surfaces, the water contact angle (CA) is about $72° \pm 1°$. According to Vogler's definition of hydrophobic surfaces, this surface should belong to the weakly hydrophobic state. After the sample is exposed to UV irradiation, water droplets spread over the surface immediately, and the water CA reaches about $0° \pm 1°$, producing a superhydrophilic state with an unstable three-phase boundary. After dark storage for a long period, the surface can revert to the original state. Thus, a transition between hydrophobicity and superhydrophilicity is realized under light stimulus, and accordingly, the three-phase boundary changes from a stable case to an unstable one. Apart from titanium dioxide, another excellent photosensitive oxide, semiconductor zinc oxide, can also be used to construct a photoresponsive surface (Liu, Feng, Zhai, Jiang, & Zhu, 2004). On a flat zinc oxide substrate, the CA can decrease from 109° to 10° through UV illumination; after long-term dark storage, the initial hydrophobic surface can be recovered. Interestingly, on microscale and nanoscale hierarchically structured zinc oxide films (see Figure 4.8(a)) prepared via CVD, both the hydrophilicity via UV illumination and the hydrophobicity under

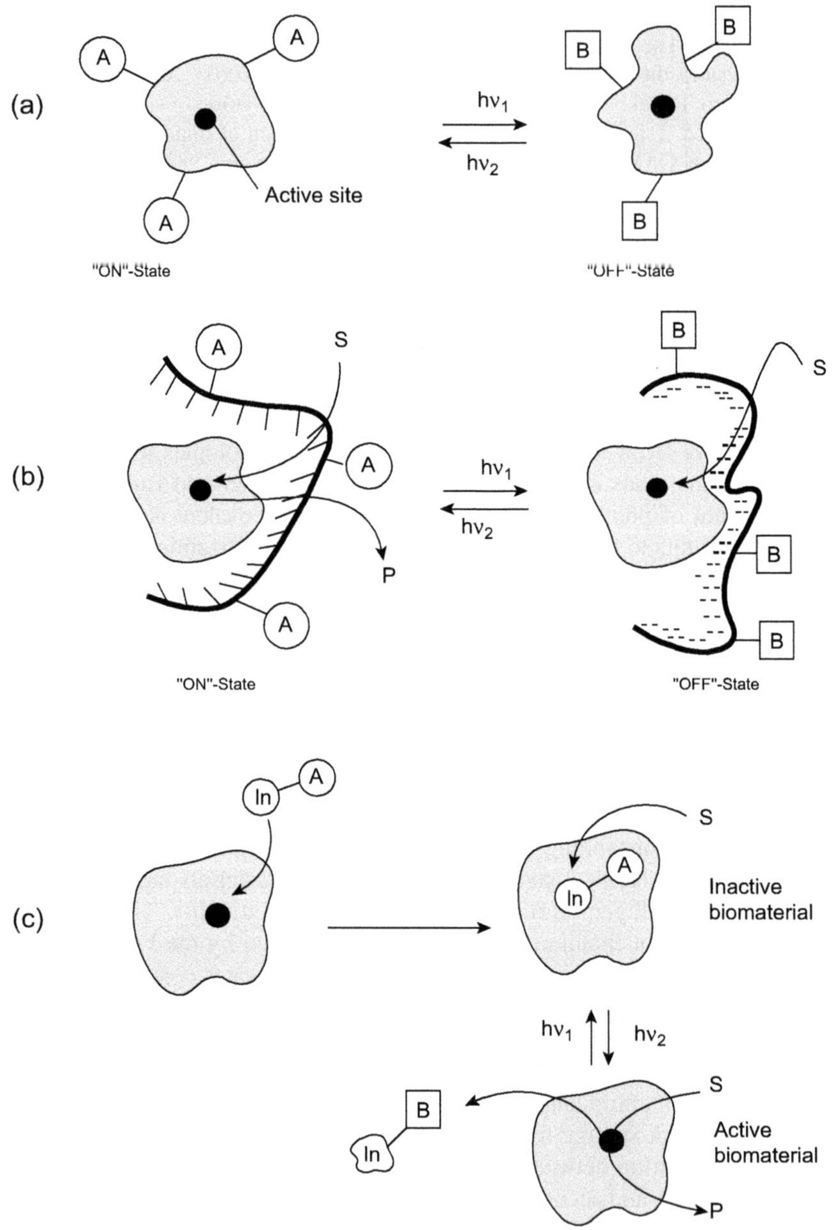

Figure 4.7 Three methods for switching the activity of a biomaterial: (a) tethering of photoisomerizable units to a protein; (b) integrating the biomaterial within a photosensitive environment; and (c) use of photoisomerizable inhibitors.
Adapted with permission from Feringa and Browne (2001).

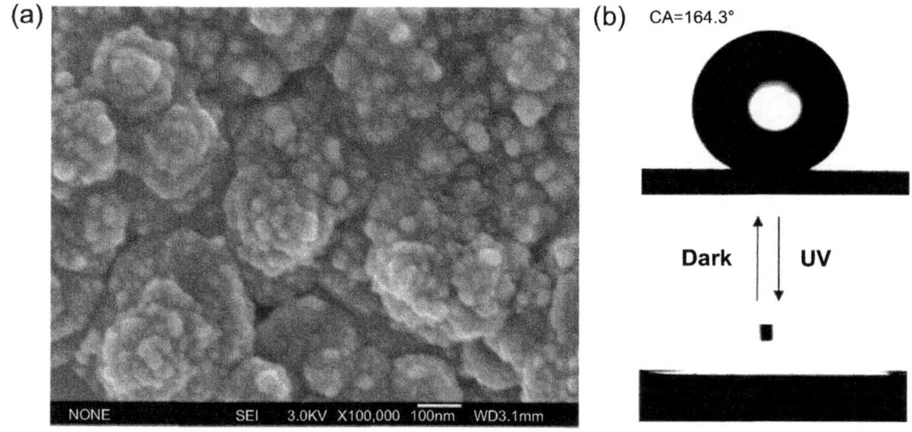

Figure 4.8 (a) Scanning electron micrograph of micro- and nanoscale hierarchically structured zinc oxide films at high magnification. (b) Photographs of water droplet shape on the zinc oxide/titanium dioxide nanorod film via dark storage and UV illumination.
Adapted with permission from Liu et al. (2004).

dark storage are amplified by the roughness. On the hierarchically structured surface, as shown in Figure 4.8(b), the CA can reach 163.4° under dark storage, whereas, via UV illumination, the CA becomes about 0° owing to the cooperation between surface 2D and 3D capillary effects (Bico, Thiele, & Quéré, 2002; Bico, Tordeux, & Quéré, 2001), suggesting that a photoresponsive switching between superhydrophobicity and superhydrophilicity has been realized. Similar remarkable surface wettability alternations were also reported on aligned zinc oxide nanorod film prepared via a solution approach (Feng et al., 2004), a vapor–solid process (Huang et al., 2005), and solution-dipping template methods (Li, He, Qin, Feng, & Zhang, 2005; Meng et al., 2005). Zhang et al. also fabricated superhydrophobic and superhydrophilic micropatterns by light lithography using the UV-induced wettability conversion of a zinc oxide nanocolumnar film (Zhang, Sato, & Fujishima, 2004).

Many unique functions are of importance in fabricating intelligent surfaces, smart windows, and microfluidic switches. It will be highly desirable to combine excellent properties with a superwettability switch on the same film. For example, an aligned tin oxide nanorod film (Figure 4.9(a)), prepared by a crystal-seed-based hydrothermal method, exhibits switchable superhydrophobicity and superhydrophilicity (Zhu, Feng, Feng, & Jiang, 2006). The CA changes from 154.1° to 0° when exposed to UV irradiation (inset in Figure 4.9(a)), and the wettability converts to its initial superhydrophobic state by storing the films in the dark for a certain time. Tin oxide is a wide band-gap semiconductor that is important for its excellent transparency and conductivity. However, transparency and superhydrophobicity are competing properties in the preparation of transparent superhydrophobic films. Nanorod structures are essential in increasing the surface roughness and forming a superhydrophobic film, but their formation decreases the transparency of the film. In this case, a film with this type of smart wettability shows 60% transmittance in the visible region and semiconducting properties. Although the transparency and conductivity

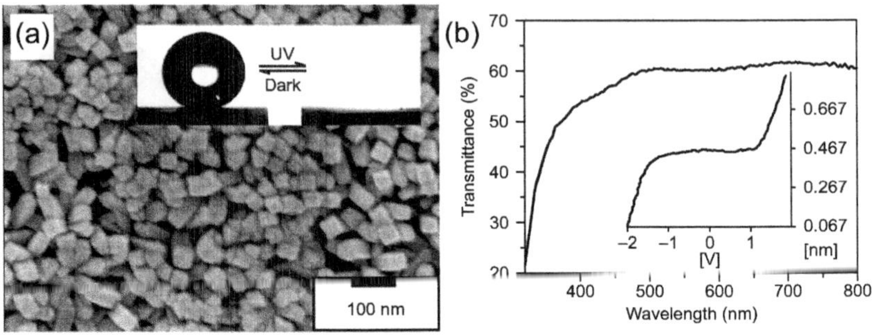

Figure 4.9 (a) Scanning electron micrograph of a tin oxide nanorod film. (b) Transmission spectrum of the as-prepared film in the visible region (insert is an *I–V* curve of the as-prepared film).
Adapted with permission from Zhu et al. (2006).

need to be improved further (Figure 4.9(b)), this study does demonstrate an embryonic combined functional switchable surface.

In the field of UV-responsive superhydrophobic/superhydrophilic switches, the use of inorganic surfaces (Feng et al., 2004; Lim, Kwak, Lee, Lee, & Cho, 2007), for example, zinc oxide, titanium dioxide, tin oxide, and silicon dioxide nanoparticles (Lim, Han, Kwak, Jin, & Cho, 2006) is often involved, whereas the use of only organic materials is less developed, as this usually results in only small changes of the surface wetting properties (Abbott, Ralston, Reynolds, & Hayes, 1999; Ichimura, Oh, & Nakagawa, 2000; Rosario et al., 2004). In 2005, based on photoresponsive properties of amphiphilic azobenzene, we fabricated a photoswitched azobenzene monolayer on rough substrates by a simple electrostatic self-assembly technique, forming an azobenzene monolayer on the laser-etched rough substrate (Jiang et al., 2005). The transition from superhydrophobic to superhydrophilic was obtained by light irradiation and a large reversible CA change of about 66° (see Figure 4.10). The surface of inverse opal was modified by a self-assembled azobenzene monolayer, and the ordered monodisperse air spheres throughout the inverse opal contribute to control both structural color and the response of wettability (Ge et al., 2006).

By using the photogenerated ionic dissociation properties of triphenylmethane leuco derivatives, Zhang et al. designed and synthesized a malachite green–terminated alkanethiol (MG–CN–SH), shown in Figure 4.11 (Jiang, Wan, Smet, Wang, & Zhang, 2008). The electrically neutral malachite green group behaves like a hydrophobic group when no extra stimulus of pH or UV is used; however, upon a stimulus by pH change or UV irradiation, the neutral malachite green can be ionized into its corresponding delocalized triphenylmethyl cation (Irie & Kunwatchakun, 1986; Jiang et al., 2007; Uda, Oue, & Kimura, 2004; Willner, Sussan, & Rubin, 1992), acting as a reversible charge generator, and a cyanide anion, which is released to the aqueous phase. Besides the malachite green as the responsive moiety, the introduction of a thiol group allows for chemisorptions onto gold-coated surfaces, leading to the formation of a self-assembled monolayer with pH- and UV-responsive wetting properties.

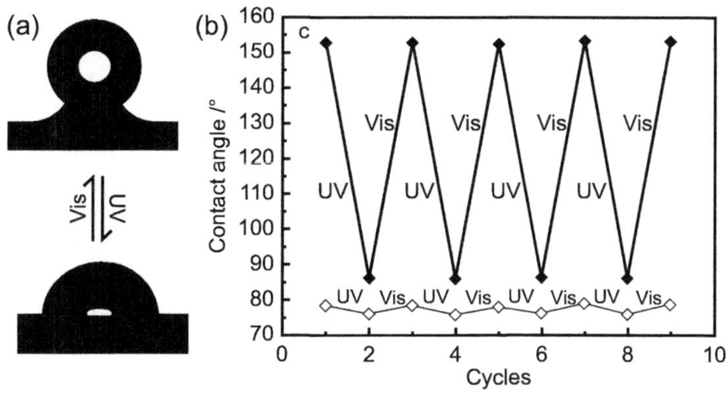

Figure 4.10 Reversible wettability transition of photoresponsive azobenzene monolayer by ultraviolet and visible light irradiation. (a) The shapes of water drops on photoresponsive monolayer with patterned substrate of 40 mm pillar spacing and flat silicon wafer upon UV and Vis irradiation. (b) Reversible wettability transition of photoresponsive monolayer by UV and Vis irradiation.

Adapted with permission from Jiang et al. (2005).

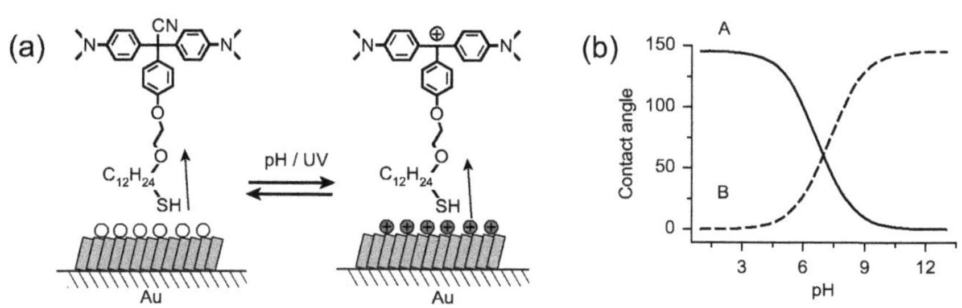

Figure 4.11 (a) Schematic illustration of a self-assembled monolayer of malachite green–terminated alkanethiol (MG–CN–SH) with pH- and ultraviolet-responsive wetting properties. (b) Curves A and B represent the ideal model for two kinds of pH-responsive surfaces with completely different pH responses.

Adapted with permission from Jiang et al. (2008).

4.3.2 Optical switches

Among addressable nanoscale devices, photoinduced molecular switches stand out for their ability to convert an optical input into a variety of useful output signals (Kumar et al., 2008). Their short response times and reversibility allow switching between different states rapidly and repeatedly. A particularly challenging and appealing application of photoswitches would be gating channels. In 2010, Driessen et al. reported reversible photo control of protein translocation by cross-linking the lateral gate of channel protein SecYEG with an azobenzene linker (Bonardi, London, Nouwen, Feringa, & Driessen, 2010). However, this kind of protein-based channels and their embedding in lipid bilayers are susceptible to deterioration in changing external environments. To overcome these difficulties, much effort has been focused on developing biomimetic analogues such as solid-state nanochannels (Han et al., 2011; Yameen et al., 2010). Light-responsive synthetic solid-state nanochannels are particularly attractive because they display several important advantages over their biological counterparts such as robustness, stability, control over channel geometry, amenable integration into devices, and tailorable surface properties (Martin & Siwy, 2007; Wanunu & Meller, 2007). For example, Zhai et al. presented a photoinduced, chemically driven, smart-gating nanochannel by photoinduced pH change in the environmental solution, which can drive the conformation response of C4-DNA motors in the inner surfaces of the nanochannel and realize the photo control (Wen, Ma, Tian, Zhai, & Jiang, 2012). Using this approach, a steady and reversible light-controlled nanochannel has been constructed (see Figure 4.12). Most recently, a kind of green carbinol base, malachite green derivative (MG–OH–NH_2) as the active element was first synthesized and then incorporated into a track-etched nanochannel in PI membrane. This successfully designed ionic gate can be driven by pH and light stimuli (see Figure 4.13) (Wen, Liu, et al., 2012).

Almost at the same time, Zhai et al. reported an artificial functional nanochannel system as a novel nanofluidic diode that displays both light-gated and pH-tunable transport properties, as well as exhibiting current rectification (Zhang et al., 2012). They constructed this light-gated nanofluidic diode system at different pH values by modifying the inner surface of the conical nanochannel with the spiropyran1′-(3-carboxypropyl)-3′,3′-dimethyl-6-nitro-spiro [2H-1] benzopyran-2,2′-indoline (SP-COOH) (Rosario et al., 2002) corresponding to the three metastable states in spiropyran that were shown in Figure 4.14.

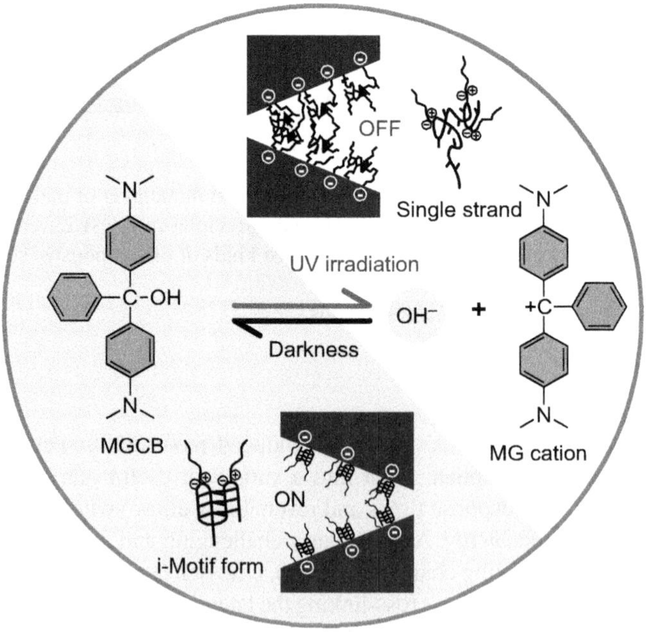

Figure 4.12 Mechanism of a photoinduced, chemical-driven, smart-gating nanochannel. Adapted with permission from Wen, Ma, et al. (2012).

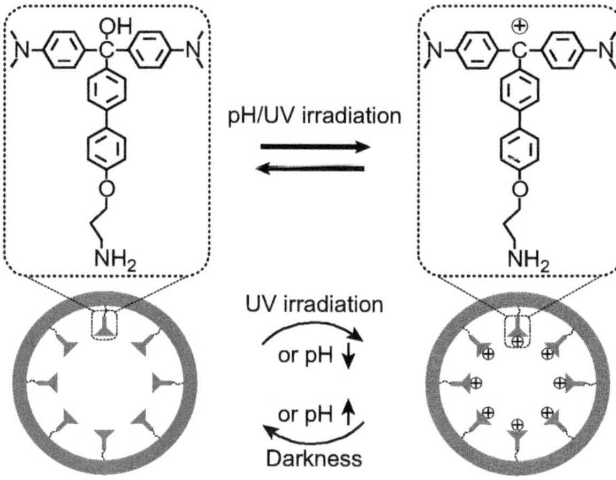

Figure 4.13 Light and pH dual-driven ionic gate: MG–OH–NH_2-functionalized nanochannel system.
Adapted with permission from Wen, Liu, et al. (2012).

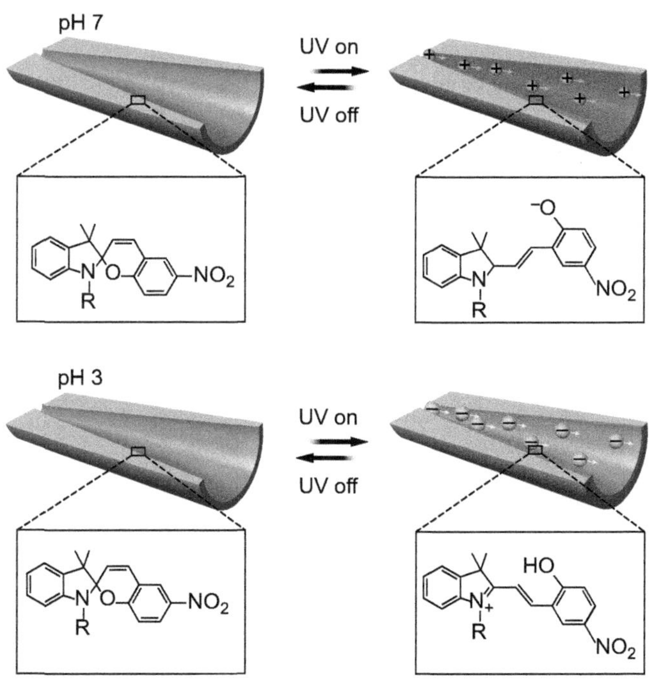

Figure 4.14 Schematic representation of the light-gated nanofluidic diode system at different pH values.
Adapted with permission from Zhang et al. (2012).

4.3.3 Optobioelectronic devices

Optobioelectronic systems provide a means for the electronic transduction of recorded optical signals. Electrons and photons play key roles in nature. Willner and co-workers described methods to harness biomaterials for the tailoring of optobioelectronic systems of new functions and applications (Wang, Liu, et al., 2012). This is schematically outlined in Figure 4.15; photoactive biomaterials might be integrated with electrode supports to yield new optobioelectronic systems. Substantial progress (e.g., photobiofuel cells) has been accomplished in the assembly of man-made photoactive biomolecular systems or natural photosystems on electrodes. In 2008, Ghirardi et al. designed a photoelectrochemical biofuel cell using a dye-sensitized nanoparticulate TiO_2 photoanode, supported on a transparent, conductive fluorine-doped tin(IV) oxide substrate (Hambourger et al., 2008). The photoanode serves to generate a charge-separated state, reminiscent of molecular reaction centers (Durrant, Haque, & Palomares, 2006). Photon absorption, by the sensitizer porphyrin (P), leads to a porphyrin-excited singlet state (P*), which relaxes via electron transfer to the TiO_2 conduction band. Willner and co-workers later reported the photonic wiring of glucose oxidase (GO*x*), a process that enables the construction of a photobiofuel cell that generates hydrogen (H_2) fuel (Tel–Vered, Yildiz, Yan, & Willner, 2010). Figure 4.16 represents the $[Ru(bpy)_3]2^+$/GO*x*|bis-aniline-cross-linked Pt–NP photobiofuel cell. They showed the photonic electron-transfer wiring of GO*x* with the electrode. The photobiofuel cell uses a photonically wired bioanode that oxidizes glucose with the concomitant evolution of hydrogen fuel at the cathode. The system represents a novel approach to harness enzymes for light-to-fuel energy conversion, and paves the way to assemble closed cycles for energy storage and use.

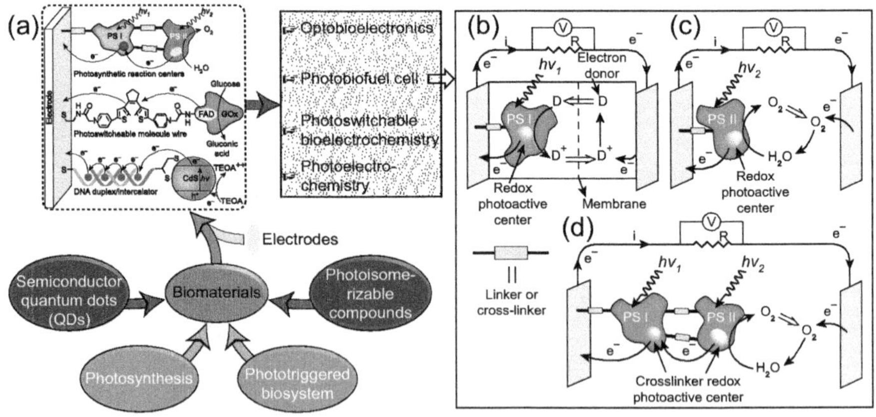

Figure 4.15 (a) Mimicking of natural photoinduced processes for designing optobioelectronic systems. Schematic configurations Photo-bioelectrochemical cells based on photoanodes consisting of photosynthetic reaction centers: (b) A PS I-based cell. (c) A PS II-based cell. (d) A PS I/PS II coupled cell.

Adapted with permission from Wang, Liu, et al. (2012).

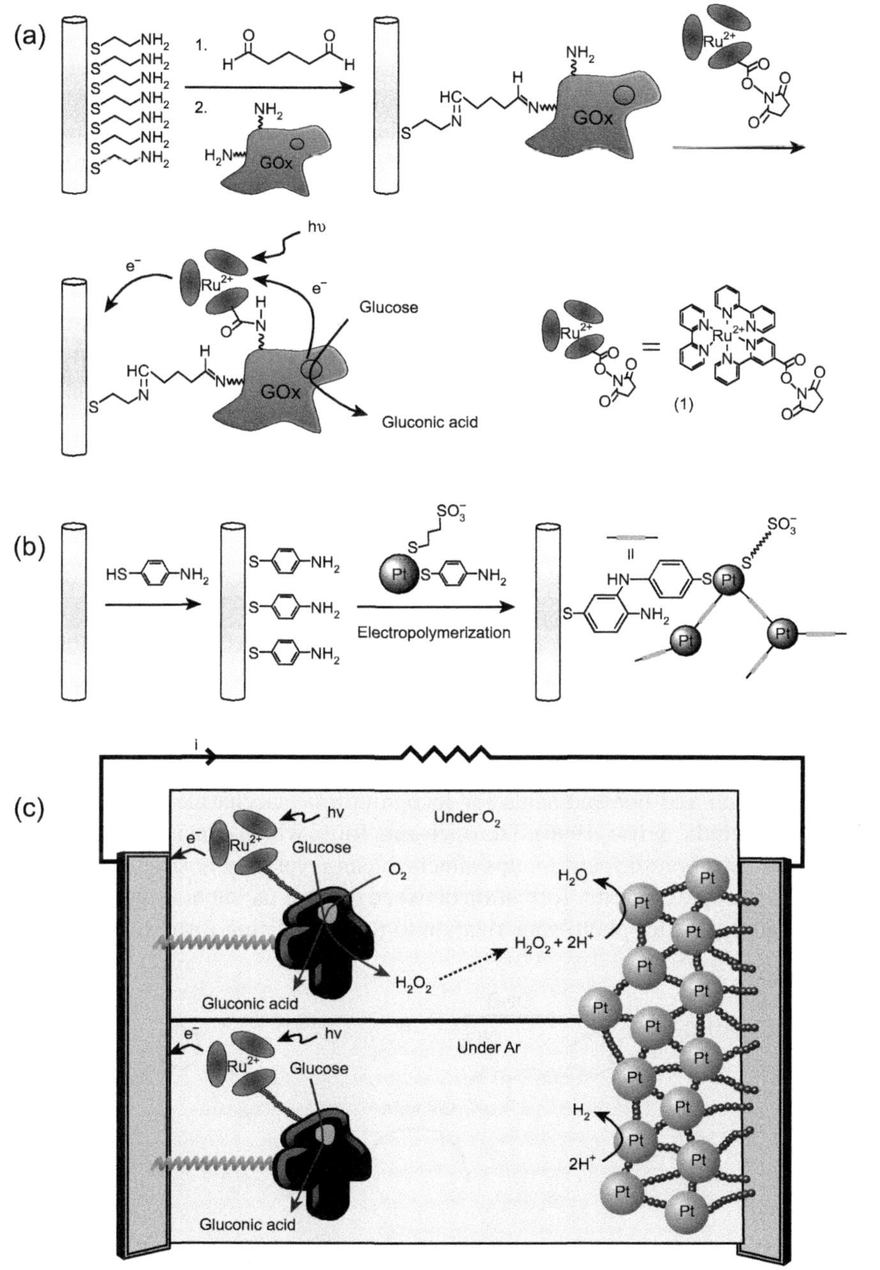

Figure 4.16 Assembly of the photobiofuel cell elements. (a) Preparation of the photobiofuel cell anode. (b) Synthesis of the photobiofuel cell cathode, which consists of a bis-aniline-cross-linked Pt-NP film. (c) Illustration of the $[Ru(bpy)_3]2^+$/GO*x*|bis-aniline-cross-linked Pt–NP photobiofuel cell.

Adapted with permission from Tel–Vered et al. (2010).

4.3.4 Biocatalysts

Enzymes are biomolecules of particular interest, as they have high substrate specificity and exhibit high catalytic efficiency. Engineering of enzymes for improved catalytic efficiency has been widely sought for industrial applications. In order to enhance enzyme catalytic rate, Agarwal's group designed a strategy that uses a photoactivated ligand bridge to modulate enzyme conformation (Agarwal, Schultz, Kalivretenos, Ghosh, & Broedel, 2012). Under activation by light of suitable wavelengths, the bridge alters the enzyme conformation by activating the dynamically important regions that promote catalysis. They described a novel chemical modification of *Candida antarctica* lipase B that allows modulation of the enzyme conformation to promote catalysis successfully. Recently, Krauss et al. reported the direct photocontrol of specific proteins by incorporating a photocaged amino acid at predetermined positions in the protein (see Figure 4.17) (Krauss, Drepper, & Jaeger, 2011). Such positions are chosen to either sterically block the substrate access to the active site (Chou, Young, & Deiters, 2010) or inhibit covalent interactions of the respective amino acid with the substrate molecule in the active site of the enzyme (Edwards, Young, & Deiters, 2009).

4.3.5 Biosensors

A biosensor is described as a compact analytical device, incorporating a biological or biomimetic sensing element, either closely connected to or integrated within a transducer system. Biosensors can be classified by their bio-recognition system or by the method of signal transduction. For example, Irie et al. showed the first research on photochromic saccharide tweezers (Irie & Kato, 1985), with a diarylethene unit as a switch and boronic acids for recognition of saccharides (see Figure 4.18) (Takeshita, Uchida, & Irie, 1996). The open-ring form, which consists of a diarylethene moiety with two boronic acid groups attached, can reversibly host glucose because it forms a 1:1 complex by ester formation between the two facing boronic acids and four hydroxyl groups. After photoisomerization to the closed-ring form, however, the two

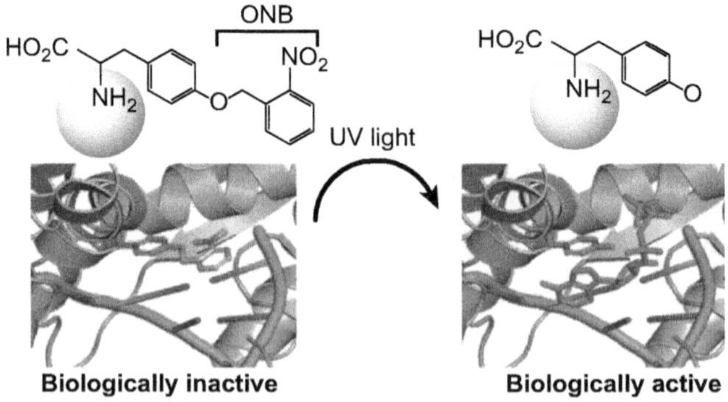

Figure 4.17 Photocaging of biological processes, showing an *o*-nitrobenzyl (ONB) moiety attached to a functionally important tyrosine residue (Y639) of the T7 RNA polymerase. Adapted with permission from Krauss et al. (2011).

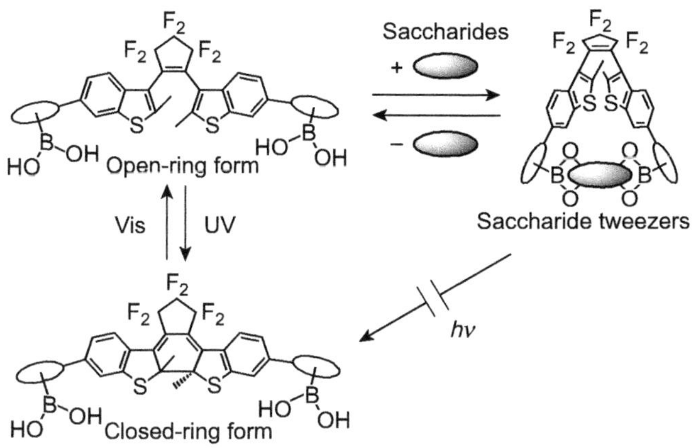

Figure 4.18 Photoswitchable recognition of saccharides by diarylethene derivative. Adapted with permission from Takeshita et al. (1996).

boronic acid groups are much more distant from each other and complexation is much weaker. The switching of the complexation can be measured by circular dichroism spectroscopy. From a "machine" viewpoint, this system can be viewed as a pair of tweezers. To emphasize the importance of defined structures on device performance and to understand and optimize sensors that can be used for the detection of hydrogen peroxide, Parak and co-workers systematically studied the QD-SAM-Au junction for the case of various dithiol SAMs adsorbed onto differently prepared gold electrodes (Khalid et al., 2011) in order to emphasize the importance of defined structures on device performance and to understand and optimize sensors. And the best device performance was obtained for a conjugated stilbenedithiol linker, which can be assembled as a structurally well-ordered mono layer by means of immersion at elevated temperature without the necessity of protection groups. Optical biosensors have been highly developed for their advantages of speed, the immunity of the signal to the electrical or magnetic interference, and the potential for higher information content.

4.3.6 Biomedical engineering

Much attention has been paid to macromolecular assemblies for drug delivery system (DDS) in therapeutics. Light-induced release of encapsulated contents within macromolecular assemblies has been explored through both chemical and physical changes to the assembly. An early study explored phospholipid vesicles (liposomes) that were sensitive to photooxidation induced by a sensitizer (Anderson & Thompson, 1992). Katz and Burdick reported on the incorporation of light-responsive moieties into macromolecular assemblies for DDS (Katz & Burdick, 2010). Considerable efforts are currently being exerted to develop more efficient and safe DDS that provide therapeutic levels of drugs in specific organs, tissues, or even cellular structures. For example, hydrogel-based DDSs are of great interest, as they can be easily modified to tune their characteristics and can lead to targeted delivery, extension of circulation time, and reduction of toxicity

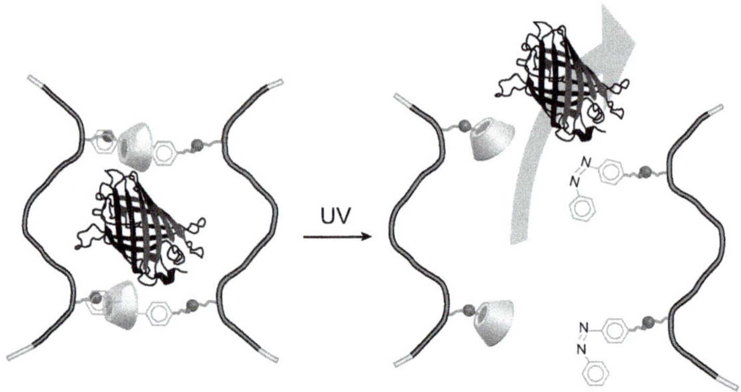

Figure 4.19 Schematic representation of photoresponsive protein release from the hydrogel composed of trans azobenzene–modified dextran and cyclodextrin-modified dextran.
Adapted with permission from Peng et al. (2010).

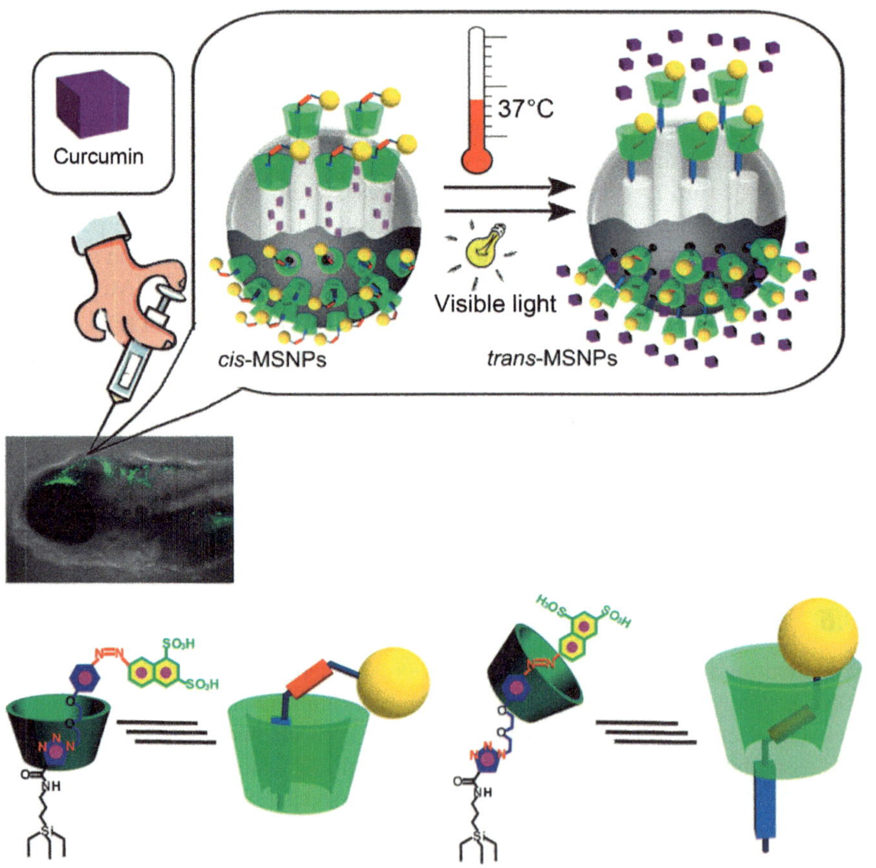

Figure 4.20 Graphical representation of the injection of drug-loaded mesoporous silica nanoparticles (MSNPs) into zebrafish larvae for in vivo drug delivery, triggered by either heating or visible light irradiation. MSNPs were functionalized with photothermal-responsive rotaxanes on the surface. The chemical structure of the rotaxane containing the α-CD ring and azobenzene unit is shown.
Adapted with permission from Yan et al. (2012).

and side effects. Drug molecules are often physically entrapped in the network and are released from the hydrogel matrix by diffusion. Kros et al. used the inclusion complex of trans azobenzene and cyclodextrin as a photoswitchable cross-linker, constructed a dextran-based photoresponsive hydrogel system, and used it for a light-controlled DDS (see Figure 4.19) (Peng, Tomatsu, & Kros, 2010). Upon UV light irradiation, azobenzene moieties isomerise from trans to cis configurations, resulting in the dissociation of cross-linking points, and allow the entrapped protein to migrate into the media. Zhao et al. developed a novel strategy (see Figure 4.20) for the preparation of a new class of photothermally responsive rotaxane-functionalized mesoporous silica nanoparticles (MSNPs) (Yan et al., 2012). The remote-controlled drug release is based on the back-and-forth movement of the a-CD ring, owing to photothermally induced reversible trans–cis isomerization of the azobenzene axle.

4.4 Conclusions and future trends

In this chapter, we have reviewed some recent progress on the developments of photonic sensitive materials and their applications, including photoinduced surface wettability switches and smart-gating nanochannels (Wen et al., 2010; Wen, Liu, et al., 2012; Zhang et al., 2012). We also listed different photonic biomaterials used in optobioelectronic devices, biocatalysts, biosensors, and the biomedical engineering field.

Reversible photoresponsive surfaces with controllable wettability have attracted enormous attention due to their importance in surface science and intelligent micro- and nano-devices (Wang, Ding, Yu, & Wang, 2011; Wang, Song, & Jiang, 2007). The photonic-sensitive materials used, such as semiconductor oxides, organic molecules, polymers, and biomolecules, have a promising future as potential responsive multifunctional surface switches in new technological applications such as smart functional windows, microfluidic devices, and bioanalysis. Photo control of recognition inside a nanoporous membrane as a biosensor is particularly attractive because of the advantages associated with ligand–receptor interactions in confined spaces. Various optical applications such as holographic pattern recording, dynamic optical filtering, and associative optical memories were developed with bR as photosensor (Bräuchle, Hampp, & Oesterhelt, 1991; Yao et al., 2005). Bacteriorhodopsin is a promising biomaterial for broad potential applications in optical and optoelectronic devices (Fábián et al., 2011; Lindvold & Lausen, 2006). Numerous applications can be envisaged for optobioelectronic systems. Redox proteins, which are made by the modification of proteins with photoisomerizable units or by the integration of proteins with photoisomerizable environments, exhibit the reversible photoswitching of the bioelectrocatalytic functions and semiconductor-biomolecule QDs hybrid systems and their immobilization on electrodes to yield photobioelectrochemical and photobiofuel cell elements have been summarized in detail by Willner and co-workers (Wang, Liu, et al., 2012). The fundamental challenge in the tailoring of optobioelectronic systems is the development of the means to electrically contact photoactive biomolecular assemblies with the electrode supports. Moreover, optobioelectronic devices hold great promise for the future development of solar energy conversion and storage devices.

The light-induced activation of biocatalytic processes might provide means to amplify weak light signals, and the control of the photocurrents by light doses would allow the fabrication of new devices for probing light intensities (actinometers). Despite substantial scientific progress in the development of biomedical applications, with photonic sensitive materials having been demonstrated, many challenges are still ahead of us. These challenges include designing new photoresponsive materials that have improved stability and sensitivity, putting these entities into complex devices that can be used in switches, and endowing photoresponsive biological activity and biocompatibility. Development of photonic sensitive switchable materials requires a scientific effort to mimic photoswitchable biological systems in nature, such as the process of photosynthesis and vision. There is no doubt that photonic sensitive materials offer exciting perspectives at the frontiers of chemistry, biology, physics, medicine, and materials science, and would bring us a bright future for their wide applications.

References

Abbott, S., Ralston, J., Reynolds, G., & Hayes, R. (1999). Reversible wettability of photoresponsive pyrimidine-coated surfaces. *Langmuir, 15*, 8923–8928. http://dx.doi.org/10.1021/la990558o.

Adams, S., Kao, J., & Tsien, R. (1989). Biologically useful chelators that take up calcium (2+) upon illumination. *Journal of the American Chemical Society, 111*, 7957–7968.

Agarwal, P. K., Schultz, C., Kalivretenos, A., Ghosh, B., & Broedel, S. (2012). Engineering hyper-catalytic enzyme by photo-activated conformation modulation. *Biophysical Journal, 102*, 273.

Aghamalyan, N. R., Hovsepyan, R. K., Poghosyan, A. R., & Lazaryan, V. G. (2004). Photodetectors on the base of ZnO thin films. *Proceedings Article*, 235–240. http://dx.doi.org/10.1117/12.556790.

Anderson, V. C., & Thompson, D. H. (1992). Triggered release of hydrophilic agents from plasmologen liposomes using visible light or acid. *Biochimica et Biophysica Acta (BBA)-Biomembranes, 1109*, 33–42.

Auernheimer, J., Dahmen, C., Hersel, U., Bausch, A., & Kessler, H. (2005). Photoswitched cell adhesion on surfaces with RGD peptides. *Journal of the American Chemical Society, 127*, 16107–16110.

Balashov, S. P., Imasheva, E. S., Govindjee, R., & Ebrey, T. G. (1996). Titration of aspartate-85 in bacteriorhodopsin: what it says about chromophore isomerization and proton release. *Biophysical Journal, 70*, 473–481. http://dx.doi.org/10.1016/S0006-3495(96)79591-7.

Becker, W. G., & Bard, A. J. (1983). Photoluminescence and photoinduced oxygen adsorption of colloidal zinc sulfide dispersions. *The Journal of Physical Chemistry, 87*, 4888–4893.

Bico, J., Thiele, U., & Quéré, D. (2002). Wetting of textured surfaces. *Colloids and Surfaces A Physicochemical Engineering Aspects, 206*, 41–46.

Bico, J., Tordeux, C., & Quéré, D. (2001). Rough wetting. *EPL (Europhysics Letters), 55*, 214.

Blankenship, R. E. (2002). *Molecular mechanisms of photosynthesis*. Blackwell Science.

Bochet, C. G. (2001). Orthogonal photolysis of protecting groups. *Angewandte Chemie International Edition, 40*, 2071–2073. http://dx.doi.org/10.1002/1521-3773(20010601).

Bonardi, F., London, G., Nouwen, N., Feringa, B. L., & Driessen, A. J. M. (2010). Light–induced control of protein translocation by the SecYEG complex. *Angewandte Chemie International Edition, 49*, 7234–7238. http://dx.doi.org/10.1002/anie.201002243.

Bouas-Laurent, H. (2001). Organic photochromism. *Pure and Applied Chemistry*, 639–665.
Bräuchle, C., Hampp, N., & Oesterhelt, D. (1991). Optical applications of bacteriorhodopsin and its mutated variants. *Advanced Materials*, *3*, 420–428. http://dx.doi.org/10.1002/adma.19910030903.
Bruchez, M., Moronne, M., Gin, P., Weiss, S., & Alivisatos, A. P. (1998). Semiconductor nanocrystals as fluorescent biological labels. *Science*, *281*, 2013–2016.
Camacho-Lopez, M., Finkelmann, H., Palffy-Muhoray, P., & Shelley, M. (2004). Fast liquid-crystal elastomer swims into the dark. *Nature Materials*, *3*, 307–310. http://dx.doi.org/10.1038/nmat1118.
Chan, W. C. W., Maxwell, D. J., Gao, X., Bailey, R. E., Han, M., & Nie, S. (2002). Luminescent quantum dots for multiplexed biological detection and imaging. *Current Opinion in Biotechnology*, *13*, 40–46.
Chan, W. C. W., & Nie, S. (1998). Quantum dot bioconjugates for ultrasensitive nonisotopic detection. *Science*, *281*, 2016–2018. http://dx.doi.org/10.1126/science.281.5385.2016.
Chechile, R. A. (2000). Vision science: photons to phenomenology. *Journal of Mathematical Psychology*, *44*, 237–237.
Chen, X., Shen, S., Guo, L., & Mao, S. S. (2010). Semiconductor-based photocatalytic hydrogen generation. *Chemical Reviews*, *110*, 6503–6570.
Choi, Y. J., & Park, H. H. (2011). Direct patterning of SnO_2 composite films prepared with various contents of Pt nanoparticles by photochemical metal-organic deposition. *Thin Solid Films*, *519*, 6214–6218. http://dx.doi.org/10.1016/j.tsf.2011.03.051.
Chou, C., Young, D. D., & Deiters, A. (2010). Photocaged T7 RNA polymerase for the light activation of transcription and gene function in pro- and eukaryotic cells. *ChemBioChem*, *11*, 972–977. http://dx.doi.org/10.1002/cbic.201000041.
Deb, S. (1963). Photoelectric properties of the alkali metal azides. *Transactions of the Faraday Society*, *59*, 1414–1422.
Durrant, J. R., Haque, S. A., & Palomares, E. (2006). Photochemical energy conversion: from molecular dyads to solar cells. *Chemical Communications*, 3279–3289. http://dx.doi.org/10.1039/B601785C.
Edwards, W. F., Young, D. D., & Deiters, A. (2009). Light-activated Cre recombinase as a tool for the spatial and temporal control of gene function in mammalian cells. *ACS Chemical Biology*, *4*, 441–445. http://dx.doi.org/10.1021/cb900041s.
Fábián, L., Heiner, Z., Mero, M., Kiss, M., Wolff, E. K., Ormos, P., et al. (2011). Protein-based ultrafast photonic switching. *Optics Express*, *19*, 18861–18870. http://dx.doi.org/10.1364/OE.19.018861.
Feng, X., Feng, L., Jin, M., Zhai, J., Jiang, L., & Zhu, D. (2004). Reversible super-hydrophobicity to super-hydrophilicity transition of aligned ZnO nanorod films. *Journal of the American Chemical Society*, *126*, 62–63. http://dx.doi.org/10.1021/ja038636o.
Feringa, B. L., & Browne, W. R. (2001). *Molecular switches*. Wiley Online Library.
Feringa, B. L., Jager, W. F., & Delange, B. (1993). Organic materials for reversible optical-data storage. *Tetrahedron*, *49*, 8267–8310.
Finkelmann, H., Nishikawa, E., Pereira, G. G., & Warner, M. (2001). A new opto-mechanical effect in solids. *Physical Review Letters*, *87*, 015501. http://dx.doi.org/10.1103/PhysRevLett.87.015501.
Ge, H., Wang, G., He, Y., Wang, X., Song, Y., Jiang, L., et al. (2006). Photoswitched wettability on inverse opal modified by a self-assembled azobenzene monolayer. *ChemPhysChem*, *7*, 575–578. http://dx.doi.org/10.1002/cphc.200500496.
Hambourger, M., Gervaldo, M., Svedruzic, D., King, P. W., Gust, D., Ghirardi, M., et al. (2008). [FeFe]-hydrogenase-catalyzed H_2 production in a photoelectrochemical biofuel cell. *Journal of the American Chemical Society*, *130*, 2015–2022. http://dx.doi.org/10.1021/ja077691k.

Han, C., Hou, X., Zhang, H., Guo, W., Li, H., & Jiang, L. (2011). Enantioselective recognition in biomimetic single artificial nanochannels. *Journal of the American Chemical Society, 133*, 7644–7647. http://dx.doi.org/10.1021/ja2004939.

Heller, H. (1983). Organic fatigue-resistant photochromic imaging materials. *Solid-State and Electron Devices, IEEE Proceedings I, 130*, 209–211. http://dx.doi.org/10.1049/ip-i-1.1983.0038.

He, T., & Yao, J. (2006). Photochromism in composite and hybrid materials based on transition-metal oxides and polyoxometalates. *Progress in Materials Science, 51*, 810–879. http://dx.doi.org/10.1016/j.pmatsci.2005.12.001.

Huang, L., Lau, S., Yang, H., Leong, E., Yu, S., & Prawer, S. (2005). Stable superhydrophobic surface via carbon nanotubes coated with a ZnO thin film. *The Journal of Physical Chemistry B, 109*, 7746–7748.

Ichimura, K., Oh, S.-K., & Nakagawa, M. (2000). Light-driven motion of liquids on a photoresponsive surface. *Science, 288*, 1624–1626. http://dx.doi.org/10.1126/science.288.5471.1624.

Ikeda, T., Nakano, M., Yu, Y., Tsutsumi, O., & Kanazawa, A. (2003). Anisotropic bending and unbending behavior of azobenzene liquid-crystalline gels by light exposure. *Advanced Materials, 15*, 201–205. http://dx.doi.org/10.1002/adma.200390045.

Irie, M. (1990). Photoresponsive polymers. *New Polymeric Materials*, 27–67.

Irie, M., & Kato, M. (1985). Photoresponsive molecular tweezers. Photoregulated ion capture and release using thioindigo derivatives having ethylenedioxy side groups. *Journal of the American Chemical Society, 107*, 1024–1028.

Irie, M., & Kunwatchakun, D. (1986). Photoresponsive polymers. 8. Reversible photostimulated dilation of polyacrylamide gels having triphenylmethane leuco derivatives. *Macromolecules, 19*, 2476–2480.

Irie, M., Sakemura, K., Okinaka, M., & Uchida, K. (1995). Photochromism of dithienylethenes with electron-donating substituents. *The Journal of Organic Chemistry, 60*, 8305–8309.

Itoga, K., Yamato, M., Kobayashi, J., Kikuchi, A., & Okano, T. (2004). Cell micropatterning using photopolymerization with a liquid crystal device commercial projector. *Biomaterials, 25*, 2047–2053.

Jiang, W., Wang, G., He, Y., Wang, X., An, Y., Song, Y., et al. (2005). Photo-switched wettability on an electrostatic self-assembly azobenzene monolayer. *Chemical Communications*, 3550–3552.

Jiang, Y., Wang, Y., Ma, N., Wang, Z., Smet, M., & Zhang, X. (2007). Reversible self-organization of a UV-responsive PEG-terminated malachite green derivative: vesicle formation and photoinduced disassembly. *Langmuir, 23*, 4029–4034. http://dx.doi.org/10.1021/la063305l.

Jiang, Y., Wan, P., Smet, M., Wang, Z., & Zhang, X. (2008). Self-assembled monolayers of a malachite green derivative: surfaces with pH- and UV-responsive wetting properties. *Advanced Materials, 20*, 1972–1977. http://dx.doi.org/10.1002/adma.200702366.

Jin, Y., Honig, T., Ron, I., Friedman, N., Sheves, M., & Cahen, D. (2008). Bacteriorhodopsin as an electronic conduction medium for biomolecular electronics. *Chemical Society Reviews, 37*, 2422–2432. http://dx.doi.org/10.1039/B806298F.

Juodkazis, S., Mukai, N., Wakaki, R., Yamaguchi, A., Matsuo, S., & Misawa, H. (2000). Reversible phase transitions in polymer gels induced by radiation forces. *Nature, 408*, 178–181. http://dx.doi.org/10.1038/35041522.

Kaplan, J. H., & Hollis, R. J. (1980). *External Na dependence of ouabain-sensitive ATP: ADP exchange initiated by photolysis of intracellular caged-ATP in human red cell ghosts.*

Katz, J. S., & Burdick, J. A. (2010). Light-responsive biomaterials: development and applications. *Macromolecular Bioscience, 10*, 339–348. http://dx.doi.org/10.1002/mabi.200900297.

Kawai, T., Nakashima, Y., & Irie, M. (2005). A novel photoresponsive π-conjugated polymer based on diarylethene and its photoswitching effect in electrical conductivity. *Advanced Materials*, *17*, 309–314. http://dx.doi.org/10.1002/adma.200400191.
Khalid, W., El Helou, M., MurböCk, T., Yue, Z., Montenegro, J. M., Schubert, K., et al. (2011). Immobilization of quantum dots via conjugated self-assembled monolayers and their application as a light-controlled sensor for the detection of hydrogen peroxide. *ACS Nano*, *5*, 9870–9876. http://dx.doi.org/10.1021/nn2035582.
Khalil, L., Rophael, M., & Mourad, W. (2002). The removal of the toxic Hg (II) salts from water by photocatalysis. *Applied Catalysis B*, *36*, 125–130. http://dx.doi.org/10.1016/S0926-3373(01)00285-5.
Kim, E., Kim, M., & Kim, K. (2006). Diarylethenes with intramolecular donor–acceptor structures for photo-induced electrochemical change. *Tetrahedron*, *62*, 6814–6821. http://dx.doi.org/10.1016/j.tet.2006.04.089.
Krauss, U., Drepper, T., & Jaeger, K. E. (2011). Enlightened enzymes: strategies to create novel photoresponsive proteins. *Chemistry – A European Journal*, *17*, 2552–2560. http://dx.doi.org/10.1002/chem.201002716.
Kubasek, W. L., Shirley, B. W., Mckillop, A., Goodman, H. M., Briggs, W., & Ausubel, F. M. (1992). Regulation of flavonoid biosynthetic genes in germinating *Arabidopsis* seedlings. *The Plant Cell Online*, *4*, 1229–1236.
Kumar, G. S., & Neckers, D. (1989). Photochemistry of azobenzene-containing polymers. *Chemical Reviews*, *89*, 1915–1925.
Kumar, A. S., Ye, T., Takami, T., Yu, B. C., Flatt, A. K., Tour, J. M., et al. (2008). Reversible photo-switching of single azobenzene molecules in controlled nanoscale environments. *Nano Letters*, *8*, 1644–1648. http://dx.doi.org/10.1039/B607187B.
Leonova, M. M., Fufina, T. Y., Vasilieva, L. G., & Shuvalov, V. A. (2011). Structure-function investigations of bacterial photosynthetic reaction centers. *Biochemistry (Moscow)*, *76*, 1465–1483. http://dx.doi.org/10.1134/S0006297911130074.
Leutwyler, W. K., Bürgi, S. L., & Burgl, H. (1996). Semiconductor clusters, nanocrystals, and quantum dots. *Science*, *271*, 933.
Li, Y., He, B., Qin, B., Feng, X., & Zhang, G. (2005). Highly enantioselective cyanosilylation of aldehydes catalyzed by novel β-amino alcohol—titanium complexes. *ChemInform*, *36*. http://dx.doi.org/10.1002/chin.200511101.
Li, M. H., Keller, P., Li, B., Wang, X., & Brunet, M. (2003). Light-driven side-on nematic elastomer actuators. *Advanced Materials*, *15*, 569–572. http://dx.doi.org/10.1002/adma.200304552.
Li, R., Li, C. M., Bao, H., Bao, Q., & Lee, V. S. (2007). Stationary current generated from photocycle of a hybrid bacteriorhodopsin/quantum dot bionanosystem. *Applied Physics Letters*, *91*. 223901–223901-3. http://dx.doi.org/10.1063/1.2801521.
Lim, H. S., Han, J. T., Kwak, D., Jin, M., & Cho, K. (2006). Photoreversibly switchable superhydrophobic surface with erasable and rewritable pattern. *Journal of the American Chemical Society*, *128*, 14458–14459. http://dx.doi.org/10.1021/ja0655901.
Lim, H. S., Kwak, D., Lee, D. Y., Lee, S. G., & Cho, K. (2007). UV-driven reversible switching of a roselike vanadium oxide film between superhydrophobicity and superhydrophilicity. *Journal of the American Chemical Society*, *129*, 4128–4129. http://dx.doi.org/10.1021/ja0692579.
Lindvold, L., & Lausen, H. (2006). A projection display based on a bacteriorhodopsin thin film. *Bionanotechnology*, 79–95.
Liu, H., Feng, L., Zhai, J., Jiang, L., & Zhu, D. (2004). Reversible wettability of a chemical vapor deposition prepared ZnO film between superhydrophobicity and superhydrophilicity. *Langmuir*, *20*, 5659–5661. http://dx.doi.org/10.1021/la036280o.

Liu, H., Matsuda, K., Gu, Z., Takahashi, K., Cui, A., Nakajima, R., et al. (2003). Reversible valence tautomerism induced by a single-shot laser pulse in a cobalt–iron Prussian blue analog. *Physical Review Letters*, *90*, 167403.

Li, J., & Zhang, J. Z. (2009). Optical properties and applications of hybrid semiconductor nanomaterials. *Coordination Chemistry Reviews*, *253*, 3015–3041. http://dx.doi.org/10.1016/j.ccr.2009.07.017.

Li, Y., Zhao, G. Y., Zhang, W. H., & Chen, Y. Q. (2006). Fine pattern fabrication on SnO_2: Sb thin films formed by the sol-gel process. *Surface Interface Analysis*, *38*, 1291–1295. http://dx.doi.org/10.1002/Sia.2442.

Martin, C. R., & Siwy, Z. S. (2007). Learning nature's way: biosensing with synthetic nanopores. *Science*, *317*, 331–332.

Matsuda, K., & Irie, M. (2000a). A diarylethene with two nitronyl nitroxides: photoswitching of intramolecular magnetic interaction. *Journal of the American Chemical Society*, *122*, 7195–7201. http://dx.doi.org/10.1021/ja000605v.

Matsuda, K., & Irie, M. (2000b). Photoswitching of intramolecular magnetic interaction: a diarylethene photochromic spin coupler. *Chemistry Letters*, *29*, 16–17.

Mattoussi, H., Mauro, J. M., Goldman, E. R., Anderson, G. P., Sundar, V. C., Mikulec, F. V., et al. (2000). Self-assembly of CdSe–ZnS quantum dot bioconjugates using an engineered recombinant protein. *Journal of the American Chemical Society*, *122*, 12142–12150. http://dx.doi.org/10.1021/ja002535y.

Mazurkevich, Y. S., & Kobasa, I. M. (1997). Photosensitive dielectric materials based on TiO_2 and SiO_2–TiO_2. *Inorganic Materials*, *33*, 1258–1260.

Mehdaoui, I., Kröner, D., Pykavy, M., Freund, H. J., & Klüner, T. (2005). Photo-induced desorption of NO from NiO (100): calculation of the four-dimensional potential energy surfaces and systematic wave packet studies. *Physical Chemistry Chemical Physics*, *8*, 1584–1592. http://dx.doi.org/10.1039/b512778e.

Meng, X., Shen, D., Zhang, J., Zhao, D., Lu, Y., Dong, L., et al. (2005). The structural and optical properties of ZnO nanorod arrays. *Solid State Communications*, *135*, 179–182.

Mitchell, G. P., Mirkin, C. A., & Letsinger, R. L. (1999). Programmed assembly of DNA functionalized quantum dots. *Journal of the American Chemical Society*, *121*, 8122–8123.

Morrison, H. (1993). *Biological applications of photochemical switches*. Wiley-Interscience.

Munekage, Y., Hashimoto, M., Miyake, C., Tomizawa, K.-I., Endo, T., Tasaka, M., et al. (2004). Cyclic electron flow around photosystem I is essential for photosynthesis. *Nature*, *429*, 579–582. http://dx.doi.org/10.1038/nature02598.

Nargeot, J., Nerbonne, J. M., Engels, J., & Lester, H. A. (1983). Time course of the increase in the myocardial slow inward current after a photochemically generated concentration jump of intracellular cAMP. *Proceedings of the National Academy of Sciences*, *80*, 2395–2399.

Nerbonne, J. M., Richard, S., Nargeot, J., & Lester, H. A. (1984). *New photoactivatable cyclic nucleotides produce intracellular jumps in cyclic AMP and cyclic GMP concentrations*.

Niemeyer, C. M. (2001). Nanoparticles, proteins, and nucleic acids: biotechnology meets materials science. *Angewandte Chemie International Edition*, *40*, 4128–4158. http://dx.doi.org/10.1002/1521-3773(20011119).

Nirmal, M., & Brus, L. (1999). Luminescence photophysics in semiconductor nanocrystals. *Accounts of Chemical Research*, *32*, 407–414.

Oesterhelt, D., & Stoeckenius, W. (1973). Functions of a new photoreceptor membrane. *Proceedings of the National Academy of Sciences*, *70*, 2853–2857.

Palmisano, G., Augugliaro, V., Pagliaro, M., & Palmisano, L. (2007). Photocatalysis: a promising route for 21st century organic chemistry. *Chemical Communications*, 3425–3437. http://dx.doi.org/10.1039/B700395C.

Pathak, S., Choi, S. K., Arnheim, N., & Thompson, M. E. (2001). Hydroxylated quantum dots as luminescent probes for in situ hybridization. *Journal of the American Chemical Society, 123*, 4103.
Peng, K., Tomatsu, I., & Kros, A. (2010). Light controlled protein release from a supramolecular hydrogel. *Chemical Communications, 46*, 4094–4096. http://dx.doi.org/10.1039/c002565h.
Renner, C., Kuscbauch, U., Löweneck, M., Milbradt, A. G., & Moroder, L. (2005). Azobenzene as photoresponsive conformational switch in cyclic peptides. *The Journal of Peptide and Protein Research, 65*, 4–14. http://dx.doi.org/10.1111/j.1399-3011.2004.00203.x.
Rosario, R., Gust, D., Garcia, A. A., Hayes, M., Taraci, J. L., Clement, T., et al. (2004). Lotus effect amplifies light-induced contact angle switching. *Journal of Physical Chemistry B, 108*, 12640–12642. http://dx.doi.org/10.1021/jp0473568.
Rosario, R., Gust, D., Hayes, M., Jahnke, F., Springer, J., & Garcia, A. A. (2002). Photon-modulated wettability changes on spiropyran-coated surfaces. *Langmuir, 18*, 8062–8069.
Roy, I., Sharma, S., & Gupta, M. (2004). Smart biocatalysts: design and applications. *Advances in Biochemical Engineering/Biotechnology, 86*, 159.
Shimoboji, T., Larenas, E., Fowler, T., Kulkarni, S., Hoffman, A. S., & Stayton, P. S. (2002). Photoresponsive polymer–enzyme switches. *Proceedings of the National Academy of Sciences, 99*, 16592–16596.
Stryer, L. (1988). Molecular-basis of visual excitation. *Cold Spring Harbor Symposia on Quantitative Biology, 53*, 283–294.
Suzuki, A., & Tanaka, T. (1990). Phase transition in polymer gels induced by visible light. *Nature, 346*, 345–347. http://dx.doi.org/10.1038/346345a0.
Takeshita, M., Uchida, K., & Irie, M. (1996). Novel saccharide tweezers with a diarylethene photoswitch. *Chemical Communications*, 1807–1808. http://dx.doi.org/10.1039/CC9960001807.
Tel-Vered, R., Yildiz, H. B., Yan, Y. M., & Willner, I. (2010). Plugging into enzymes with light: photonic "wiring" of enzymes with electrodes for photobiofuel cells. *Small, 6*, 1593–1597. http://dx.doi.org/10.1002/smll.201000296.
Turner, A. D., Pizzo, S. V., Rozakis, G., & Porter, N. A. (1988). Photoreactivation of irreversibly inhibited serine proteinases. *Journal of the American Chemical Society, 110*, 244–250. http://dx.doi.org/10.1021/ja00209a040.
Uchida, K., Saito, M., Murakami, A., Nakamura, S., & Irie, M. (2003). Non–destructive readout of the photochromic reactions of diarylethene derivatives using infrared light. *Advanced Materials, 15*, 121–125.
Uda, R. M., Oue, M., & Kimura, K. (2004). Effective photocontrol of micelle formation by malachite green derivative carrying a long alkyl chain. *Chemical Letters, 33*, 586–587. http://dx.doi.org/10.1246/cl.2004.586.
Von Wettstein, D., Gough, S., & Kannangara, C. G. (1995). Chlorophyll biosynthesis. *The Plant Cell, 7*, 1039.
Wang, X., Ding, B., Yu, J., & Wang, M. (2011). Engineering biomimetic superhydrophobic surfaces of electrospun nanomaterials. *Nano Today, 6*, 510–530.
Wang, Q., Fang, T., Liu, P., Deng, B., Min, X., & Li, X. (2012). Direct synthesis of high-quality water-soluble CdTe: Zn^{2+} quantum dots. *Inorganic Chemistry, 51*, 9208–9213. http://dx.doi.org/10.1021/ic300473u.
Wang, S., Feng, X., Yao, J., & Jiang, L. (2006). Controlling wettability and photochromism in a dual-responsive tungsten oxide film. *Angewandte Chemie International Edition, 45*, 1264–1267. http://dx.doi.org/10.1002/anie.200502061.
Wang, R., Hashimoto, K., Fujishima, A., Chikuni, M., Kojima, E., Kitamura, A., et al. (1997). Light-induced amphiphilic surfaces. *Nature, 388*, 431–432.

Wang, F., Liu, X., & Willner, I. (2012). Integration of photoswitchable proteins, photosynthetic reaction centers and semiconductor/biomolecule hybrids with electrode supports for optobioelectronic applications. *Advanced Materials, 25,* 349–377. http://dx.doi.org/10.1002/adma.201201772.
Wang, S. T., Song, Y. L., & Jiang, L. (2007). Photoresponsive surfaces with controllable wettability. *Journal of Photochemistry and Photobiology C, 8,* 18–29. http://dx.doi.org/10.1016/j.jphotochemrev.2007.03.001.
Wanunu, M., & Meller, A. (2007). Chemically modified solid-state nanopores. *Nano Letters, 7,* 1580–1585.
Wei, M., Xu, X., Wang, X., Li, F., Zhang, H., Lu, Y., et al. (2006). Study on the photochromism of Ni–Al layered double hydroxides containing nitrate anions. *European Journal of Inorganic Chemistry, 2006,* 2831–2838. http://dx.doi.org/10.1002/ejic.200600058.
Wen, L., Hou, X., Tian, Y., Nie, F. Q., Song, Y., Zhai, J., et al. (2010). Bioinspired smart gating of nanochannels toward photoelectric–conversion systems. *Advanced Materials, 22,* 1021–1024. http://dx.doi.org/10.1002/adma.200903161.
Wen, L., Liu, Q., Ma, J., Tian, Y., Li, C., Bo, Z., et al. (2012). Malachite green derivative–functionalized single nanochannel: light–and–pH dual–driven ionic gating. *Advanced Materials, 24,* 6193–6198. http://dx.doi.org/10.1002/adma.201202673.
Wen, L., Ma, J., Tian, Y., Zhai, J., & Jiang, L. (2012). A photo–induced, and chemical–driven, smart–gating nanochannel. *Small, 8,* 838–842. http://dx.doi.org/10.1002/smll.201101661.
Whaley, S. R., English, D., Hu, E. L., Barbara, P. F., & Belcher, A. M. (2000). Selection of peptides with semiconductor binding specificity for directed nanocrystal assembly. *Nature, 405,* 665–668.
Williams, E., Hodgson, W., & Brinen, J. (2006). Synthetic photochromic sodalite. *Journal of the American Chemical Society, 52,* 139–144.
Willner, I., & Rubin, S. (1996). Control of the structure and functions of biomaterials by light. *Angewandte Chemie International Edition in English, 35,* 367–385. http://dx.doi.org/10.1002/anie.199603671.
Willner, I., Sussan, S., & Rubin, S. (1992). Photostimulated transport of carboxylate and phenolate anions across a liquid-liquid membrane using a photochromic cationic copolymer as carrier. *Journal of the Chemical Society Chemical Communications,* 100–101. http://dx.doi.org/10.1039/C39920000100.
Yameen, B., Ali, M., Neumann, R., Ensinger, W., Knoll, W., & Azzaroni, O. (2010). Proton-regulated rectified ionic transport through solid-state conical nanopores modified with phosphate-bearing polymer brushes. *Chemical Communications, 46,* 1908–1910.
Yan, H., Teh, C., Sreejith, S., Zhu, L., Kwok, A., Fang, W., et al. (2012). Functional mesoporous silica nanoparticles for photothermal-controlled drug delivery in vivo. *Angewandte Chemie International Edition, 51,* 8373–8377. http://dx.doi.org/10.1002/anie.201203993.
Yao, B., Ren, Z., Menke, N., Wang, Y., Zheng, Y., Lei, M., et al. (2005). Polarization holographic high-density optical data storage in bacteriorhodopsin film. *Applied Optics, 44,* 7344–7348.
Yu, Y., Nakano, M., & Ikeda, T. (2003). Photomechanics: directed bending of a polymer film by light. *Nature, 425,* 145–145. http://dx.doi.org/10.1038/425145a.
Zhang, M., Hou, X., Wang, J., Tian, Y., Fan, X., Zhai, J., et al. (2012). Light and pH cooperative nanofluidic diode using a spiropyran–functionalized single nanochannel. *Advanced Materials, 24,* 2424–2428. http://dx.doi.org/10.1002/adma.201104536.
Zhang, X.-T., Sato, O., & Fujishima, A. (2004). Water ultrarepellency induced by nanocolumnar ZnO surface. *Langmuir, 20,* 6065–6067. http://dx.doi.org/10.1021/la049471f.
Zhu, W., Feng, X., Feng, L., & Jiang, L. (2006). UV-manipulated wettability between superhydrophobicity and superhydrophilicity on a transparent and conductive SnO_2 nanorod film. *Chemical Communications,* 2753–2755. http://dx.doi.org/10.1039/b603634a.

Responsive polymer brushes for biomedical applications

5

N. Akkilic, W.M. de Vos
University of Twente, Enschede, The Netherlands

5.1 Introduction

Polymer brushes can easily be considered one of the most powerful tools to control interfacial properties. A good definition of a polymer brush is that it is a dense layer of polymers, end-attached to an interface, that stretch out into the surrounding solution (Advincula, Brittain, Caster, & Rühe, 2004; Currie, Norde, & Cohen Stuart, 2003; Halperin, 1999; Milner, 1991; Szleifer & Carignano, 2007; Zhao & Brittain, 2000). The stretching is a direct result of the high polymer density within the brush, and it is also this high polymer density that provides it with its characteristic properties and large potential for applications. The effect of density is schematically shown in Figure 5.1. If only a few polymers are end-grafted to an interface (and the polymer does not adsorb to the surface), the polymer will adapt its preferred conformation and form a polymer coil (Figure 5.1(a), mushroom regime). However, if the density of chains is increased, the polymer coils will come into contact and start to overlap. To minimize contact between the polymer segments, and to retain as much as possible of their entropic freedom, the polymer chains will stretch away from the surface as shown in Figure 5.1(b). This is the so-called brush regime.

As mentioned, many investigations over the years have shown the brush to be very useful for a significant number of applications. The most well-known and heavily studied brush application is that of an antifouling coating (Chen, Ferris, Zhang, Ducker, & Zauscher, 2010; Currie et al., 2003). The high polymer density in the brush allows its use as a barrier, keeping fouling particles away from the surface. If a fouling particle, for example a protein molecule, comes into the brush, it needs to deform the brush equilibrium structure. The brush responds to this with a restoring force pushing the particle out of the brush, away from the interface. This same effect can also be used to stabilize particle suspensions (Napper, 1983), with the high polymer density around the particles preventing aggregation. Furthermore, the brush has been shown to be able to substantially reduce the friction between two interfaces (Klein, Kumacheva, Mahalu, Perahia, & Fetters, 1994). Finally, the high surface area of the polymer brush can also be used, not to prevent, but rather to accommodate the adsorption of protein molecules (Wittemann & Ballauff, 2006).

A much more recent development, however, is the drive to create responsive polymer brushes (Chen et al., 2010; Stuart et al., 2010). These brushes respond with certain behavior, such as swelling, collapsing, charging to an external trigger, and allowing

Switchable and Responsive Surfaces and Materials for Biomedical Applications. http://dx.doi.org/10.1016/B978-0-85709-713-2.00005-5

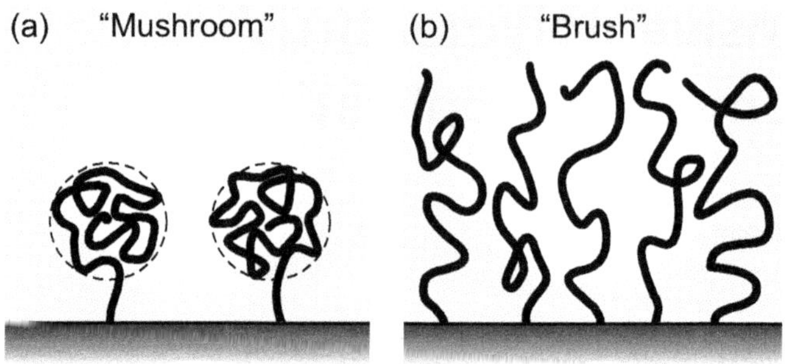

Figure 5.1 Schematic depiction of polymers grafted to an interface in a sparse fashion. (a) (mushroom regime) and (b) a dense fashion (brush regime).

an even greater control over surface properties than was previously possible. It has been known for quite some time that polymer brushes respond to solvent quality (Karim et al., 1995) and that polyelectrolyte brushes can react strongly to changes in salt concentration and (for weak polyelectrolytes) pH (Balastre et al., 2002; Borisov, Zhulina, & Birshtein, 1994). Still, the field of responsive polymer brushes has really expanded with the introduction of polymer brushes consisting of two chemically different species. This includes brushes based on diblock copolymers, brushes from polymers bearing chemically different side chains, and the so-called mixed polymer brush where chemically different homopolymers are grafted side by side. Combining polymers with different properties and different responsiveness, in combination with the various available brush architectures, provides an enormous amount of design opportunities. Examples of responsive brushes include a brush that "remembers" the solvent it was in contact with last (Kumar vyas, Schneider, Nandan, & Stamm, 2008), and brushes that allow switching between a very hydrophilic and an ultra-hydrophobic interface (Draper, Luzinov, Minko, Tokarev, & Stamm, 2004; Motornov et al., 2003). Other examples include brushes that could change between antifouling and protein accommodating by a change in pH (Uhlmann et al., 2006), and colloidal particles coated with a brush where the pH could be used to switch between aggregation and dissociation of the particles (Motornov et al., 2007).

This chapter encompasses a review on the system of responsive polymer brushes, while specifically focusing on the large potential that these systems can have in the field of biomedical applications.

5.1.1 Brush preparation

By now, there are several well understood and well-established methods to produce polymer brushes, all of which have their own advantages and disadvantages. There are a number of good reviews that discuss these methods in great detail (Advincula et al., 2004; Currie et al., 2003; Edmondson, Osborne, & Huck, 2004; Rühe et al., 2004; Tsujii, Ohno, Yamamoto, Goto, & Fukuda, 2006; Zhao & Brittain, 2000), but a short summary is given here. The major techniques for the production of polymer brushes can be divided into three different categories: adsorption, "grafting to," and "grafting from" (Figure 5.2).

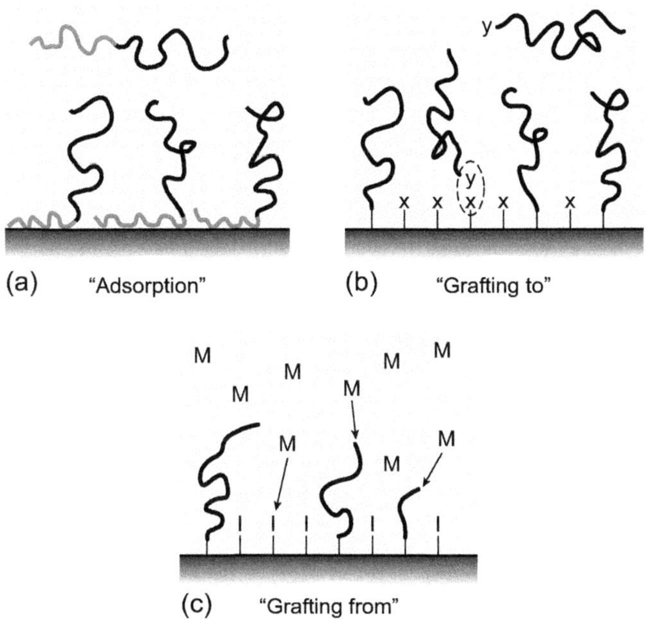

Figure 5.2 The main methods to produce polymer brushes: (a) adsorption from solution, (b) grafting to the interface, (c) grafting from the interface.

The most simple and the first approach to be used for the formation of polymer brushes is based on adsorption (Figure 5.2(a)) (Marra & Hair, 1988; Watanabe & Tirrell, 1993). For this approach one need a diblock copolymer, of which one block strongly adsorbs to the interface (anchor block) and the other block is preferably noninteracting with the interface. This method has the major advantage that a polymer brush is spontaneously formed when using the right combination of diblock copolymer, surface, and solvent and can thus quickly and easily be used on large surfaces. However, the disadvantages of this technique are also large. One is that a solvent is needed in which both blocks of the diblock copolymer are soluble or else one will get micelle formation, usually leading to an inhomogeneous adsorption. The most serious disadvantage, however, is that this technique can only form polymer brushes with very low grafting densities. During the adsorption process, the polymer brush formed acts as a barrier not only for fouling agents (such as proteins) but also for the adsorption of additional diblock copolymers that attempt to enter the polymer brush.

Both in the "grafting to" approach and the "adsorption" approach preformed polymers are used. An obvious advantage of this is that one can determine properties beforehand such as the chain length and the polydispersity. The "grafting to" approach (Figure 5.2(b)) uses polymers with a reactive chain end that are connected to the surface using a suitable reaction (Auroy, Auvray, & Leger, 1991; Maas, Cohen Stuart, Sieval, Zuilhof, & Sudhölter, 2003). High grafting densities can be achieved, provided this reaction can be done without solvent (in melt) (Maas et al., 2003). This approach is easier and more suitable for large surfaces than the "grafting from" approach. It is even possible to use this approach to produce mixed polymer brushes, brushes where chemically different polymer chains are combined into a single polymer brush. Disadvantages are that the grafting density is hard to control and that a reactive surface is needed.

Of these approaches, the "grafting from" approach (Figure 5.2(c)) is typically seen as the most powerful and versatile method. In the approach, polymer chains are polymerized from the grafting interface. For this, one first prepares a surface with covalently or physically attached monomers from which polymerization is initiated. With this approach, high degrees of polymerization and grafting densities (Tsujii et al., 2006) can be achieved, typically up to 0.8 chains per nm^2. The achieved polydispersity depends on the type of polymerization reaction and it is relatively easy to create different brush architectures, such as branched brushes and diblock-copolymer brushes. This approach has the disadvantage that the grafting density is hard to control, and that it is difficult to separately determine grafting density and chain length.

5.1.2 Biomedical applications for nonresponsive brushes

While the focus in this chapter is on the field of responsive polymer brushes, it is key to understand that some brushes that do not fall into that category, especially poly(ethylene oxide) (PEO) brushes, already play a big role in biomedical applications. Many years of research have demonstrated that especially PEO brushes show excellent antifouling properties against biofouling. If a fouling agent (e.g., a protein molecule) penetrates into a polymer brush, this will lead to an increase in the local polymer density and thus to an increase in local osmotic pressure. This osmotic pressure will force the particle out of the brush to restore the brush equilibrium. Indeed, in many experiments, grafted PEO chains have been found to prevent or strongly reduce the adsorption of proteins to surfaces (Du, Chandaroy, & Hui, 1997; Halperin, 1999; Roosjen, Kaper, Van Der Mei, Norde, & Busscher, 2003). For example, Du et al. (1997) showed that for three different types of protein, grafted PEO chains clearly reduce their adsorbed amount. In their case, especially the brush grafting density was found to be important, with very low protein adsorptions obtained at high graft densities. PEO is by far the most studied polymer, because it is very soluble in water, known to be nontoxic (Herold, Keil, & Bruns, 1989), very flexible (Pattanayek & Juvekar, 2002), and compatible with living cells (Albertsson, 1986). Other polymers such as poly(dextrans) and poly(acrylamide) have also been studied and demonstrated similar antifouling effects.

Although most antifouling work on PEO brushes has been performed with the intention of preventing protein adsorption, brushes can also function to keep away much larger objects. A nice example of the latter is the work by Roosjen, Kaper, Van Der Mei, Norde, and Busscher (2003), Roosjen, van der Mei, Busscher, and Norde (2004), and Roosjen, de Vries, Van Der Mei, Norde, and Busscher (2005) on the use of PEO brushes to prevent the attachment of bacteria to a glass interface. In Figure 5.3, we show a graphical conclusion of this work. An untreated glass slide (left) shows large

Figure 5.3 Optical microscopy photographs of the adherence of the bacteria *Staphylococcus epidermidis* HBH 276 onto an uncoated glass slide (left) and a glass slide coated with a PEO polymer brush on the right side. PEO, poly(ethylene oxide).
Reprinted with permission from Roosjen et al. (2005), copyright (2005) John Wiley and Sons.

amounts of attached bacteria. However, on a glass slide that been treated by attaching a polymer brush (right), bacterial adhesion is almost completely prevented. Clearly, the antifouling effect of polymer brushes is also relevant for larger foulants such as bacteria.

An important biomedical application for the antifouling properties of the PEO polymer brush is its use in medical implants. One problem on the implant surface is that microorganisms might grow as a so-called biofilm (Costerton, Stewart, & Greenberg, 1999; Gristina, 1987). Biofilms on medical implants can lead to serious infections that would make it necessary to remove them from the body (Costerton et al., 1999; Gristina, 1987). By preventing the adsorption of protein molecules and bacteria, the brush helps to prevent the growth of these biofilms and thus possible infections. Another biomedical application is the use of polymer brushes, such as PEO, in biosensors. For biosensors, one wants the interaction between surface and the molecules that one wants to detect to be as specific as possible. Coating a biosensor surface with PEO brushes can prevent aspecific protein adsorption, so that only adsorption of the protein under investigation is a possibility.

5.2 Brush architecture

It can be said that the work on responsive polymer brush systems really intensified with the development of more varied brush architectures. Homopolymers can have very interesting and useful switching properties, but combining chemically different polymers into a single brush can result in a much richer switching behavior. In Figure 5.4, we schematically compare the switching behavior of a homopolymer brush with that of a mixed polymer brush, a brush consisting of two chemically different polymers grafted in the same brush. For the homopolymer brush (Figure 5.4(a)), we

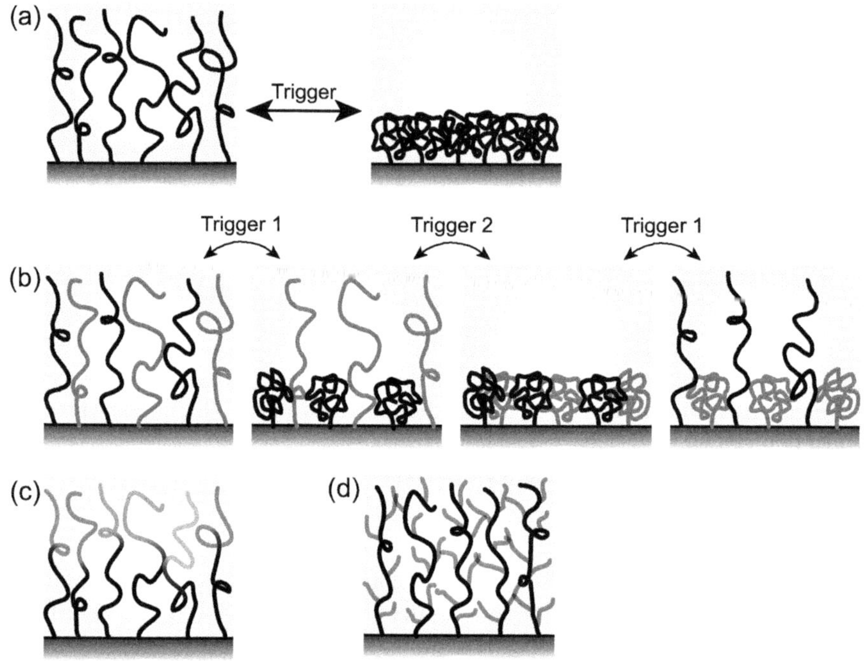

Figure 5.4 Schematic illustration of conformational changes of polymer brushes of (a) homopolymers and (b) mixed polymer brushes. Furthermore, the schematic structure of (c) block copolymers and (d) branched polymer brushes is shown.

can use a trigger to switch between two states: for example, collapsed and swollen or charged and uncharged. However, in the mixed brush, there is the potential for more switching possibilities. For the mixed brush this can, for example, result in the four different states, as shown in Figure 5.4(b). Clearly, the availability of additional brush architectures significantly enhanced the systems that can be designed to suit various biomedical applications. We want to stress that much more brush architectures are available; for example, brushes based on diblock copolymers (Figure 5.4(c)) or brushes based on polymers that have chemically different side chains (branched brush, Figure 5.4(d)). Clearly, there are even opportunities to design brushes that have not just two, but three or more chemically different polymers.

5.3 Types of responsive polymer brushes

Several types of responsive behavior have already been integrated with polymer brushes. In Table 5.1, we provide a selection of references including the used polymer and the trigger related to the responsive behavior. Of the triggers presented in the table temperature, pH, ionic strength, and solvent have received a lot of attention over the years and are already related to quite interesting applications. Voltage and light responsive

Table 5.1 A selection of literature on the integration of responsive behavior in polymer brushes (the literature is categorized based on the used trigger)

Trigger	Responsive polymer	References
Temperature	• Poly(*N*-isopropylacrylamide)s	Bittrich et al. (2012), Heinz et al. (2008), Klein Gunnewiek et al. (2012), Kooij, Sui, Hempenius, Zandvliet, and Vancso (2012), Montagne, Polesel-Maris, Pugin, and Heinzelmann (2008), Plunkett, Zhu, Moore, and Leckband (2006), Sui, Chen, Hempenius, and Vancso (2011), Abu-Lail, Kaholek, Lamattina, Clark, and Zauscher (2006), and Sun et al. (2004)
pH	• Poly(methacrylic acid)s	Santonicola, De Groot, Memesa, Meszyńska, and Vancso (2010), and Schüwer and Klok (2011)
	• Poly(acrylic acid)s	Aulich et al. (2010), Van Camp et al. (2010), Bittrich et al. (2010), Wu et al. (2007), Akkahat and Hoven (2011), Wang and Somasundaran (2006), Chiang, Dong, Ober, and Baird (2011), and Lee, Boyce, Nese, Sheiko, and Matyjaszewski (2008)
Ions	• 2-(Methacryloyloxy)ethyl dimethyl-(3-sulfopropyl)ammonium hydroxide	Azzaroni, Brown, and Huck (2006) Quintana, Gosa, Jańczewski, Kutnyanszky, and Vancso (2013)
	• Sulfobetaine methacrylamide	Nguyen, Baggerman, Paulusse, Zuilhof, and Van Rijn (2011)
	• Sulfobetaine methacrylate	Nguyen, Baggerman, Paulusse, Van Rijn, and Zuilhof (2011)
	• Poly(sulfobetaine acrylamide)	
Solvent	• Polystyrene–poly(2-vinylpyridine)	Kumar vyas et al. (2008)
	• Poly(methyl methacrylate)–polystyrene	Zhao, Haasch, and Maclaren (2004) Han, Ryu, and Park (2013) Yu, Wang, and Han (2007)
Voltage	• Ferrocene functional polymethacrylate	Kim, Ratcliff, Armstrong, Kowalewski, and Pyun (2009), Xu et al. (2010) Mazurowski et al. (2012), Elbert et al. (2013)
	• Poly(2-(methacryloyloxy)ethyl ferrocenecarboxylate)	Kutnyanszky, Hempenius, and Vancso (2014), Sui, Feng, Song, Hempenius, and Vancso (2012), and Sui et al. (2013)
	• Poly(ferrocenylsilane)–poly (*N*-isopropylacrylamide)	Weir et al. (2011)
Light	• 4,5-Dimethoxy-2-nitrobenzyl methacrylate	Brown, Azzaroni, and Huck (2009)
	• 6-Nitroveratryloxycarbonyl	Erath et al. (2013)
	• Azo polymer brushes	Wang, Ye, Wang, and Wang (2011), Cui et al. (2012), Salierno, García, and Del Campo (2013)

polymer brushes have received less attention, but have already demonstrated a large degree of potential. In this section, we will discuss these triggers in detail.

5.3.1 Temperature-responsive brushes

For a temperature-responsive polymer brush, the focus is on polymers for which the solvent quality depends strongly on the temperature. By far the most studied polymer for this purpose is poly(*N*-isopropylacrylamide) (PNIPAM) because it switches from a hydrophilic to a hydrophobic state when the temperature is increased to above 32 °C (Sun et al., 2004) (Figure 5.5(a) and (b)). The structure of a temperature-responsive PNIPAM brush highly depends on grafting density (Plunkett et al., 2006; Zhu, Yan, Winnik, & Leckband, 2006). For example, an octopus-like micelle formation of end-grafted PNIPAM was observed by atomic force microscopy (AFM), at low grafting densities and low molecular weights (Choi, Choi, & Leckband, 2013). On the other hand, at high grafting densities and molecular weights, collapsing of NIPAM chains led to the formation of continuous thin films (Choi et al., 2013). Bittrich et al. (2012) showed conformational changes of very thin (dry thickness of 3–24 nm) PNIPAM brushes with low molecular weight and grafting density (Figure 5.5(c)). Although PNIPAM brushes with various grafting densities (0.04–0.33 nm^{-2}) were swollen in buffer solution, a true collapse was not found to occur resulting from a transition from the brush regime to an intermediate and mushroom conformation.

5.3.2 pH-responsive brushes

Poly(acrylic acid) (PAA) and poly(methacrylic acid) (PMAA) brushes are the most studied weak polyelectrolytes, among many others (Currie, Sieval, Fleer, & Stuart, 2000). These polyelectrolyte brushes can undergo conformational changes (Figure 5.6(a)) upon changing the pH and/or ionic strength. Because the charge density along the polymer chains depends on the pH of the solution, the thickness of, for example, a PMAA brush can swell by about a factor of four when going from acidic to basic conditions. The swelling degree is dependent on the grafting density (Parnell et al., 2009; Schüwer & Klok, 2011) and the chain architecture of a brush system. When the PMAA brush is collapsed at a low pH, it behaves like a rigid film, which leads to a low damping of the quartz crystal microbalance resonator (Figure 5.6(b)). Schüwer and Klok (Schüwer & Klok, 2011) showed that pK_a of the PMAA brushes depends on the grafting density, when they compared a thin PMAA brush (~15 nm dry layer thickness) with high and less densely covered surfaces. With increasing grafting density, the degree of dissociation of PMAA brushes decreased, resulting in a shift in pK_a value.

5.3.3 Ionic strength responsive brushes

All polyelectrolyte brushes, both weakly and strongly charged, are responsive to the ionic strength in (aqueous) solution (Balastre et al., 2002; Borisov et al., 1994; Israels, Leermakers, & Fleer, 1994). At a low salt concentration (the osmotic brush regime),

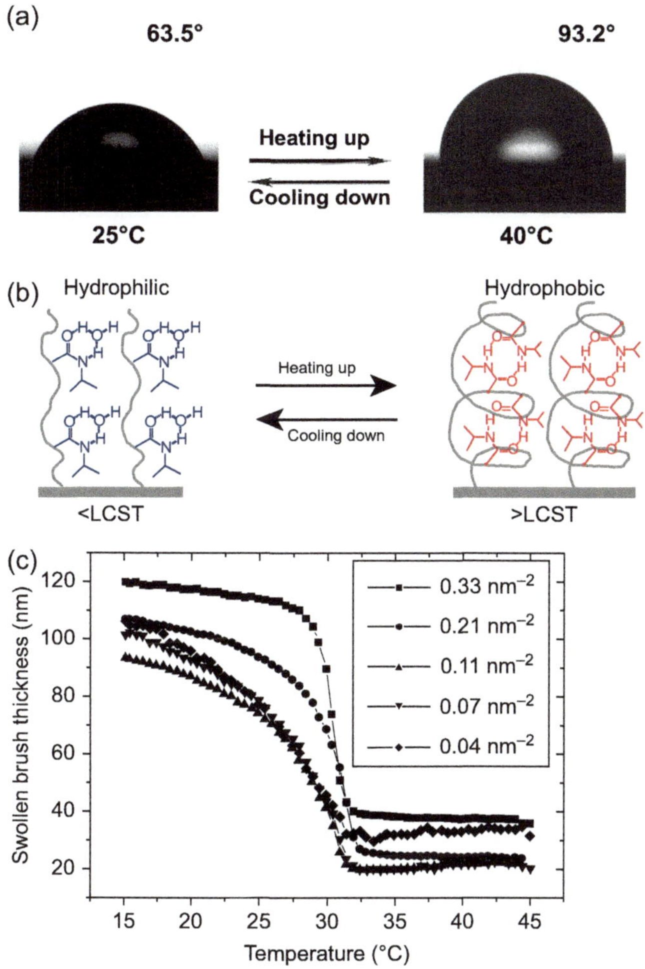

Figure 5.5 (a) Water contact angles of thermally responsive PNIPAAm-modified surface at 25 °C and 40 °C; (b) depiction of the reversibility of hydrogen bonding between PNIPAAm chains and water molecules below lower critical solution temperature (LCST; left) and intramolecular hydrogen bonding between C=O and N–H groups in PNIPAAm chains above the LCST (right). Data in (a) and (b) are reprinted with permission from Sun et al. (2004), copyright (2004) John Wiley and Sons. (c) Swollen brush thicknesses as a function of temperature for PNIPAM (47 k) brushes with different grafting densities.

Reprinted with permission from Bittrich et al. (2012), copyright (2012) American Chemical Society.

the brushes will be highly swollen, with swelling mainly depended on the degree of dissociation of the charged groups. However, at high salt concentrations (the salted brush regime), ionic interactions are screened, and swelling will be significantly less. The degree of swelling is in that case also more determined by the graft density of the brush than the degree of dissociation.

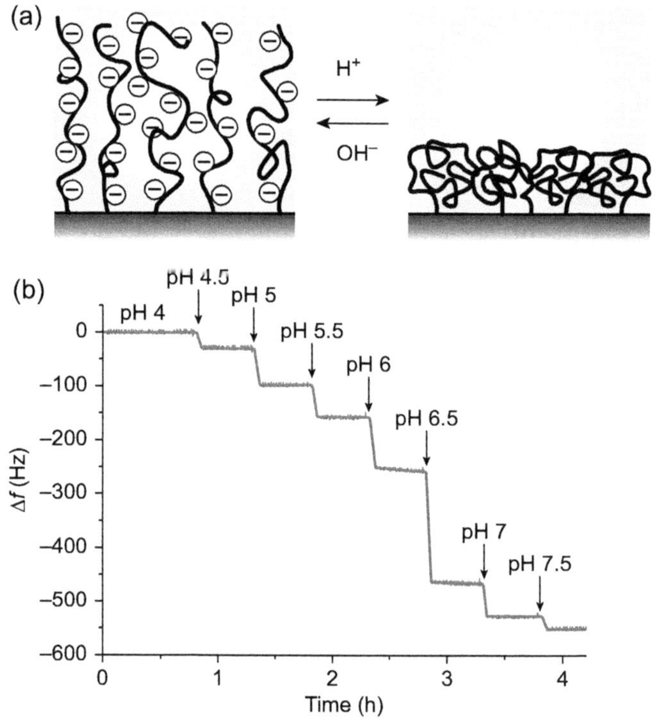

Figure 5.6 (a) Schematic illustration of conformation changes of a pH-responsive polymer brush. (b) Response of a poly(methacrylic acid) brush (11 nm thick) modified quartz crystal microbalance chip upon exposure to increasing pH values.
Reprinted with permission from Schüwer and Klok (2011), copyright (2011) American Chemical Society.

Still, there are other examples of ionic strength responsive brushes. A zwitterionic brush contains on its monomers, both a positively and a negatively moiety (Nguyen, Baggerman, Paulusse, Van Rijn, et al., 2011; Nguyen, Baggerman, Paulusse, Zuilhof, et al., 2011; Quintana et al., 2013). At a low salt concentration, these oppositely charged moieties interact with each other leading to a low degree of swelling or even a collapse of the zwitterionic polymer (Figure 5.7). However, by increasing the ionic strength, the charged groups begin to interact more with salt ions in the solution, leading to a very substantial increase in swelling. The collapse of zwitterionic polymers at a low salt concentration, and swelling at a high salt concentration, is also referred to as the anti-polyelectrolyte effect (Schulz et al., 1986).

5.3.4 Solvent-responsive brushes

In a sense, all polymer brushes are responsive to solvent. For example, a simple PEO polymer brush will swell in water, but will be collapsed when exposed to an apolar solvent such as hexane. Still, what we discuss here as solvent-resistant brushes

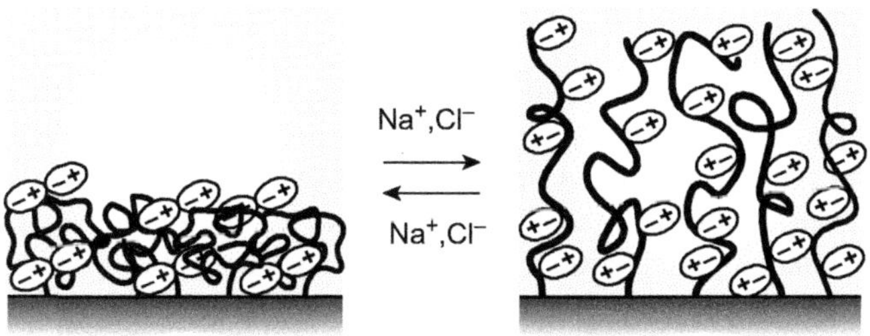

Figure 5.7 Schematic illustration of conformation changes of an ionic strength-responsive polymer brush.

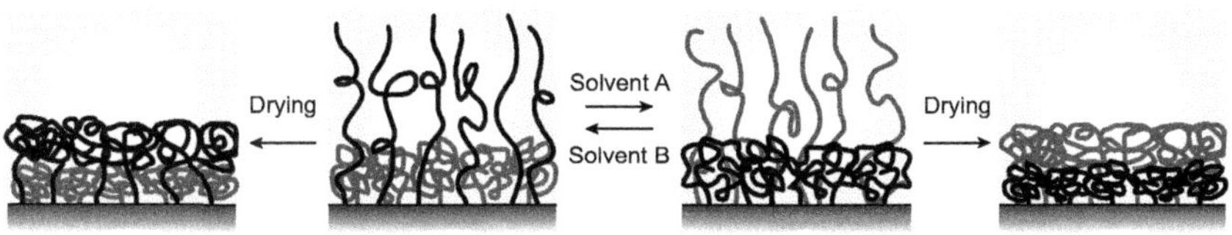

Figure 5.8 Schematic diagram showing switching behavior of mixed polymer brushes upon treatment with different selective solvents.

are more complicated systems where solvent exposure have more lasting effect. For example, Kumar vyas et al. (2008) designed a gradient-mixed brush system of polystyrene (PS): poly(2-vinylpyridine) (P2VP). The authors show that the wetting behavior of the PS-P2VP brushes can be tuned by preexposure to selective solvents. PS chains were swollen and stretched away from the surface when exposed to toluene, whereas the P2VP chains remained collapsed. On the other hand, ethanol (or acidic water) treatment forced to P2VP chains stretch away from the surface and collapsing of the PS chains (Figure 5.8). Drying the samples leads to the top layer being formed by the polymer chains that were most recently swollen. As such, the surface "remembers" the solvent it was last exposed to, with a related effect on the wetting of that surface with water.

Another interesting design of mixed polymer brushes was obtained by sequential grafting of asymmetric nonsticky/sticky diblock copolymers of poly(styrene-b-3-(trimethoxysilyl)propylmethacrylate) and monomethoxypoly(ethylene glycol)-trimethoxysilane onto silicon wafers (Han et al., 2013). The nanoscopic morphology of these mixed polymer brushes exhibited stimuli responsive behavior to various temperatures and solvents.

Yu et al. (2007) developed a phase separation technique to move the CdS nanoparticles reversibly in the perpendicular direction. The CdS nanoparticles were chemically bonded to polystyrene-b-(poly(methyl methacrylate)-*co*-poly(-methacrylic acid) (CdS)) (PS-b-(PMMA-*co*-PMAA(CdS))) brushes and could be vertically lowered into the brush by exposure to toluene, whereas they were lifted out of the brush by exposure to acetone (or ethanol). Hence, authors show the possible extension of movement by manipulating the thickness of two blocks of the polymer brushes.

5.3.5 Electro-responsive brushes

By coating a polymer brush on a conductive surface, it becomes possible to use an electric field as a trigger for certain types of polymer brushes. In an applied electric field, a force acts on positively charged moieties along the direction of the field, whereas a force with an opposite direction acts on negatively charged moieties. Initially shown by molecular dynamics simulations (Hui, Zhenhai, & Jiang, 2009), an electric field can thus also affect the swelling of a charged brush, by "pulling" on its charges. Depending on the charge of the brush, one direction of the electric field will lead to an increased swelling, whereas the other leads to an opposite effect. This effect was investigated experimentally (Weir et al., 2011) with a positively charged poly(2-dimethylamino ethyl methacrylate) (PDMAEMA) and by use of neutron reflection. It was found that application of a positive voltage (3 V) between the silica surface and a gold electrode placed in parallel at a 3 mm distance led to an enhanced swelling of the PDMAEMA brush on a silicon surface. When a higher voltage (5 V) was applied, the brush was damaged, indicating that chains were pulled off the surface by the generated electric field.

Another approach for responsive behavior is the use of electrochemistry. Redox responsive brushes were, for example, prepared (Mazurowski et al., 2012) from poly2-(methacryloyloxy)ethylferrocenecarboxylate). Oxidation of the ferrocene moiety led to a charging of the brush and subsequent swelling (see Figure 5.9), whereas subsequent reduction had the opposite effect.

5.3.6 Light-induced brushes

Only a little work has been done on light-responsive brushes, but a few examples do exist (Cui et al., 2012; Salierno et al., 2013; Wang et al., 2011). An example encompasses a hydrophobic polymer brush based on poly(4,5-dimethoxy-2-nitrobenzyl methacrylate) (Brown et al., 2009). By exposure to ultraviolet light, this brush can be gradually turned into a charged PMAA brush because of the photoremoval of the chromophore and generation of free –COOH groups. Although this type of response is not reversible, it does show the possibility to create photoresponsive brushes to control the wettability of a surface.

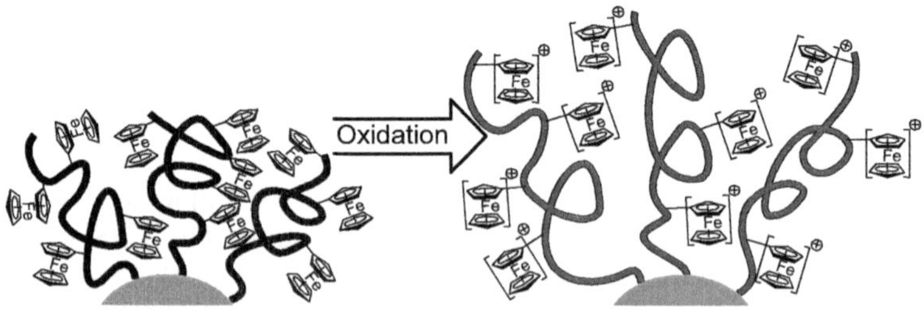

Figure 5.9 Electrochemically responsive brushes: oxidation of a brush containing ferrocene groups leads to charging and subsequent swelling of the polymer brush.
Reprinted with permission from Mazurowski et al. (2012), copyright (2012) American Chemical Society.

5.4 Biomedical applications

5.4.1 Homopolymers

In this section, we will point out various types of stimuli-responsive homopolymer brushes that show practical biomedical applications. Xue, Choi, Choi, Braun, and Leckband (2012) showed how the grafting density of PNIPAM brushes may affect protein adsorption and reversible cell attachment (Figure 5.10). Under poor solvent conditions, T > LCST, protein adsorption is negligible at high grafting densities. However, at intermediate densities protein adsorption was found to occur through ternary adsorption. Figure 5.10 shows the fibroblast attachment on different PNIPAM grafting densities at 23 °C and 37 °C. Cells are firmly attached and well distributed on the surface with a lower grafting density of PNIPAM (0.01 chain nm^{-2}), whereas on dense brushes (0.11 and 0.21 chain nm^{-2}), fewer and more rounded fibroblasts were seen. Interestingly, at intermediate chain densities reversible adsorption was observed.

pH-responsive PAA, in combination with certain biomolecules, may serve as a highly efficient biosensor. For instance, biotin-attached PAA brushes were used for the detection of streptavidin (SA) binding (Akkahat & Hoven, 2011). Self-assembled monolayer of carboxyl-terminated alkanethiol (11-mercaptoundecanoic acid) was used as a comparison and showed an 11-fold lower specificity to biotin–SA interaction. On the other hand, an enhanced binding affinity of an analyte (e.g., SA, bovine serum albumin) to PAA brushes which are covalently immobilized with a biomolecule (e.g., biotin and anti-bovine serum albumin) was observed with the increase in grafting density (Akkahat, Mekboonsonglarp, Kiatkamjornwong, & Hoven, 2012). However, the latter also causes a resistance for the accessibility of the target analyte to the probe which is immobilized on the PAA brush.

Another application of pH-responsive polymers was shown by de Groot et al. (2013), as shown in Figure 5.11. Surface-initiated atom transfer radical polymerization synthesized PMAA brushes were used to create pH-responsive nanoporous platforms. It was shown that at pH 4, gating of ions was allowed through the nano channels, whereas at pH 8, the pores were closed. The authors predict that their pH-responsive channels have potential as electrochemical biosensors and in bioseperation technology.

Tam, Ornatska, Pita, Minko, and Katz (2008) reported on an electrochemically responsive interface based on poly(4-vinyl pyridine) (P4VP) brush functionalized with Os-complex redox units (Figure 5.12). At acidic pH (pH = 3.0), the pyridine functional groups become positively charged thus yielding a hydrophilic swollen state allowing direct contact between the conductive surface and the Os-complex redox units (on-state). When the pH was increased to neutral pH values (pH = 7.0), the pyridine groups showed an enhanced hydrophobicity that resulted in brush collapsing (off-state). Oxidation of glucose biocatalyzed by glucose oxidase occurs by the mediation of the Os-functionalized P4VP brush in the on-state conditions. When the pH was increased from pH 4.0 to 7.0, a sharp inhibition of the bioelectrocatalytic process was observed due to the collapsed brushes. Similarly, depending on the on-off state of P4VP brushes grafted on indium-tin oxide electrode, anionic species such as

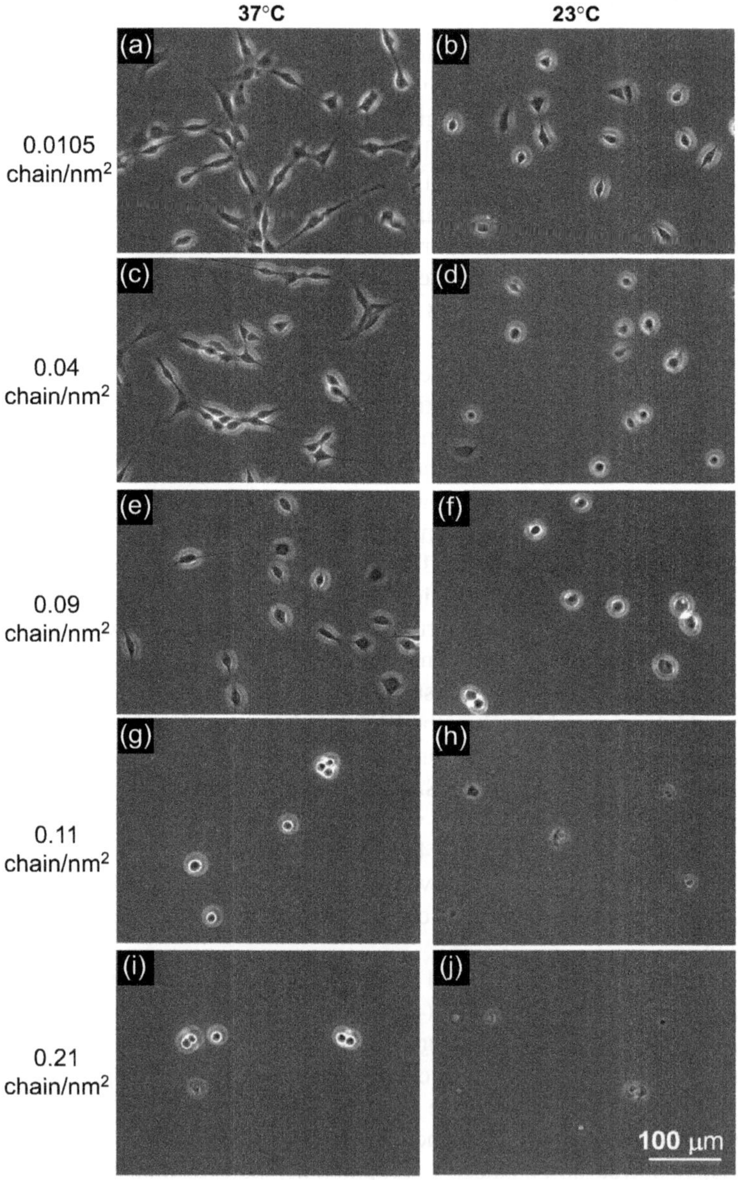

Figure 5.10 (a–j) 3T3 fibroblast attachment at 37 °C and 23 °C on PNIPAM brushes with different grafting densities.
Reprinted with permission from Xue et al. (2012), copyright (2012) John Wiley and Sons.

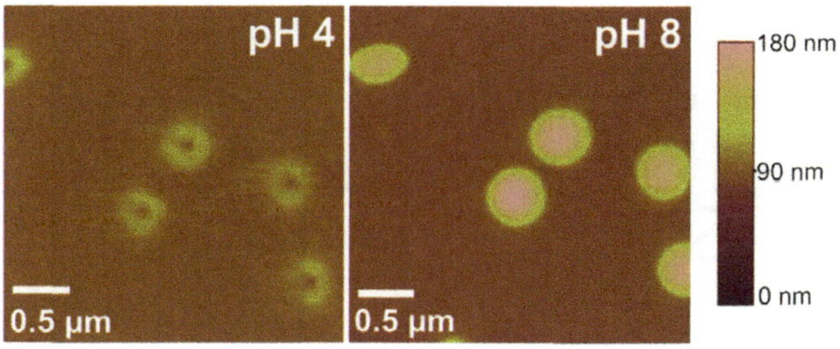

Figure 5.11 Contact mode atomic force microscopy (CM-AFM) images of PMAA brush functionalized nanoporous at pH 4 or pH 8.
Reprinted with permission from de Groot et al. (2013), copyright (2013) American Chemical Society.

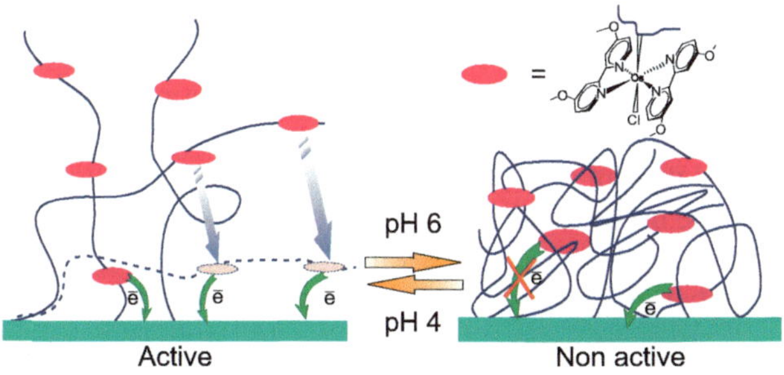

Figure 5.12 Reversible pH-controlled transformation of the redox-polymer brush on the electrode surface between electrochemically active and inactive states.
Reprinted with permission from Tam et al. (2008), copyright (2008) American Chemical Society.

ferrocyanide were shown to be electrochemically active or inactive. Tam et al. suggest their electrochemically controlled electrode surface with tunable and switchable properties is a good candidate for the development of electrochemical biosensors, immune-controlled biofuel cells (Tam, Strack, Pita, & Katz, 2009), or artificial bioelectronics systems (Privman, Tam, Pita, & Katz, 2008).

Responsive polymer brushes are not only investigated on planar substrates, but can also be used to decorate nanoparticles (Figure 5.13). Here, the key biomedical application is the accommodation (immobilize) of biofunctional moieties such as proteins or enzymes (Wittemann & Ballauff, 2006; Wittemann, Haupt, & Ballauff, 2003). If there is sufficient attraction between a protein and a polymer brush to overcome excluded volume effects (the antifouling properties of the brush), much larger amounts of protein can accumulate inside a brush layer than can be adsorbed onto just the particles surface. The role of the attached polymer chains is thus to strongly increase the available surface area. In Figure 5.13, we schematically show a spherical polymer brush filled with adsorbed nanoparticles. Because, protein molecules immobilized by polymers are found to be relatively weakly bound, they keep their conformation and (enzymatic)

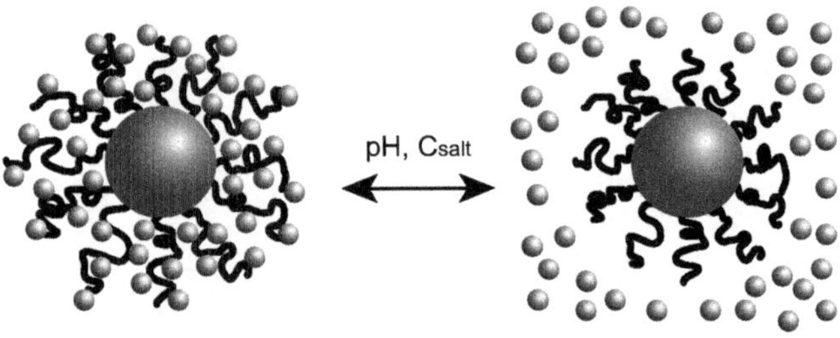

Figure 5.13 Schematic depiction of the uptake and release of enzyme molecules into and from a responsive polymer brush.

activity largely intact (Caruso & Schüler, 2000; Xia, Mattison, Romano, Dubin, & Muhoberac, 1997), whereas proteins adsorbing on a smooth, hard surface often change their conformation to adjust to the hard surface, resulting in a loss of activity (Norde & Giacomelli, 1999). Protein uptake is best achieved with polyelectrolyte brushes as strong (electrostatic) interaction is necessary to overcome the excluded volume effects of the brush. Another advantage of electrostatic interaction is that changes in the pH and addition of salt can be used as triggers to decrease the electrostatic interaction and release the adsorbed protein molecules. To conclude, enzyme-filled brushes on colloidal particles might be used for drug delivery purposes, or as nano-reactors in solution. As the used polyelectrolyte brushes are by nature responsive, the uptake and release of the enzymes and other biofunctional moieties can be controlled.

5.4.2 Block copolymer

Copolymer architecture such as graft, branch, or linear offers variety of applications in responsive biomaterial design (Idota, Ebara, Kotsuchibashi, Narain, & Aoyagi, 2012; Kelley, Albert, Sullivan, & Epps, 2013; Laloyaux et al., 2010). Tuning the composition of co-monomer sequence (e.g., random, gradient, block) and molecular weight, it is furthermore possible to control and manipulate of the copolymer brush response (Gunkel, Weinhart, Becherer, Haag, & Huck, 2011).

A nice example is given by Lei et al. (2013), in which they studied adsorption of lysozyme on poly(2-(dimethylamino ethyl) methacrylate)-block-poly(methacrylic acid) (PDMAEMA-b-PMAA) diblock-copolymer brushes (Figure 5.14). Lysozyme has a relatively high isoelectric point ($I_p = 11.1$) thus controlling its adsorption (on Si) by pH is too difficult. It has been shown that lysozyme adsorption decreases on weak base polyelectrolyte, PDMAEMA, and increases on weak acid polyelectrolyte, PMAA brushes. Authors show that the thickness of the PMAA block in PDMAEMA-b-PMAA is critical: when the outer PMAA block is less than 10 nm, adsorption of the lysozyme increases with pH. Lysozyme adsorption reached to 16.4-fold more at pH 10 compared with its value at pH 4. At thicker regimes (>10 nm), diblock-copolymer brush exhibited similar behavior to that of PMAA homopolymer brush.

Depending on the film thickness and coating conditions, structure and wettability of the polymers may vary and this can affect cell attachment. For example, Yang et al. (2010) reported adhesion, growth, and detachment of HeLa cells seeded on

Figure 5.14 Illustration of the swelling behavior of PDMAEMA-b-PMAA diblock-copolymer brushes when varying the pH from 4 to 7 to 10.
Reprinted with permission from Lei et al. (2013), copyright (2014) American Chemical Society.

poly(*N*-isopropylacrylamide-*co*-hydroxypropylmethacrylate-*co*-3-(trimethoxysilyl) propyl methacrylate) copolymer at different concentrations (Figure 5.15). At 0.1–0.2 mg/ml, cell attachment/detachment was not different then the positive control (glass coverslip). Partial and complete detachment of the cells upon cooling to ambient temperature was observed at the PNIPAAm copolymer concentrations of 0.5–3 mg/ml (Figure 5.15(a) and (b)) and at 5 mg/ml (Figure 5.15(c) and (d)), respectively. However, cells showed much less preference to the cells that are seeded on thicker (8–10 mg/ml) polymer films (Figure 5.15(e)). This important finding showed that an optimal film thickness is needed to accommodate adequate cell attachment (or detachment).

The interaction of ligands and cell membrane receptors plays multiple important roles for inducing cell spreading, proliferation, differentiation, and signal transduction. Ebara et al. (2004) grafted poly(*N*-isopropylacrylamide-*co*-2-carboxyisopropylacrylamide) (P(IPAAm-*co*-CIPAAm)) copolymer to investigate interactions between cell adhesion peptides (RGDS, (Arg-Gly-Asp-Ser)) and cell integrin receptors located on cell membranes. (P(IPAAm-*co*-CIPAAm)) copolymer grafted on tissue culture grade PS dishes discussed previously, the LCST facilitates the spreading of the human umbilical vein endothelial cells in serum-free conditions. Human umbilical vein endothelial cells attached to the surface at 37 °C can dissociate the immobilized RGDS from integrin cells below the LCST, at a temperature that the grafted copolymer chains start to hydrate. This on-off control over the specific integrin-peptide binding is not only important for the cell function regulation but also can serve several applications in tissue engineering.

Poly(*N*-isopropylacrylamide)-block polystyrene (PNIPAAm-b-PS) brushes were shown to be responsive both to temperature and solvent (Yu et al., 2010) (Figure 5.16). Interestingly, PNIPAAm-b-PS brushes showed protein resistance similar to the homo-PNIPAM (Cheng, Canavan, Graham, Castner, & Ratner, 2006; Huber, Manginell, Samara, Kim, & Bunker, 2003) despite the protein-adsorptive, hydrophobic PS block at the outer part of the brush. Still, authors show adsorption of fibrinogen on this block copolymer grafted surface was $<8\,ng/cm^2$ at both room temperature and 37 °C, suggesting good blood compatibility. Furthermore, a significant influence of the hydrophobic PS block on cellular attachment was clearly demonstrated.

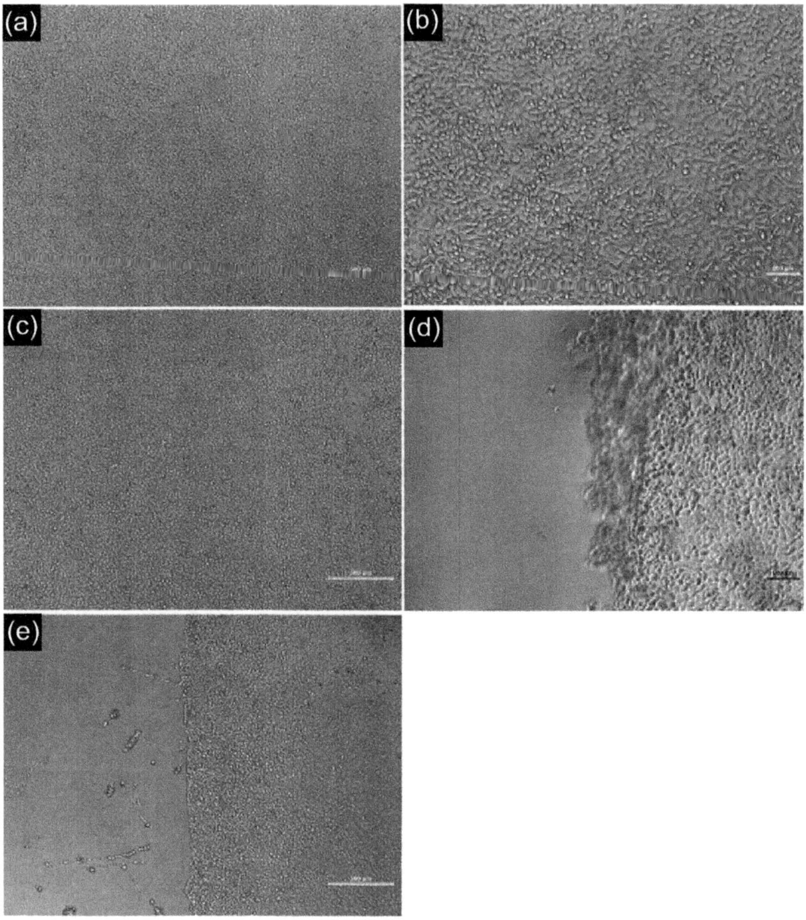

Figure 5.15 Adhesion (a, c, e; 37 °C) and detachment (b, d; ambient temperature) of HeLa cells seeded on three different PNIPAAm copolymer films coated at 1 mg/ml (a, b), 5 mg/ml (c, d), and 10 mg/ml (e) copolymer concentrations.
Reprinted with permission from Yang et al. (2010), copyright (2010) American Chemical Society.

5.4.3 Mixed brushes

Mixed brushes have enormous potential to be used in biomedical applications because of their multiresponsiveness in one system.

To tune protein adsorption, brushes of PEO and PAA were either grafted simultaneously or sequentially using the "grafting to" approach (Delcroix et al., 2012). It was concluded that simultaneous grafting of the polymers gives a much larger range of composition compared to sequential grafted brushes. The mixed brush could adsorb high amounts of albumin at pH 5 and low ionic strength (I) from swollen chains of PAA, but 86% of the adsorbed protein could then be desorbed by changing pH and the ionic strength. This is because at high pH and ionic strength, PAA chains collapse and thus PEO chains are more exposed to the outermost surface of the mixed brush system.

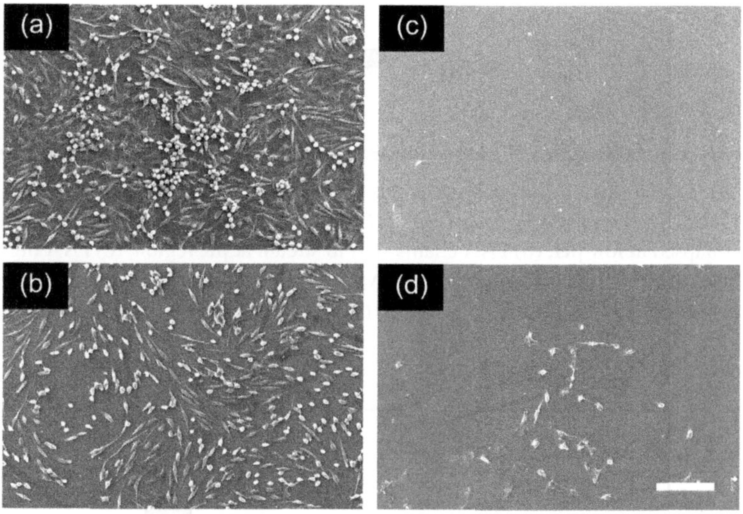

Figure 5.16 Scanning electron microscopic images of L929-cells on (a) an unmodified Si surface at 37 °C, (b) a Si-PNIPAAm surface at 37 °C, (c) a Si-PNIPAAm-b-PS surface at 37 °C, and (d) a Si-PNIPAAm-b-PS surface at 23 °C. The scale bar in all images is 100 μm. Reprinted with permission from Yu et al. (2010), copyright (2010) American Chemical Society.

Hoy et al. (2010) designed a mixed brush system composed of a hydrophilic homopolymer, poly(ethylene glycol) (PEG), and an amphiphilic block copolymer, poly(acrylic acid-*b*-styrene) (PAA-PS), where the hydrophobic block, PS, is immiscible with PAA and PEG (Figure 5.17). The surface properties were controlled by the grafted PEG amount on solid surface. By applying the external triggers such as pH, ionic strength, and Ca^{2+} concentration, hydrophobic domains which are embedded in the hydrophilic polymeric matrix are exposed. As a control, the adsorption of fibrinogen was investigated on homopolymer PS, PAA, PEG, and PAA-PS block copolymer brushes. Protein adsorption was observed for all brushes; however, the mixed polymer system treated with $CaCl_2$ showed the lower adsorption ($<1.5 \times 10^{-9}$ mol/m^2). To remove the Ca ions bound to brush, the same mixed brush was treated with HCl. As a result the adsorption process recovered and a protein monolayer of 10^{-8} mol/m^2 were obtained.

Motornov exploited the pH sensitivity of PAA and P2VP mixed polyelectrolyte brush for selective gating ion transport (Motornov et al., 2009). The PAA-P2VP brush was shown to have a bipolar permselectivity behavior: The system blocks transport of positively charged species below pH 4.0, whereas it allows permeation of the negatively charged species. However, above pH 5, the system shows an opposite behavior, where the permeation of positively charged species is allowed and negatively charged species are blocked. Authors predict that their system is a good candidate to be used as a smart drug-delivery capsule that can work in acidic and neutral pH conditions.

To regulate microbial adhesion, mixed brushes of PEG and a positively charged polymer, P2VP or quarternized poly(2-vynil pyridine) (QP2VP) were investigated and compared with homo polymer brushes of the respective components (Zdyrko et al., 2009). Only with PEG brush, almost all the bacteria (*Staphylococcus aureus*)

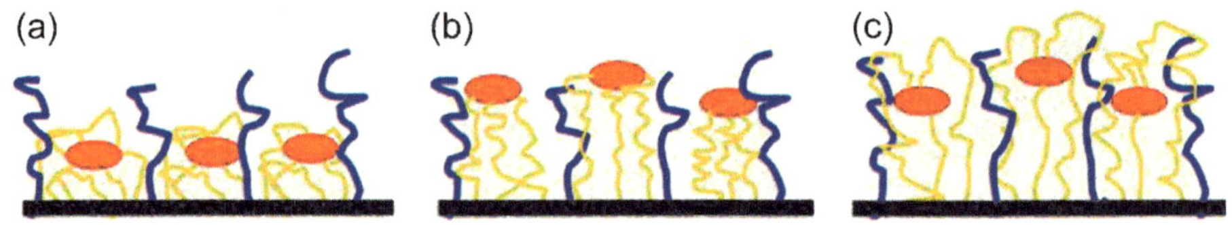

Figure 5.17 Depiction of mixed polymer brush system composed of PEG (blue line) and PAA-PS block copolymer (PAA, yellow curve; PS, orange ellipse) at three different states. (a) PAA collapsed at low pH; (b) PAA chains start to stretch at pK region; (c) PAA highly extended at high pH or high ionic strength. PAA, Poly(acrylic acid); PEG, poly(ethylene glycol). Reprinted with permission from Hoy et al. (2010), copyright (2010) John Wiley and Sons.

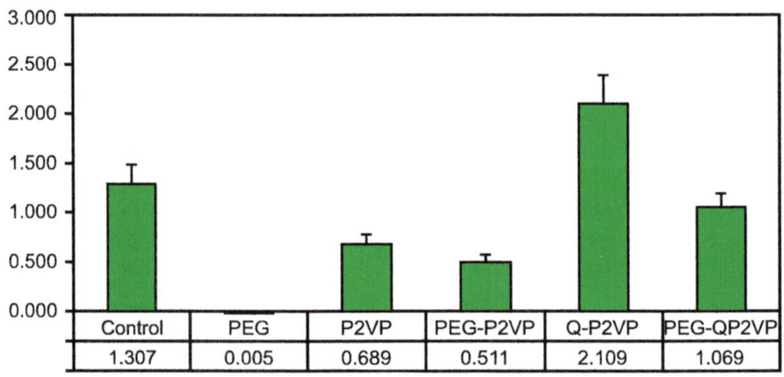

Figure 5.18 *Staphylococcus aureus* adsorption to the polymer brushes. Control surface: Si wafer. *Y*-axis is in relative fluorescence of the samples after *S. aureus* adsorption. Numbers below the bars indicating relative amount of *S. aureus* adsorbed to different brushes. Reprinted with permission from Zdyrko et al. (2009), copyright (2009) John Wiley and Sons.

was repelled from the surface, whereas QP2VP brush showed 160% higher bacterial adsorption compare to the control (Si wafer surface). Interestingly, P2VP brush reduced the adsorption about 50%. To conclude (Figure 5.18), PEG and QP2VP covered surfaces resulted in a 400-fold difference in the adsorption of bacteria (from 0.005 to 2.109 au). Thus, mixing these two polymer brushes allows designing surfaces with tunable bacteria adsorption properties.

5.5 Summary and future trends

It is clear that there is a bright future for responsive polymer brushes in the field of biomedical applications. They provide a unique control over surfaces that is difficult to match by any other surface coating. They have been shown be able to control wettability, cell adhesion, bacterial adhesion, and protein adsorption, which provide many opportunities for functional coatings of medical implants, protein carriers, biosensors, etc. In addition, the chemistry to produce them is well understood and allows a versatile range of brush chemistries that allow responsiveness to a large variation of possible triggers. Combining different chemistries in a single brush system, for example, by use of mixed brushes, branched brushes, and diblock-copolymer brushes, further enhanced the versatility and thus applicability of the responsive polymer brush. Still there are also drawbacks to the current state-of-the-art systems that limit widespread applications. In this outlook, we give two examples of responsive brush systems with current limitations but great potential. In addition, we include a discussion on possible ways to overcome these limitations.

An application that has received very large amounts of research attention is the use of spherical polymer brushes (SPBs), as discussed under the homopolymers section. The weakness of the current generation of SPBs is that the particles lose their colloidal stability when oppositely charged proteins adsorb to the brush. The oppositely charged proteins neutralize the charge in the brush to a large extent, thus creating an almost neutral particle. Without the charge which gave the stabilization the particles aggregate. However, proteins are also found to adsorb to like-charged polymer brushes (at the wrong side of the isoelectric point of the protein). In that case, the adsorption does not lead to neutralization of the polyelectrolyte brush, and the particle dispersion remains stable. As adsorption above the isoelectric point is only found a few (about 1.5) pH units above the isoelectric point (de Vos, Biesheuvel, De Keizer, Kleijn, & Cohen Stuart, 2008), the pH range in which the particles are stable and also carry proteins is severely limited. To some extent, one could widen this pH range by choosing a polyelectrolyte that has some (chemical) affinity to the protein. However, to strongly increase the pH range in which they can be applied, we suggest incorporating an extra built-in stabilization mechanism into the "next generation" of SPB protein carriers. One way would be to graft diblock copolymers on the particle. Such diblock copolymers should consist of a polyelectrolyte part for the storage of the proteins, and a neutral polymer (such as PEO) for the particle stabilization. This idea is schematically shown in Figure 5.19.

The biggest potential to improve the biomedical applicability of responsive polymer brushes lies in the specificity of their interactions. Currently, a brush might switch between the adsorption and nonadsorption of protein molecules, but it is not specific to a single type of protein. There are also no specific interactions yet with specific cells or bacteria. Here we propose a responsive polymer brush design that would combine responsive and specific behavior. The key is to combine the specific interactions of protein molecules such as enzymes or antibodies with the responsive behavior of brush. Imagine a polymer brush consisting of two chemically different polymers. For one polymer the stretching depends strongly on parameters such as pH, temperature and/or ionic strength. This could be a PNIPAM chain that collapses above a temperature of 32 °C or a PMAA chain that becomes uncharged and because of its more hydrophobic nature, collapses at low pH. The other polymer is chosen so that its stretching is quite independent of these parameters, for example, the commonly used PEO. This combination provides us with a brush that with a certain trigger can be switched between a brush with two types of swollen chains and a brush with one swollen and one collapsed component. To functionalize this brush, we end-attach a

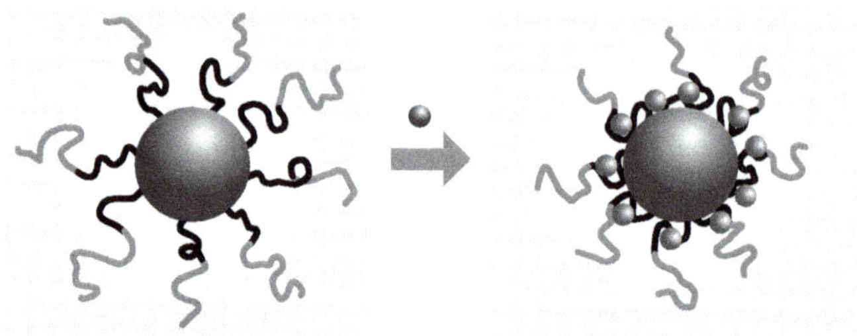

Figure 5.19 Schematic depiction of a particle covered with a brush consisting of a polyelectrolyte block (dark line) and a neutral (water soluble) polymer block (gray line). Upon the addition of oppositely charged protein molecules (gray spheres), the polyelectrolyte block complexes with the protein, forming a net neutral phase. The neutral polymer block stabilizes this particle.

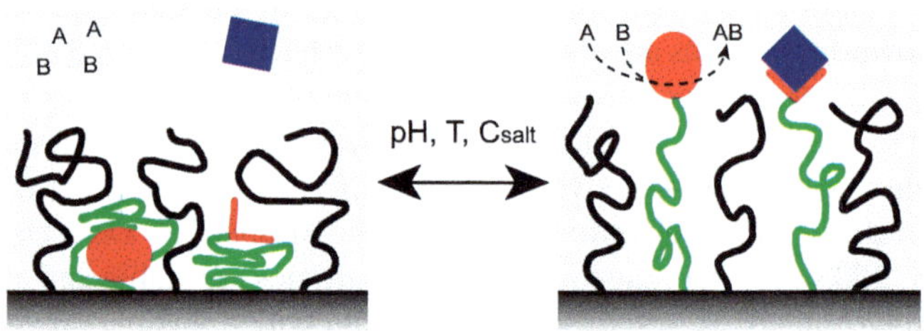

Figure 5.20 Schematic representation of a responsive polymer brush combined with biological molecules. A change in pH, temperature, or salt concentration leads to a change between a protective state in which an enzyme and a receptor molecule are hidden away deep inside of a brush layer, and an active state in which they are exposed to the solution.

biological molecule, such as an enzyme, antibody, and/or a receptor molecule to the switching polymer. With that in place, we have a responsive polymer brush that with certain trigger switches between a protective state, in which the biological component is hidden deep inside a protective brush layer, and an active state in which the biological component is exposed to the solution. This idea is schematically shown in Figure 5.20.

Although this is a relatively simple concept, its true strength lies in its many possible applications, depending on the attached biological component. The most direct example would be to use such layers for catalytic particles. Colloidal particles could be coated with these layers, using a (catalytic) enzyme as the biological component. These particles could then switch between an "on" mode in which the enzyme is exposed to the solution and an "off" mode in which the enzyme is protected inside the brush. But the same approach can be used to create biosensors, which can be switched on and off. It is even possible to combine a responsive brush with an antibacterial protein such as lysozyme allowing switching between a bacterial adhesive and bacterial deactivating interface. We can only conclude that the combination of responsive surface layers with the specific functionality of biomolecules would make for very interesting and powerful systems, with a large potential for biomedical applications.

References

Abu-Lail, N. I., Kaholek, M., Lamattina, B., Clark, R. L., & Zauscher, S. (2006). Micro-cantilevers with end-grafted stimulus-responsive polymer brushes for actuation and sensing. *Sensors and Actuators B: Chemical*, *114*, 371–378.

Advincula, R. C., Brittain, W. J., Caster, K. C., & Rühe, J. (2004). *Polymer brushes*. Weinheim: Wiley-VHC.

Akkahat, P., & Hoven, V. P. (2011). Introducing surface-tethered poly(acrylic acid) brushes as 3D functional thin film for biosensing applications. *Colloids and Surfaces B: Biointerfaces*, *86*, 198–205.

Akkahat, P., Mekboonsonglarp, W., Kiatkamjornwong, S., & Hoven, V. P. (2012). Surface-grafted Poly(acrylic acid) brushes as a precursor layer for biosensing applications: effect of graft density and swellability on the detection efficiency. *Langmuir*, *28*, 5302–5311.

Albertsson, P.-Å. (1986). *Partition of cell particles and macromolecules*. New York: Wiley.

Aulich, D., Hoy, O., Luzinov, I., Brücher, M., Hergenröder, R., Bittrich, E., et al. (2010). In situ studies on the switching behavior of ultrathin poly(acrylic acid) polyelectrolyte brushes in different aqueous environments. *Langmuir: the ACS Journal of Surfaces and Colloids*, *26*, 12926–12932.

Auroy, P., Auvray, L., & Leger, L. (1991). Structures of end-grafted polymer layers: a small-angle neutron scattering study. *Macromolecules*, *24*, 2523–2528.

Azzaroni, O., Brown, A. A., & Huck, W. T. S. (2006). UCST wetting transitions of polyzwitterionic brushes driven by self-association. *Angewandte Chemie International Edition*, *45*, 1770–1774.

Balastre, M., Li, F., Schorr, P., Yang, J., Mays, J. W., & Tirrell, M. V. (2002). A study of polyelectrolyte brushes formed from adsorption of amphiphilic diblock copolymers using the surface forces apparatus. *Macromolecules*, *35*, 9480–9486.

Bittrich, E., Burkert, S., Müller, M., Eichhorn, K.-J., Stamm, M., & Uhlmann, P. (2012). Temperature-sensitive swelling of poly(N-isopropylacrylamide) brushes with low molecular weight and grafting density. *Langmuir: The ACS Journal of Surfaces and Colloids*, *28*, 3439–3448.

Bittrich, E., Rodenhausen, K. B., Eichhorn, K. J., Hofmann, T., Schubert, M., Stamm, M., et al. (2010). Protein adsorption on and swelling of polyelectrolyte brushes: a simultaneous ellipsometry-quartz crystal microbalance study. *Biointerphases*, *5*, 159–167.

Borisov, O. V., Zhulina, E. B., & Birshtein, T. M. (1994). Diagram of the states of a grafted polyelectrolyte layer. *Macromolecules*, *27*, 4795–4803.

Brown, A. A., Azzaroni, O., & Huck, W. T. S. (2009). Photoresponsive polymer brushes for hydrophilic patterning. *Langmuir*, *25*, 1744–1749.

Caruso, F., & Schüler, C. (2000). Enzyme multilayers on colloid particles: assembly, stability, and enzymatic activity. *Langmuir*, *16*, 9595–9603.

Chen, T., Ferris, R., Zhang, J., Ducker, R., & Zauscher, S. (2010). Stimulus-responsive polymer brushes on surfaces: transduction mechanisms and applications. *Progress in Polymer Science*, *35*, 94–112.

Cheng, X., Canavan, H., Graham, D., Castner, D., & Ratner, B. (2006). Temperature dependent activity and structure of adsorbed proteins on plasma polymerized N-isopropyl acrylamide. *Biointerphases*, *1*, 61–72.

Chiang, E. N., Dong, R., Ober, C. K., & Baird, B. A. (2011). Cellular responses to patterned poly(acrylic acid) brushes. *Langmuir*, *27*, 7016–7023.

Choi, B. C., Choi, S., & Leckband, D. E. (2013). Poly(N-isopropyl acrylamide) brush topography: dependence on grafting conditions and temperature. *Langmuir*, *29*, 5841–5850.

Costerton, J. W., Stewart, P. S., & Greenberg, E. P. (1999). Bacterial biofilms: a common cause of persistent infections. *Science, 284*, 1318–1322.

Cui, J., Nguyen, T.-H., Ceolín, M., Berger, R., Azzaroni, O., & Del Campo, A. (2012). Phototunable response in caged polymer brushes. *Macromolecules, 45*, 3213–3220.

Currie, E. P. K., Norde, W., & Cohen Stuart, M. A. (2003). Tethered polymer chains: surface chemistry and their impact on colloidal and surface properties. *Advances in Colloid and Interface Science, 100–102*, 205–265.

Currie, E. P. K., Sieval, A. B., Fleer, G. J., & Stuart, M. A. C. (2000). Polyacrylic acid Brushes: surface pressure and salt induced swelling. *Langmuir, 16*, 8324–8333.

Delcroix, M. F., Huet, G. L., Conard, T., Demoustier-Champagne, S., Du Prez, F. E., et al. (2012). Design of mixed PEO/PAA brushes with switchable properties toward protein adsorption. *Biomacromolecules, 14*, 215–225.

Draper, J., Luzinov, I., Minko, S., Tokarev, I., & Stamm, M. (2004). Mixed polymer brushes by sequential polymer Addition: anchoring layer effect. *Langmuir, 20*, 4064–4075.

Du, H., Chandaroy, P., & Hui, S. W. (1997). Grafted poly-(ethylene glycol) on lipid surfaces inhibits protein adsorption and cell adhesion. *Biochimica et Biophysica Acta (BBA) – Biomembranes, 1326*, 236–248.

Ebara, M., Yamato, M., Aoyagi, T., Kikuchi, A., Sakai, K., & Okano, T. (2004). Temperature-responsive cell culture surfaces enable "On–Off" affinity control between cell integrins and RGDS ligands. *Biomacromolecules, 5*, 505–510.

Edmondson, S., Osborne, V. L., & Huck, W. T. S. (2004). Polymer brushes via surface-initiated polymerizations. *Chemical Society Reviews, 33*, 14–22.

Elbert, J., Gallei, M., Rüttiger, C., Brunsen, A., Didzoleit, H., Stühn, B., et al. (2013). Ferrocene polymers for switchable surface wettability. *Organometallics, 32*, 5873–5878.

Erath, J., Cui, J., Schmid, J., Kappl, M., Campo, A. D., & Fery, A. (2013). Phototunable surface interactions. *Langmuir, 29*, 12138–12144.

Gristina, A. (1987). Biomaterial-centered infection: microbial adhesion versus tissue integration. *Science, 237*, 1588–1595.

de Groot, G. W., Santonicola, M. G., Sugihara, K., Zambelli, T., Reimhult, E., Vörös, J., et al. (2013). Switching transport through nanopores with pH-responsive polymer brushes for controlled ion permeability. *ACS Applied Materials & Interfaces, 5*, 1400–1407.

Gunkel, G., Weinhart, M., Becherer, T., Haag, R., & Huck, W. T. S. (2011). Effect of polymer Brush architecture on antibiofouling properties. *Biomacromolecules, 12*, 4169–4172.

Halperin, A. (1999). Polymer brushes that resist adsorption of model proteins: design parameters. *Langmuir, 15*, 2525–2533.

Han, M., Ryu, J.-S., & Park, J.-W. (2013). Homo- and mixed polymer brushes prepared by surface-grafting of asymmetric non-sticky/sticky diblock copolymers and their stimuli–responsive behaviors. *Reactive and Functional Polymers, 73*, 66–72.

Heinz, P., Brétagnol, F., Mannelli, I., Sirghi, L., Valsesia, A., Ceccone, G., et al. (2008). Poly(N-isopropylacrylamide) grafted on plasma-activated poly(ethylene oxide): thermal response and interaction with proteins. *Langmuir, 24*, 6166–6175.

Herold, D. A., Keil, K., & Bruns, D. E. (1989). Oxidation of polyethylene glycols by alcohol dehydrogenase. *Biochemical Pharmacology, 38*, 73–76.

Hoy, O., Zdyrko, B., Lupitskyy, R., Sheparovych, R., Aulich, D., Wang, J., et al. (2010). Synthetic hydrophilic materials with tunable strength and a range of hydrophobic interactions. *Advanced Functional Materials, 20*, 2240–2247.

Huber, D. L., Manginell, R. P., Samara, M. A., Kim, B.-I., & Bunker, B. C. (2003). Programmed adsorption and release of proteins in a microfluidic device. *Science, 301*, 352–354.

Hui, O., Zhenhai, X., & Jiang, Z. (2009). Static and dynamic responses of polyelectrolyte brushes under external electric field. *Nanotechnology, 20*, 195703.

Idota, N., Ebara, M., Kotsuchibashi, Y., Narain, R., & Aoyagi, T. (2012). Novel temperature-responsive polymer brushes with carbohydrate residues facilitate selective adhesion and collection of hepatocytes. *Science and Technology of Advanced Materials*, *13*, 064206.

Israels, R., Leermakers, F. A. M., & Fleer, G. J. (1994). On the theory of grafted weak polyacids. *Macromolecules*, *27*, 3087–3093.

Karim, A., Tsukruk, V. V., Douglas, J. F., Satija, S. K., Fetters, L. J., Reneker, D. H., et al. (1995). Self-organization of polymer brush layers in a poor solvent. *Journal de Physique II France*, *5*, 1441–1456.

Kelley, E. G., Albert, J. N. L., Sullivan, M. O., & Epps, I. I. I. T. H. (2013). Stimuli-responsive copolymer solution and surface assemblies for biomedical applications. *Chemical Society Reviews*, *42*, 7057–7071.

Kim, B. Y., Ratcliff, E. L., Armstrong, N. R., Kowalewski, T., & Pyun, J. (2009). Ferrocene functional polymer brushes on indium tin oxide via surface-initiated atom transfer radical polymerization. *Langmuir*, *26*, 2083–2092.

Klein, J., Kumacheva, E., Mahalu, D., Perahia, D., & Fetters, L. J. (1994). Reduction of frictional forces between solid surfaces bearing polymer brushes. *Nature*, *370*, 634–636.

Klein Gunnewiek, M., Di luca, A., Sui, X., Van blitterswijk, C. A., Moroni, L., & Vancso, G. J. (2012). Controlled surface initiated polymerization of N-isopropylacrylamide from polycaprolactone substrates for regulating cell attachment and detachment. *Israel Journal of Chemistry*, *52*, 339–346.

Kooij, E. S., Sui, X., Hempenius, M. A., Zandvliet, H. J. W., & Vancso, G. J. (2012). Probing the Thermal collapse of poly(N-isopropylacrylamide) grafts by quantitative in situ ellipsometry. *The Journal of Physical Chemistry B*, *116*, 9261–9268.

Kumar vyas, M., Schneider, K., Nandan, B., & Stamm, M. (2008). Switching of friction by binary polymer brushes. *Soft Matter*, *4*, 1024–1032.

Kutnyanszky, E., Hempenius, M. A., & Vancso, G. J. (2014). Polymer bottlebrushes with a redox responsive backbone feel the heat: synthesis and characterization of dual responsive poly(ferrocenylsilane)s with PNIPAM side chains. *Polymer Chemistry*.

Laloyaux, X., Fautré, E., Blin, T., Purohit, V., Leprince, J., Jouenne, T., et al. (2010). Temperature-responsive polymer brushes switching from bactericidal to cell-repellent. *Advanced Materials*, *22*, 5024–5028.

Lee, H.-I., Boyce, J. R., Nese, A., Sheiko, S. S., & Matyjaszewski, K. (2008). pH-induced conformational changes of loosely grafted molecular brushes containing poly(acrylic acid) side chains. *Polymer*, *49*, 5490–5496.

Lei, H., Wang, M., Tang, Z., Luan, Y., Liu, W., Song, B., et al. (2013). Control of lysozyme adsorption by pH on surfaces modified with polyampholyte brushes. *Langmuir*.

Maas, J. H., Cohen Stuart, M. A., Sieval, A. B., Zuilhof, H., & Sudhölter, E. J. R. (2003). Preparation of polystyrene brushes by reaction of terminal vinyl groups on silicon and silica surfaces. *Thin Solid Films*, *426*, 135–139.

Marra, J., & Hair, M. L. (1988). Interactions between two adsorbed layers of poly(ethylene oxide)/polystyrene diblock copolymers in heptane—toluene mixtures. *Colloids and Surfaces*, *34*, 215–226.

Mazurowski, M., Gallei, M., Li, J., Didzoleit, H., Stühn, B., & Rehahn, M. (2012). Redox-responsive polymer brushes grafted from polystyrene nanoparticles by means of surface initiated atom Transfer radical polymerization. *Macromolecules*, *45*, 8970–8981.

Milner, S. T. (1991). Polymer brushes. *Science*, *251*, 905–914.

Montagne, F., Polesel-Maris, J. R., Pugin, R., & Heinzelmann, H. (2008). Poly (N-isopropylacrylamide) thin films densely grafted onto gold surface: preparation, characterization, and dynamic AFM study of temperature-induced chain conformational changes. *Langmuir*, *25*, 983–991.

Motornov, M., Minko, S., Eichhorn, K.-J., Nitschke, M., Simon, F., & Stamm, M. (2003). Reversible tuning of wetting behavior of polymer surface with responsive polymer brushes. *Langmuir*, *19*, 8077–8085.

Motornov, M., Sheparovych, R., Lupitskyy, R., Macwilliams, E., Hoy, O., Luzinov, I., et al. (2007). Stimuli-responsive colloidal systems from mixed brush-coated nanoparticles. *Advanced Functional Materials*, *17*, 2307–2314.

Motornov, M., Tam, T. K., Pita, M., Tokarev, I., Katz, E., & Minko, S. (2009). Switchable selectivity for gating ion transport with mixed polyelectrolyte brushes: approaching 'smart' drug delivery systems. *Nanotechnology*, *20*, 434006.

Napper, D. H. (1983). *Polymeric stabilisation of colloidal particles*. London: Academic Press.

Nguyen, A. T., Baggerman, J., Paulusse, J. M. J., Van Rijn, C. J. M., & Zuilhof, H. (2011). Stable protein-repellent zwitterionic polymer brushes grafted from silicon nitride. *Langmuir*, *27*, 2587–2594.

Nguyen, A. T., Baggerman, J., Paulusse, J. M. J., Zuilhof, H., & Van Rijn, C. J. M. (2011). Bioconjugation of protein-repellent zwitterionic polymer brushes grafted from silicon nitride. *Langmuir*, *28*, 604–610.

Norde, W., & Giacomelli, C. E. (1999). Conformational changes in proteins at interfaces: from solution to the interface, and back. *Macromolecular Symposia*, *145*, 125–136.

Parnell, A. J., Martin, S. J., Dang, C. C., Geoghegan, M., Jones, R. A. L., Crook, C. J., et al. (2009). Synthesis, characterization and swelling behaviour of poly(methacrylic acid) brushes synthesized using atom transfer radical polymerization. *Polymer*, *50*, 1005–1014.

Pattanayek, S. K., & Juvekar, V. A. (2002). Prediction of adsorption of nonionic polymers from aqueous solutions to solid surfaces. *Macromolecules*, *35*, 9574–9585.

Plunkett, K. N., Zhu, X., Moore, J. S., & Leckband, D. E. (2006). PNIPAM chain collapse depends on the molecular weight and grafting density. *Langmuir*, *22*, 4259–4266.

Privman, M., Tam, T. K., Pita, M., & Katz, E. (2008). Switchable electrode controlled by enzyme logic network system: approaching physiologically regulated bioelectronics. *Journal of the American Chemical Society*, *131*, 1314–1321.

Quintana, R., Gosa, M., Jańczewski, D., Kutnyanszky, E., & Vancso, G. J. (2013). Enhanced stability of low fouling zwitterionic polymer brushes in seawater with diblock architecture. *Langmuir*, *29*, 10859–10867.

Roosjen, A., Kaper, H. J., Van Der Mei, H. C., Norde, W., & Busscher, H. J. (2003). Inhibition of adhesion of yeasts and bacteria by poly(ethylene oxide)-brushes on glass in a parallel plate flow chamber. *Microbiology*, *149*, 3239–3246.

Roosjen, A., van der Mei, H. C., Busscher, H. J., & Norde, W. (2004). Microbial adhesion to poly(ethylene oxide) brushes: influence of polymer chain length and temperature. *Langmuir*, *20*, 10949–10955.

Roosjen, A., de Vries, J., Van Der Mei, H. C., Norde, W., & Busscher, H. J. (2005). Stability and effectiveness against bacterial adhesion of poly(ethylene oxide) coatings in biological fluids. *Journal of Biomedical Materials Research Part B: Applied Biomaterials*, *73B*, 347–354.

Rühe, J., Ballauff, M., Biesalski, M., Dziezok, P., Gröhn, F., Johannsmann, D., et al. (2004). Polyelectrolyte brushes. In M. Schmidt (Ed.), *Polyelectrolytes with defined molecular architecture I*. Berlin Heidelberg: Springer.

Salierno, M. J., García, A. J., & Del Campo, A. (2013). Photo-activatable surfaces for cell migration assays. *Advanced Functional Materials* (n/a-n/a).

Santonicola, M. G., De Groot, G. W., Memesa, M., Meszyńska, A., & Vancso, G. J. (2010). Reversible pH-controlled switching of poly(methacrylic acid) grafts for functional Biointerfaces. *Langmuir*, *26*, 17513–17519.

Schulz, D. N., Peiffer, D. G., Agarwal, P. K., Larabee, J., Kaladas, J. J., Soni, L., et al. (1986). Phase behaviour and solution properties of sulphobetaine polymers. *Polymer, 27*, 1734–1742.

Schüwer, N., & Klok, H.-A. (2011). Tuning the pH sensitivity of poly(methacrylic acid) brushes. *Langmuir, 27*, 4789–4796.

Stuart, M. A. C., Huck, W. T. S., Genzer, J., Muller, M., Ober, C., Stamm, M., et al. (2010). Emerging applications of stimuli-responsive polymer materials. *Nature Materials, 9*, 101–113.

Sui, X., Chen, Q., Hempenius, M. A., & Vancso, G. J. (2011). Probing the collapse dynamics of poly(N-isopropylacrylamide) brushes by AFM: effects of co-nonsolvency and grafting densities. *Small, 7*, 1440–1447.

Sui, X., Feng, X., Di Luca, A., Van Blitterswijk, C. A., Moroni, L., Hempenius, M. A., et al. (2013). Poly(N-isopropylacrylamide)-poly(ferrocenylsilane) dual-responsive hydrogels: synthesis, characterization and antimicrobial applications. *Polymer Chemistry, 4*, 337–342.

Sui, X., Feng, X., Song, J., Hempenius, M. A., & Vancso, G. J. (2012). Electrochemical sensing by surface-immobilized poly(ferrocenylsilane) grafts. *Journal of Materials Chemistry, 22*, 11261–11267.

Sun, T., Wang, G., Feng, L., Liu, B., Ma, Y., Jiang, L., et al. (2004). Reversible switching between Superhydrophilicity and Superhydrophobicity. *Angewandte Chemie International Edition, 43*, 357–360.

Szleifer, I., & Carignano, M. A. (2007). Tethered polymer layers. In I. Prigogine & S. A. Rice (Eds.), *Advances in Chemical Physics: Polymeric Systems*. John Wiley & Sons, Inc.

Tam, T. K., Ornatska, M., Pita, M., Minko, S., & Katz, E. (2008). Polymer brush-modified electrode with switchable and tunable redox activity for bioelectronic applications. *The Journal of Physical Chemistry C, 112*, 8438–8445.

Tam, T. K., Strack, G., Pita, M., & Katz, E. (2009). Biofuel cell logically controlled by antigen–antibody recognition: towards immune-regulated bioelectronic devices. *Journal of the American Chemical Society, 131*, 11670–11671.

Tsujii, Y., Ohno, K., Yamamoto, S., Goto, A., & Fukuda, T. (2006). Structure and properties of high-density polymer brushes prepared by surface-initiated living radical polymerization. In R. Jordan (Ed.), *Surface-initiated polymerization I*. Berlin Heidelberg: Springer.

Uhlmann, P., Houbenov, N., Brenner, N., Grundke, K., Burkert, S., & Stamm, M. (2006). In-situ investigation of the adsorption of globular model proteins on stimuli-responsive binary polyelectrolyte brushes. *Langmuir, 23*, 57–64.

Van Camp, W., Du Prez, F. E., Alem, H., Demoustier-Champagne, S., Willet, N., Grancharov, G., et al. (2010). Poly(acrylic acid) with disulfide bond for the elaboration of pH-responsive brush surfaces. *European Polymer Journal, 46*, 195–201.

de Vos, W. M., Biesheuvel, P. M., De Keizer, A., Kleijn, J. M., & Cohen Stuart, M. A. (2008). Adsorption of the protein bovine serum albumin in a planar poly(acrylic acid) brush layer as measured by optical reflectometry. *Langmuir, 24*, 6575–6584.

Wang, J., & Somasundaran, P. (2006). Reversible conformational behavior of poly(acrylic acid) LB film with changes in pH, ionic strength and time. *Colloids and Surfaces A: Physicochemical and Engineering Aspects, 273*, 63–69.

Wang, D., Ye, G., Wang, X., & Wang, X. (2011). Graphene functionalized with Azo polymer brushes: surface-initiated polymerization and photoresponsive properties. *Advanced Materials, 23*, 1122–1125.

Watanabe, H., & Tirrell, M. (1993). Measurement of forces in symmetric and asymmetric interactions between diblock copolymer layers adsorbed on mica. *Macromolecules, 26*, 6455–6466.

Weir, M. P., Heriot, S. Y., Martin, S. J., Parnell, A. J., Holt, S. A., Webster, J. R. P., et al. (2011). Voltage-Induced swelling and deswelling of weak polybase brushes. *Langmuir*, *27*, 11000–11007.

Wittemann, A., & Ballauff, M. (2006). Interaction of proteins with linear polyelectrolytes and spherical polyelectrolyte brushes in aqueous solution. *Physical Chemistry Chemical Physics*, *8*, 5269–5275.

Wittemann, A., Haupt, B., & Ballauff, M. (2003). Adsorption of proteins on spherical polyelectrolyte brushes in aqueous solution. *Physical Chemistry Chemical Physics*, *5*, 1671–1677.

Wu, T., Gong, P., Szleifer, I., Vlček, P., Šubr, V., & Genzer, J. (2007). Behavior of surface-anchored poly(acrylic acid) brushes with grafting density gradients on solid Substrates: 1. Experiment. *Macromolecules*, *40*, 8756–8764.

Xia, J., Mattison, K., Romano, V., Dubin, P. L., & Muhoberac, B. B. (1997). Complexation of trypsin and alcohol dehydrogenase with poly(diallyldimethylammonium chloride). *Biopolymers*, *41*, 359–365.

Xu, L. Q., Wan, D., Gong, H. F., Neoh, K.-G., Kang, E.-T., & Fu, G. D. (2010). One-pot preparation of ferrocene-functionalized polymer brushes on gold substrates by combined surface-initiated atom Transfer radical polymerization and "Click chemistry". *Langmuir*, *26*, 15376–15382.

Xue, C., Choi, B.-C., Choi, S., Braun, P. V., & Leckband, D. E. (2012). Protein adsorption modes determine reversible cell attachment on poly(N-isopropyl acrylamide) brushes. *Advanced Functional Materials*, *22*, 2394–2401.

Yang, L., Pan, F., Zhao, X., Yaseen, M., Padia, F., Coffey, P., et al. (2010). Thermoresponsive copolymer nanofilms for controlling cell adhesion, growth, and detachment. *Langmuir*, *26*, 17304–17314.

Yu, K., Wang, H., & Han, Y. (2007). Motion of integrated CdS nanoparticles by phase separation of block copolymer brushes. *Langmuir*, *23*, 8957–8964.

Yu, Q., Zhang, Y., Chen, H., Zhou, F., Wu, Z., Huang, H., et al. (2010). Protein adsorption and cell Adhesion/Detachment behavior on dual-responsive silicon surfaces modified with poly(N-isopropylacrylamide)-block-polystyrene copolymer. *Langmuir*, *26*, 8582–8588.

Zdyrko, B., Klep, V., Li, X., Kang, Q., Minko, S., Wen, X., et al. (2009). Polymer brushes as active nanolayers for tunable bacteria adhesion. *Materials Science and Engineering: C*, *29*, 680–684.

Zhao, B., & Brittain, W. J. (2000). Polymer brushes: surface-immobilized Macromolecules. *Progress in Polymer Science*, *25*, 677–710.

Zhao, B., Haasch, R. T., & Maclaren, S. (2004). Solvent-induced self-assembly of mixed poly(methyl methacrylate)/polystyrene brushes on planar silica Substrates: molecular weight effect. *Journal of the American Chemical Society*, *126*, 6124–6134.

Zhu, X., Yan, C., Winnik, F. M., & Leckband, D. (2006). End-grafted low-molecular-weight PNIPAM does not collapse above the LCST. *Langmuir*, *23*, 162–169.

Preparation and analysis of switchable copolymers for biomedical application

6

M. Nitschke, S. Gramm
Max Bergmann Center of Biomaterials, Dresden, Germany

6.1 Introduction

Switchable polymer coatings can be prepared on a wide variety of biomaterials. To implement a particular switching effect, a stimuli-responsive layer is attached to the surface. Its molecular structure introduces the desired physicochemical mechanism. In case of copolymers (also polymer blends, mixed brushes, etc.), the stimuli-responsive unit can be combined with a second unit that allows to set other material properties in a particular way.

The chapter illustrates this concept and introduces a choice of preparation techniques and anchoring mechanisms to fabricate such systems. Selected examples, focused on thermo-responsive materials, demonstrate how switchable copolymer coatings can be designed to meet the specific requirements of demanding applications. However, to ensure or to optimise the proper function of a switchable coating in contact with an aqueous environment or a biofluid requires a comprehensive knowledge on the system behaviour. Depending on the intended use, this includes the change of values like surface energy, charge distribution, mechanical properties or optical properties upon switching. Accordingly, the second part of the chapter introduces a range of advanced analytical techniques as applied to switchable coatings on solid surfaces.

6.2 Switchable copolymer coatings

6.2.1 Anchorage of the coating on the surface

When the material of a switchable coating is dissolvable in aqueous environments after phase transition, an effective anchorage mechanism toward the surface is essential. For that purpose, different routes that use the second, non-switching, comonomer unit were proposed.

First, a reactive comonomer unit can be introduced for a direct covalent immobilisation. When the material surface to be coated contains hydroxyl groups (e.g., glass, silicon, titanium alloys), the silyl–hydroxyl reaction can be used. Following this route, Yang et al. (2010) combine *N*-isopropyl acrylamide (NiPAAm) with a 3(trimethoxysilyl)propyl methacrylate unit to provide a stable thermo-responsive cell culture surface on glass carriers. On an amine-modified surface, a randomly distributed copolymer of

Switchable and Responsive Surfaces and Materials for Biomedical Applications. http://dx.doi.org/10.1016/B978-0-85709-713-2.00006-7

NiPAAm and 10% acrylic acid was immobilised as reported by Lee and von Recum (2010). However, a prior activation of the acid groups by carbodiimide is required. In case of an epoxy functionalised substrate, carboxyl-containing comonomer units can be used for immobilisation directly. This was demonstrated by Kaufmann et al. (2010) for NiPAAm-based carboxyalkylacrylamide copolymers that bind to the substrate upon heat treatment at 170 °C.

Instead of covalent bonds, electrostatic interactions can be employed to immobilise a copolymer coating. For that purpose, the thermo-responsive unit is combined with a charged unit. This concept was demonstrated by Martins, Mano, and Alves (2011) for chitosan-graft-NiPAAm. For the formation of multilayer structures according to the layer-by-layer technique, negatively charged alginate was prepared in between positively charged chitosan-graft-NiPAAm layers.

A purely physical approach based on an entrapment procedure was developed by Guo and Ulbricht (2011) to fabricate thermo-responsive polypropylene membranes. The polypropylene surface was exposed to solutions of an amphiphilic block copolymer of polyNiPAAm (PNiPAAm) with a hydrophobic poly(butylacrylate) block. After treatment under optimised conditions (i.e., solvent, concentration, temperature, exposure time), a hydrophilisation as well as a switchable wettability were achieved.

A remarkably simple and straightforward procedure toward switchable coatings for cell culture purposes was proposed by Nakayama et al. (2012). A diblock copolymer combining a thermo-responsive PNiPAAm segment with a hydrophobic poly(butyl methacrylate) segment was sythesised and subsequently prepared by spin coating on commercial tissue culture polystyrene. The physical adsorption provided by the unsoluble poly(butyl methacrylate) block was sufficient to stabilise the coating under cell culture conditions and to ensure growth and stimulated detachment of cell sheets.

6.2.2 Cross-linking the coating

Most of the approaches mentioned in the previous section lead to the immobilisation of a copolymer monolayer. However, the thickness of the coating can be increased by adding a comonomer unit that allows cross-linking. According to this concept, different chemical mechanisms were reported.

Nash et al. (2012) used a copolymer of NiPAAm and acrylamidobenzophenone that allows to cross-link the material based on a radical mechanism. Upon ultraviolet radiation, the benzophenone unit attaches to C–H bonds in the thermo-responsive copolymer itself, but also on an appropriate carrier material like polystyrene. Surface immobilised coatings with different thicknesses up to 200 nm were fabricated by cross-linking spin-coated films obtained from different copolymer concentrations in the solution.

A copolymer of NiPAAm with 1 mol% of a vinyl monomer carrying a protected isocyanate group was used by Hirata, Okazaki, and Iwata (2004) to fabricate cross-linked coatings. Upon heat treatment at 150 °C for 1.5 h, the isocyanate group is deprotected and cross-links to the amide group of the NiPAAm unit. A dry layer thickness of 35 nm was obtained, but higher values are possible. Given that the substrate carries appropriate functional groups like OH or NH_2, the technique can cross-link and immobilise the coating material simultaneously.

Kuckling, Harmon, and Frank (2002) prepared a copolymer of NiPAAm and 2-(dimethylmaleimido)-*N*-ethylacrylamide that undergoes a [2 + 2] cyclodimerization under ultraviolet irradiation. This leads to cross-linked layers of several hundred nanometres in the dry state while the cross-linking density can be adjusted easily by the comonomer content.

6.2.3 Control over phase transition behaviour

The phase transition behaviour of NiPAAm-based thermo-responsive copolymers can be tuned by the comonomer unit. The introduction of hydrophobic or hydrophilic units shifts the transition temperature toward lower or higher values, respectively. The amount of this shift as well as the characteristics of the transition (e.g., appearing over a wide range or well defined at a particular temperature) depend on the molar fraction of the comonomer and the molecular architecture. A range of examples covering different cases is given below.

For applications aiming at the control of marine biofouling, a significantly reduced phase transition temperature compared to the PNiPAAm homopolymer is required. For that purpose, Cordeiro et al. (2009) sythesised copolymers of NiPAAm with *N*-(1-phenyl ethyl)acrylamide; 10–15 mol% of this hydrophobic unit leads to a shift of more than 10 K. Another example for lowering the phase transition temperature is found in the work of Tsuda et al. (2004). NiPAAm-based copolymers with different amounts of the hydrophobic unit *n*-butyl methacrylate were studied to fabricate tunable thermo-responsive cell culture carriers. As shown in Figure 6.1, a pronounced shift of the phase

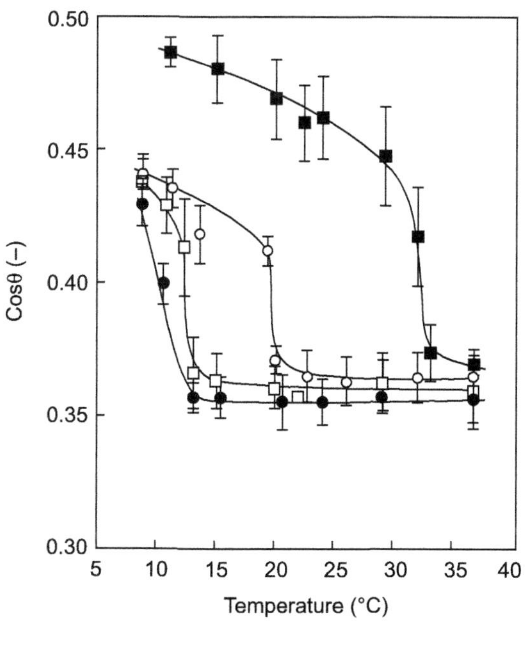

Figure 6.1 Temperature-dependent water contact angle for surfaces grafted with PNiPAAm (filled square) and with copolymers with an increasing molar content of *n*-butyl methacrylate (open circle, open square, filled circle).

Reprinted from Tsuda et al. (2004) by permission of John Wiley & Sons, Inc.

transition up to 20 K toward lower temperatures was obtained. As expected, reduced amplitude of the wettability switching occurs upon comonomer addition. Rochev et al. (2004) investigated copolymers of NiPAAm with up to 50 mol% of the more hydro phobic *N*-tert-butylacrylamide leading to phase transition temperatures as low as 10 °C.

The introduction of hydrophilic units into NiPAAm-based copolymers was investigated by Nitschke et al. (2007). It was demonstrated by *in situ* ellipsometry for diethylene glycol methyl ether methacrylate that the addition of a permanently hydrated unit to the copolymer structure promotes the onset of the swelling at higher temperature. A molar fraction as low as 0.8 mol% shifts the phase transition temperature of a surface immobilised thermo-responsive thin film by several K toward higher values (Figure 6.2).

Kaufmann et al. (2010) combine both options (i.e., to increase or to decrease the phase transition temperature) in a single molecular architecture and propose a copolymer of NiPAAm and carboxyalkylacrylamides. A different number of alkyl units in the side chain (in this study 2, 5 and 10) gradually changes the balance between hydrophilic and hydrophobic interactions of the comonomer unit. This allows to adjust the phase transition temperature of the copolymer above or below the value of the NiPAAm homopolymer.

6.2.4 Tuning cell adhesion and cell detachment

Stimulated detachment of adherent cells from thermo-responsive cell culture carriers is a distinguished application of switchable copolymer coatings. While NiPAAm is the most common unit that provides a temperature-dependent swelling and collapsing with a phase transition close to cell cultivation temperature of 37 °C, appropriate comonomers can be used to modify the system behaviour compared to a pure PNiPAAm coating. Depending on the particular cell type, it may be necessary to improve the initial cell attachment in order to ensure efficient formation of a confluent layer. On the other hand, the stimulus provided by a pure PNiPAAm coating may be insufficient to promptly detach a grown sheet of strongly adherent cells. Strategies to improve initial cell attachment or to accelerate cell detachment include

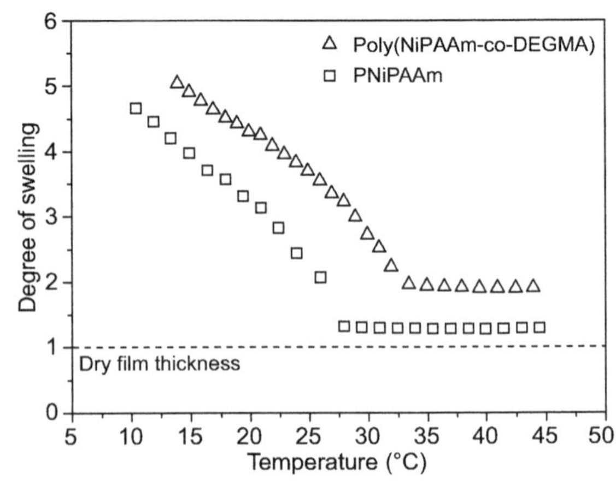

Figure 6.2 Temperature-dependent swelling behaviour in water of PNiPAAm and poly(NiPAAm-co-DEGMA) thin films.
Reprinted from Nitschke et al. (2007) by permission of John Wiley & Sons, Inc.

the introduction of hydrophobic or hydrophilic units as well as the adjustment of the mechanical properties and the switching amplitude of the thermo-responsive coating.

To adjust initial cell adhesion and proliferation above the phase transition temperature, Rochev et al. (2004) investigated copolymers of NiPAAm with up to 50 mol% of the hydrophobic *N*-tert-butylacrylamide. For two different cell lines, it was shown that cell adhesion and growth are significantly improved with an increasing comonomer content.

To accelerate cell detachment below the phase transition temperature, Kwon, Kikuchi, Yamato, and Okano (2003) co-grafted NiPAAm with small amounts of the hydrophilic macromonomer polyethylene glycol methacrylate (MW 4000) on microporous membranes. Compared to pure NiPAAm, the PEG-containing coating allowed a more rapid detachment of single cells and cell sheets upon temperature decrease. This was attributed to an improved diffusion of water molecules in the grafted layer and a more rapid hydration of PNiPAAm chains. However, the introduction of PEG chains promotes the hydration of the thermo-responsive coating also above the phase transition temperature, which leads to impaired cell adhesion. To overcome this problem, Ebara et al. (2003) proposed a copolymer of NiPAAm with 2-carboxyisopropylacrylamide (Figure 6.3). This material possesses similar hydrophobicity and similar cell attachment properties compared to the homopolymer above the phase transition temperature but improved hydrophilicity and cell detachment below.

Chang et al. (2010) suggested copolymeric hydrogels of non-ionic NiPAAm with zwitterionic sulfobetaine methacrylate. The copolymer composition allows to tune the protein adsorption in a wide range. In particular, for copolymers with rather small

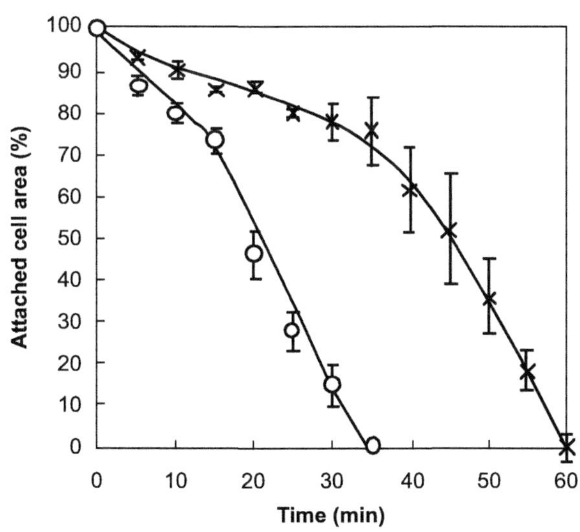

Figure 6.3 Cell sheet recovery from thermo-responsive carriers as a function of incubation time at 20 °C: TCPS grafted with pure NiPAAm (x) compared to TCPS co-grafted with NiPAAm and 1 mol% 2-carboxyisopropylacrylamide (o).

fractions of sulfobetaine methacrylate, a pronounced temperature-responsive swelling behaviour along with an accelerated hydration and a more rapid cell detachment compared to the NiPAAm homopolymer were found.

6.2.5 Biomolecular functionalisation

Beyond physicochemical approaches, switchable cell culture carriers can be provided with a biomolecular functionalisation to more effectively tune attachment, growth and proliferation as well as detachment of adherent cells. To allow covalent protein or peptide immobilisation, reactive units are introduced into the thermo-responsive material.

Following to this strategy, Hatakeyama, Kikuchi, Yamato, and Okano (2006) prepared carboxyl-functionalised thermo-responsive coatings on tissue culture polystyrene by electron beam–induced copolymerisation of *N*-isopropylacrylamide and 2-carboxyisopropylacrylamide. The subsequent co-immobilization of RGDS, a cell adhesive peptide, and insulin was found to promote initial cell adhesion and growth while the cultured tissue monolayers still detach spontaneously upon temperature decrease.

Another route towards biomolecular functionalisation of switchable cell culture carriers was proposed by Gramm et al. (2011). The thermo-responsive polymer poly (vinyl methyl ether) was blended with a small amount of the alternating copolymer of vinyl methyl ether and maleic anhydride. After electron beam cross-linking, stable films with a dry thickness of up to 70 nm were obtained. The introduced anhydride moieties allow for subsequent protein or peptide immobilization without toxic coupling agents. This concept was applied and further developed by Teichmann et al. (2013). The impact of different adhesion-promoting molecules immobilised in different concentrations to the thermo-responsive coating was evaluated and compared.

6.2.6 Dual responsive coatings

Dual responsive copolymers that complement the thermo-responsive behaviour with a second stimuli-responsive switching mechanism represents a promising class of materials. However, keeping both effects independent of each other is a crucial point. For example, when combining NiPAAm with acrylic acid as a pH responsive comonomer, the thermo-sensitivity is significantly degraded with increasing comonomer content.

To decouple thermo-sensitivity and pH sensitivity in a dual responsive system, Xu, Zhang, Cheng, Zhuo, and Kennedy (2007) prepared a copolymer of NiPAAm and *N*-isopropyl-maleamic acid. As a result, a pronounced pH sensitivity was added without weakening the thermo-responsive switching (Figure 6.4). The effect was attributed to the continuous isopropylamide sequence in the copolymer chain.

Another approach to a dual-responsive material that combines thermo-sensitivity with pH sensitivity was proposed by Kaufmann et al. (2010). Copolymers of NiPAAm and carboxyalkylacrylamides with molar fractions up to 10 mol% were synthesised. As shown in a quartz crystal microbalance (QCM) study, the introduction of pH-sensitive

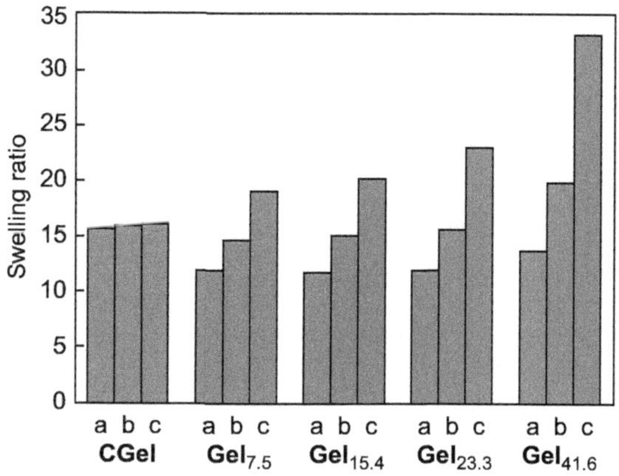

Figure 6.4 pH sensitivity of NiPAAm (CGel) and copolymers with an increasing content of *N*-isopropyl-maleamic acid at room temperature (a: pH 1; b: pH 6; c: pH 13).
Reprinted with permission from Xu et al. (2007). Copyright 2007 Elsevier.

carboxylic groups spacially separated from the NiPAAm chain does not impair the thermo-sensitivity of the material.

6.3 Advanced analytical techniques

6.3.1 Contact angle measurements by axisymmetric drop shape analysis

The most common approach to characterise the wetting behaviour of a flat surface is the sessile drop (or captive bubble) technique. The contact angle between the liquid drop (or the gas bubble) and the solid surface is determined using a tangent aligned with the profile of the drop (or bubble) at the contact point with the solid surface. Measurements can be performed dynamically by increasing and decreasing the droplet (or bubble) volume to determine the advancing and receding contact angle, respectively. These values are usually different (contact angle hysteresis) for various reasons, e.g., in case of inhomogeneous or rough surfaces. The accuracy of the method as described above is about ±2°.

To improve the significance of the technique, drop profile analysis can be applied instead of measuring the contact angle directly. Axisymmetric drop shape analysis (ADSA, Figure 6.5) extracts experimental drop profiles from video images while slowly increasing or decreasing the droplet volume (Grundke, 2008; Kwok et al., 1998). The best fit with theoretical assumptions based on the Laplace equation of capillarity allows to calculate the surface/interfacial tension and subsequently the contact angle for each particular image. In addition, droplet radius, droplet volume and the contact area are computed. In this way, ADSA reveals phenomena like a slip/stick behaviour or any time-dependent changes of the contact angle. Furthermore, irregular or inconsistent values can be identified easily. Finally, an accuracy of ±0.2° or better can be achieved.

The ADSA technique in the captive bubble mode was used by Cordeiro et al. (2009) to study the thermo-responsive switching of poly(*N*-isopropylacrylamide-co-*N*-(1-phenylethyl) acrylamide) and to investigate the possible impact of electrolytes on wetting behaviour and transition temperature.

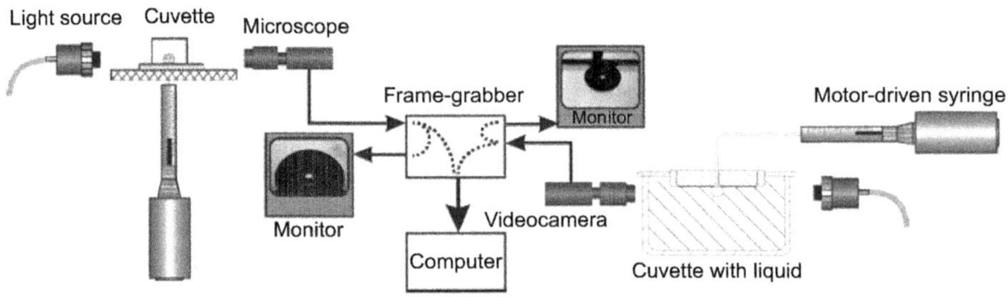

Figure 6.5 Scheme of the experimental set-up for sessile drop experiments (left) and for captive air bubble measurements (right) using ADSA.
Reprinted with permission from Uhlmann, Skorupa, Werner, and Grundke (2005). Copyright 2005 American Chemical Society.

6.3.2 Monitoring of switching layer systems by spectroscopic ellipsometry

Ellipsometry measures the change of the polarization state when light is reflected from a surface. In the simple case of a transparent film on a reflecting substrate, the film thickness and the index of refraction can be determined with high accuracy. With a sufficient set of independent experimental data obtained for different angles of incidence and/or different wavelengths (spectroscopic ellipsometry), more complicated samples can be characterised. This includes, e.g., multilayer structures or layers with a graded index of refraction. Ellipsometry measurements are possible in air but also in environments like water or aqueous solutions using a liquid cell. Altogether, this makes ellipsometry a powerful tool to investigate the physicochemical properties of surface-immobilised swellable hydrogel films (Hinrichs & Eichhorn, 2014). In particular, spectroscopic ellipsometry can characterise the temperature-dependent switching behaviour of a thermo-responsive coating on a reflecting substrate. From the ellipsometric raw data, physical values like the thickness or the optical properties of the swollen polymer film can be calculated using fit procedures and certain model assumptions (Kooij, Sui, Hempenius, Zandvliet, & Vancso, 2012).

A plot of the film thickness versus temperature as illustrated in Figure 6.6 reveals many details. The degree of swelling in the expanded and the collapsed state can be evaluated. After heating the system above the transition temperature once, the switching behaviour becomes fully reversible. The hysteresis between corresponding heating and cooling curves is usually attributed to the kinetics of the interaction mechanisms and may be less or more pronounced depending on the applied heating/cooling rate. Furthermore, the phase transition temperature can be estimated from the graph. However, the transition typically occurs over a wider range for the immobilised polymer compared to the same polymer in solution as shown by Schmaljohann, Beyerlein, Nitschke, and Werner (2004). It becomes obvious whether or not the solvent is fully expelled from the layer above the transition temperature. While this is the case for pure PNiPAAm, copolymers with hydrophilic side chains can keep a certain amount of water independent of

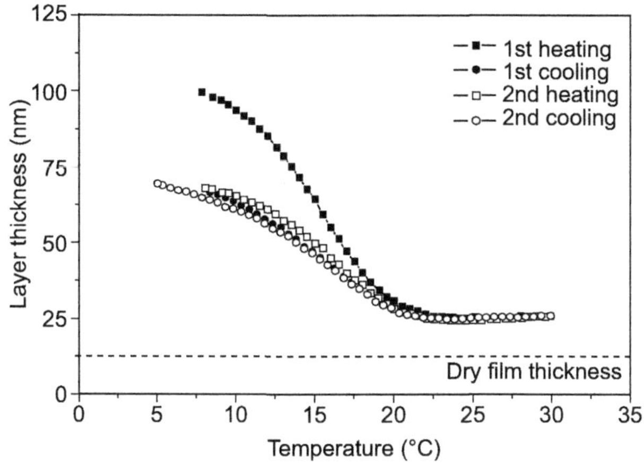

Figure 6.6 Poly(*N*-isopropylacrylamide-co-*N*-(1-phenylethyl) acrylamide) film thickness versus temperature during first and second heating/cooling cycles in distilled water (heating/cooling rate 1 K/min).

Reproduced by permission of the Royal Society of Chemistry from Cordeiro et al. (2009).

the actual temperature as shown for diethylene glycol methacrylate as a comonomer unit (Figure 6.2).

Spectroscopic ellipsometry allows to detect small shifts of the phase transition temperature introduced either by a variation of the thermo-responsive polymer or the liquid environment. The latter dependency was demonstrated by Schmaljohann et al. (2003) for buffer solutions and cell culture medium and by Cordeiro et al. (2009) for artificial sea water. In both cases, the electrolyte shifts the transition temperature toward lower values.

For some thermo-responsive materials, the switching characteristics can be largely influenced by the preparation procedure. As shown by Gramm et al. (2011) for electron beam–immobilised poly(vinyl methyl ether) thin films (Figure 6.7), the degree of swelling (i.e., the switching amplitude) can be adjusted in a wide range by the absorbed energy dose while keeping the phase transition temperature the same.

Moreover, ellipsometric studies allow to unravel peculiarities in the switching kinetics of thermo-responsive copolymers. For example, the variety of interaction mechanisms in a thermo-responsive polymer may lead to the coexistence of different contributions to the swelling process acting on different time scales (Schmaljohann et al., 2004).

Laterally resolved information on the micrometre scale can be obtained directly without a scanning process by imaging ellipsometry. This was illustrated by Schmaljohann et al. (2005) who demonstrated the advantages of this technique for a micropatterned NiPAAm-based copolymer coating.

6.3.3 Soft surfaces characterised by electrokinetic measurements

Electrokinetic effects streaming potential and streaming current are widely used to characterise charge formation processes at the interface between planar polymer coatings

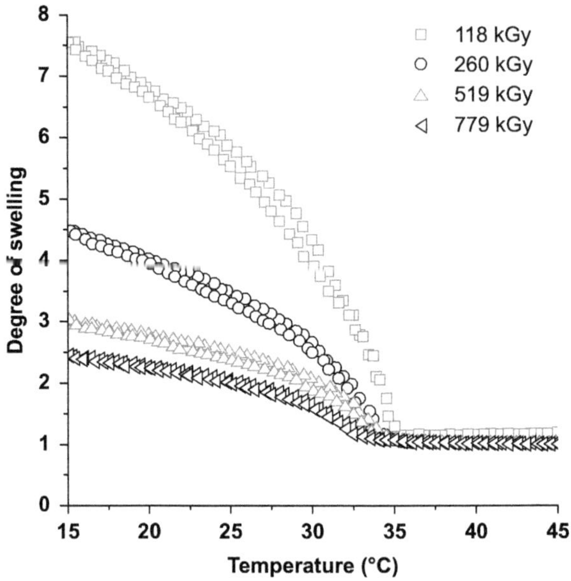

Figure 6.7 Temperature-dependent swelling behaviour of electron beam–immobilised poly(vinyl methyl ether) thin films in water during heating and cooling.
Reprinted with permission from Gramm et al. (2011).

and aqueous solutions. In case of hard (impermeable) films (volume fraction = 1) without any roughness and lateral inhomogeneity, the results can be converted into the electrokinetic or zeta potential (ζ) applying the Smoluchowski theory (Delgado, Gonzalez-Caballero, Hunter, Koopal, & Lyklema, 2005). In this theory, the zeta potential is the electrical potential of the imaginary hydrodynamic shear plane separating the inner region (in the direct vicinity of the surface) without any fluid motion and the outer region where fluid motion occurs. In case of soft (permeable) polymer films, no sharp boundary exists (volume fraction <1). Liquids as well as ions can penetrate into the coating, and the charge distribution is strongly influenced by the distribution of ionisable groups along the polymer chains.

The appropriate treatment of streaming potential/current data obtained for soft (permeable) surfaces requires more comprehensive models beyond the classical concept of zeta potential. The theoretical formalism reported by Duval, Zimmermann, Cordeiro, Rein, and Werner (2009) covers diffuse, permeable and charged hydrogel layers immobilised on charged rigid surfaces. Streaming current data on surface-immobilised poly(*N*-isopropylacrylamide)-co-*N*-(1-phenylethyl) acrylamide thin films above and below the transition temperature were successfully reproduced by the model for the whole range of pH and ionic strength investigated. Calculations revealed more details on the behaviour of the thermo-responsive copolymer coating like a homogeneous segment distribution above and below the transition temperature along with an increase of the hydrodynamic penetration length upon expanding the film. A similar approach developed by Zimmermann, Kuckling, Kaufmann, Werner, and Duval (2010), that also takes into account surface conductivity, was applied to a negatively charged thermo-responsive coating of poly(*N*-isopropylacrylamide-co-carboxyacrylamide). Here, clear evidence for a heterogeneous swelling of the copolymer film below the transition temperature, i.e., a gradient in the polymer segment density, was found.

The concept for a consistent derivation of electrohydrodynamic and structural properties of surface immobilised swollen films was further developed and applied to materials with high densities of dissociable groups like poly(acrylic acid) films (Duval, Küttner, Nitschke, Werner, & Zimmermann, 2011) and polyelectrolyte multilayers (Duval, Küttner, Werner, & Zimmermann, 2011).

In summary, electrokinetic measurements combined with an advanced theory for soft surfaces is an effective tool for the comprehensive characterisation of switchable hydrogel coatings. The accessible physicochemical parameters (distribution of polymer segment density, charge density, hydrodynamic softness, interphasial diffuseness) complement the results as obtained by techniques like ellipsometry or indenter measurements.

6.3.4 Scanning force techniques

Scanning force microscopy (SFM) is frequently used to image the morphology of solid surfaces with a lateral resolution in the nanometre range, which is possible in vacuum or air as well as in liquid media. Furthermore, an SFM instrument allows for the acquisition of force distance curves to determine the local mechanical properties of the sample. With the colloidal probe technique, well-defined indentation experiments on rather soft surfaces like swollen hydrogels in aqueous media can be performed. For quantitative evaluation of the elastic properties, raw data are fitted, e.g., according to the Hertz model, assuming an infinitely hard sphere indenting a flat surface. To account for the properties of the experimental set-up, a hard mica surface can be used as a reference. However, the indentation should be small compared to the total film thickness when the material under investigation is a hydrogel layer.

The approach described above is illustrated by the work of Matzelle et al. (2002). Force distance curves were obtained for the indentation of a colloidal probe in PNiPAAm coatings immersed in water above and below the phase transition temperature (Figure 6.8). The calculation of Young's modulus using the Hertz model revealed a variation of more than two orders of magnitude between the swollen and the collapsed state at 10 °C and 35 °C, respectively.

Harmon, Kucking, and Frank (2003) used a similar technique to investigate the impact of the cross-linking degree of PNiPAAm on the temperature-dependent variation of the elastic modulus (Figure 6.9). Crossing the phase transition temperature for a given degree of cross-linking changes the modulus by two orders of magnitude. Changing the cross-linking degree shifts the modulus about one order of magnitude above as well as below the phase transition temperature.

Beyond the elastic modulus, SFM-based colloidal probe measurements were successfully applied to study interaction forces between thermo-responsive polymers and protein-coated surfaces. This was shown by Cole, Voelcker, Thissen, Horn, and Griesser (2010) with bovine serum albumin–coated probes on PNiPAAm surfaces. Furthermore, time-dependent colloidal probe measurements on thermo-responsive polymer coatings

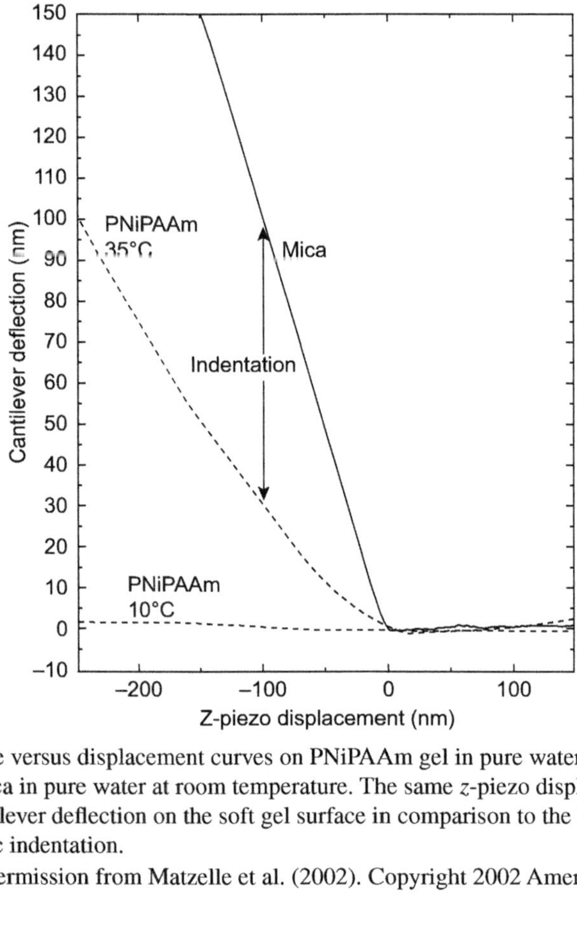

Figure 6.8 Force versus displacement curves on PNiPAAm gel in pure water at 10 °C and 35 °C and on mica in pure water at room temperature. The same *z*-piezo displacement results in a smaller cantilever deflection on the soft gel surface in comparison to the hard mica sample because of elastic indentation.

Reprinted with permission from Matzelle et al. (2002). Copyright 2002 American Chemical Society.

can characterise the kinetics of the switching process in terms of interaction forces as demonstrated by Kessel et al. (2010) for a switchable cell culture surface.

6.3.5 Quartz crystal microbalance with dissipation monitoring

A QCM uses an oscillating quartz crystal at its resonance frequency. Measuring the shift of the resonance frequency Δf caused by a thin film deposited on the quartz surface allows to calculate the mass per unit area. Adding a dissipation monitoring to this concept (QCM-D), i.e., also monitoring the decay of the quartz oscillation when switching off the ac excitation, provides additional information on the viscoelastic properties of the thin film material. In particular, the change of dissipation ΔD introduced by the thin film can be calculated.

The technique also works in liquid environments. In case of hydrogel thin films immersed in an aqueous solution, QCM-D also senses the mass of water molecules that are bound to the hydrogel due to hydration. Combined with the capability to monitor viscoelastic properties of the hydrated layer, this makes QCM-D a highly

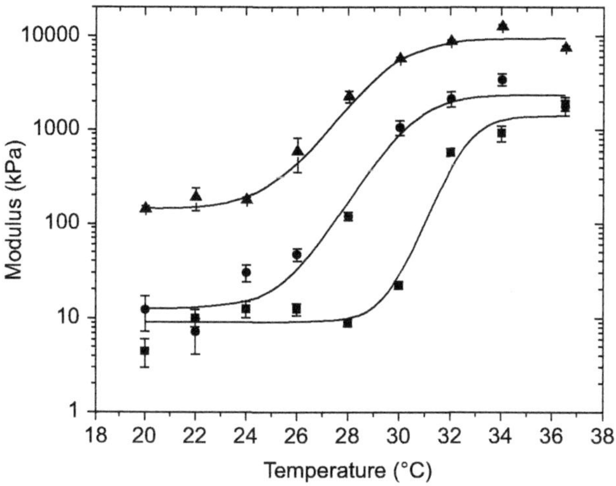

Figure 6.9 Temperature-dependent elastic modulus in de-ionised water for PNiPAAm hydrogel layers with different degrees of cross-linking. The solid lines are to guide the eye. Reprinted with permission from Harmon et al. (2003). Copyright 2003 American Chemical Society.

advantageous technique to characterise the phase transition behaviour of thermo-responsive polymer coatings. However, some technical issues have to be considered. Since the effective range of the oscillation is limited in a soft film, the thickness should not exceed a few 100 nm to get a response also from the outer part of the swollen hydrogel layer. To obtain the true signal of the hydrogel film, some effects of the experimental set-up have to be compensated for: Δf and ΔD are first measured for bare quartz crystals immersed in the liquid environment at various temperatures. These values are subtracted from the corresponding Δf and ΔD of the coated quartz crystals, which provides corrected values for the surface-bound film.

Figure 6.10 shows a typical QCM-D plot illustrating the behaviour of Δf and ΔD upon temperature variation for a PNiPAAm-based copolymer film (Alf, Godfrin, Hatton, & Gleason, 2010). For increasing temperature, the decrease of the compatibility between polymer and water molecules leads to a gradual dehydration of the hydrogel and a collapse of the film. Due to the expelled amount of water, the observed frequency increases. The collapse of the hydrogel causes a change of the viscoelastic properties. The dense and compact layer above the phase transition temperature results in less energy dissipation (ΔD) relative to the extended and flexible one below.

Based on this measurement strategy, the phase transition temperature of a responsive system can be determined as a function of the liquid environment or the material parameters. This is demonstrated by Jhon et al. (2006) for grafted PNiPAAm films and different salt concentrations in the medium. For an increasing NaCl concentration up to 1 M, a depression of the phase transition temperature of several K is observed. The study of Kaufmann et al. (2010) investigates copolymers of NiPAAm and carboxyalkylacrylamide. QCM-D reveals the shift of the phase transition temperature depending on different alkyl chain length and comonomer fractions.

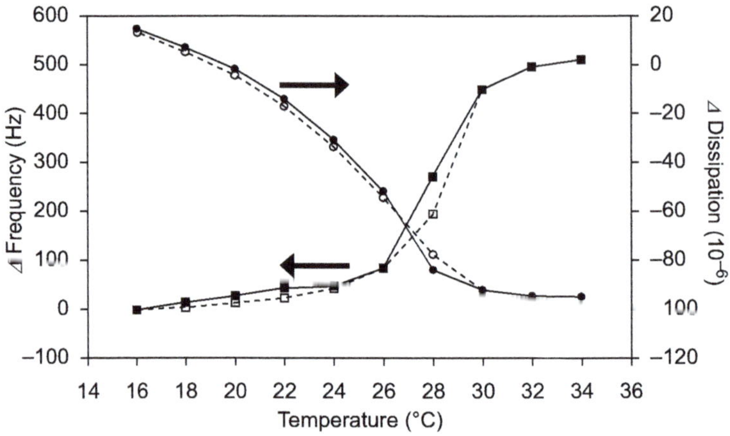

Figure 6.10 QCM-D data of a PNiPAAm-based copolymer film at temperatures around the phase transition. Solid lines correspond to data obtained ramping up temperature, while dashed lines correspond to ramping down temperature.

Reprinted from Alf et al. (2010) by permission of John Wiley & Sons, Inc.

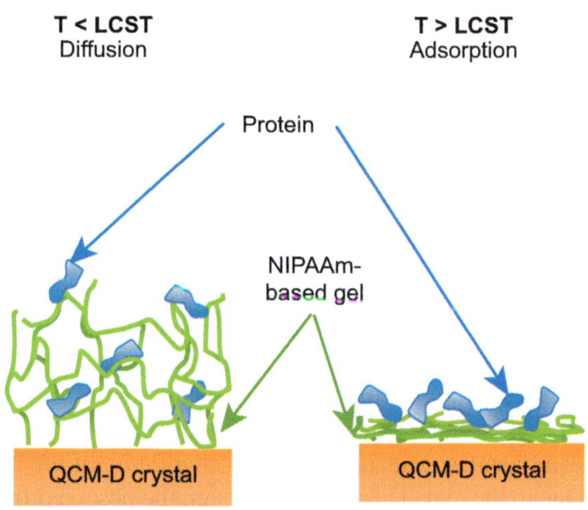

Figure 6.11 Schematic representation of the interaction between a PNiPAAm-based copolymer film and bovine serum albumin. Above the LCST of the film, simple monolayer adsorption was seen, while below the LCST, QCM-D results indicated protein diffusion into the swollen hydrogel.

Reprinted with permission from Alf et al. (2011a). Copyright 2011 American Chemical Society.

In another study, QCM-D provides insight into how a protein interacts with a hydrogel film, e.g., monolayer or multilayer adsorption onto the surface versus penetration into the material. Alf et al. (2011a) show how bovine serum albumin penetrates an expanded thermo-responsive copolymer but forms an adsorbed monolayer when the material is collapsed above the phase transition temperature (Figure 6.11).

Finally, QCM-D can provide information on the swelling kinetics of thermo-responsive thin films. The system behaviour upon temperature variation can be followed in real time, and an immediate response can be clearly distinguished from a delayed one (Alf, Hatton, & Gleason, 2011b).

6.4 Future trends

Diagnostic techniques that were originally intended for hard and impermeable materials are further developed to cover the case of soft and permeable interfaces and coatings. This applies in particular to the way of data processing and interpretation. Recent developments of this kind comprise electrokinetic measurements (consistent treatment of streaming potential, streaming current and surface conductivity), spectroscopic ellipsometry (sophisticated model approaches for swelling layers), as well as the ADSA technique.

Advanced analytical techniques as well as special accessories evolve from custom-made laboratory units into commercially available devices. Toward a more comprehensive characterisation of swelling and switching layer systems, different analytical techniques are combined in a single set-up. Examples closely related to the topic of this chapter are a unit for simultaneous measurements by QCM-D and ellipsometry (see application notes at www.q-sense.com) as well as a combined set-up for microslit electrokinetic measurements and reflectometric interference spectroscopy or Fourier transform infrared spectroscopy, respectively (see www.zetascience.info).

6.5 Sources of further information

With the focus on bioengineering applications, Rzaev, Dincer, and Piskin's (2007) review article summarises a wide range of studies on copolymers of NiPAAm with various comonomers including synthesis routes, structure, properties and applications.

References

Alf, M. E., Godfrin, P. D., Hatton, T. A., & Gleason, K. K. (2010). Sharp hydrophilicity switching and conformality on nanostructured surfaces prepared via initiated chemical vapor deposition (iCVD) of a novel thermally responsive copolymer. *Macromolecular Rapid Communications*, *31*, 2166–2172.

Alf, M. E., Hatton, T. A., & Gleason, K. K. (2011a). Insights into thin, thermally responsive polymer layers through quartz crystal microbalance with dissipation. *Langmuir*, *27*, 10691–10698.

Alf, M. E., Hatton, T. A., & Gleason, K. K. (2011b). Novel *N*-isopropylacrylamide based polymer architecture for faster LCST transition kinetics. *Polymer*, *52*, 4429–4434.

Chang, Y., Yandi, W., Chen, W. Y., Shih, Y. J., Yang, C. C., Ling, Q. D., et al. (2010). Tunable bioadhesive copolymer hydrogels of thermoresponsive poly(*N*-isopropyl acrylamide) containing zwitterionic polysulfobetaine. *Biomacromolecules*, *11*, 1101–1110.

Cole, M. A., Voelcker, N. H., Thissen, H., Horn, R. G., & Griesser, H. J. (2010). Colloid probe AFM study of thermal collapse and protein interactions of poly(*N*-isopropylacrylamide) coatings. *Soft Matter*, *6*, 2657–2667.

Cordeiro, A. L., Zimmermann, R., Gramm, S., Nitschke, M., Janke, A., Schäfer, N., et al. (2009). Temperature dependent physicochemical properties of poly(*N*-isopropyl acrylamide-co-*N* (1-phenylethyl) acrylamide) thin films. *Soft Matter*, *5*, 1367–1377.

Delgado, A. V., Gonzalez-Caballero, E., Hunter, R. J., Koopal, L. K., & Lyklema, J. (2005). Measurement and interpretation of electrokinetic phenomena – (IUPAC technical report). *Pure and Applied Chemistry*, *77*, 1753–1805.

Duval, J. F. L., Küttner, D., Nitschke, M., Werner, C., & Zimmermann, R. (2011). Interrelations between charging, structure and electrokinetics of nanometric polyelectrolyte films. *Journal of Colloid and Interface Science*, *362*, 439–449.

Duval, J. F. L., Küttner, D., Werner, C., & Zimmermann, R. (2011). Electrohydrodynamics of soft polyelectrolyte multilayers: point of zero-streaming current. *Langmuir*, *27*, 10739–10752.

Duval, J. F. L., Zimmermann, R., Cordeiro, A. L., Rein, N., & Werner, C. (2009). Electrokinetics of diffuse soft interfaces. IV. Analysis of streaming current measurements at thermoresponsive thin films. *Langmuir*, *25*, 10691–10703.

Ebara, M., Yamato, M., Hirose, M., Aoyagi, T., Kikuchi, A., Sakai, K., et al. (2003). Copolymerization of 2-carboxyisopropylacrylamide with *N*-isopropylacrylamide accelerates cell detachment from grafted surfaces by reducing temperature. *Biomacromolecules*, *4*, 344–349.

Gramm, S., Teichmann, J., Nitschke, M., Gohs, U., Eichhorn, K.-J., & Werner, C. (2011). Electron beam immobilization of functionalized poly(vinyl methyl ether) thin films on polymer surfaces – towards stimuli responsive coatings for biomedical purposes. *Express Polymer Letters*, *5*, 970–976.

Grundke, K. (2008). Characterization of polymer surfaces by wetting and electrokinetic measurements – contact angle, interfacial tension, zeta potential. In M. Stamm (Ed.), *Polymer surfaces and interfaces: Characterization, modification and applications* (pp. 103–138). Berlin-Heidelberg, Springer.

Guo, H. F., & Ulbricht, M. (2011). Preparation of thermo-responsive polypropylene membranes via surface entrapment of poly(*N*-isopropylacrylamide)-containing macromolecules. *Journal of Membrane Science*, *372*, 331–339.

Harmon, M. E., Kucking, D., & Frank, C. W. (2003). Photo-cross-linkable PNIPAAm copolymers. 5. Mechanical properties of hydrogel layers. *Langmuir*, *19*, 10660–10665.

Hatakeyama, H., Kikuchi, A., Yamato, M., & Okano, T. (2006). Bio-functionalized thermoresponsive interfaces facilitating cell adhesion and proliferation. *Biomaterials*, *27*, 5069–5078.

Hinrichs, K. & Eichhorn, K. J. (2014). *Ellipsometry of functional organic surfaces and films*. Berlin-Heidelberg, Springer.

Hirata, I., Okazaki, M., & Iwata, H. (2004). Simple method for preparation of ultra-thin poly (*N*-isopropylacrylamide) hydrogel layers and characterization of their thermo-responsive properties. *Polymer*, *45*, 5569–5578.

Jhon, Y. K., Bhat, R. R., Jeong, C., Rojas, O. J., Szleifer, I., & Genzer, J. (2006). Salt-induced depression of lower critical solution temperature in a surface-grafted neutral thermoresponsive polymer. *Macromolecular Rapid Communucations*, *27*, 697–701.

Kaufmann, M., Jia, Y. F., Renner, L., Gupta, S., Kuckling, D., Werner, C., et al. (2010). Tuneable swelling of thermo- and pH-responsive copolymer films. *Soft Matter, 6*, 937–944.

Kessel, S., Schmidt, S., Muller, R., Wischerhoff, E., Laschewsky, A., Lutz, J. F., et al. (2010). Thermoresponsive PEG-based polymer layers: surface characterization with AFM force measurements. *Langmuir, 26*, 3462–3467.

Kooij, E. S., Sui, X. F., Hempenius, M. A., Zandvliet, H. J. W., & Vancso, G. J. (2012). Probing the thermal collapse of poly(*N*-isopropylacrylamide) grafts by quantitative in situ ellipsometry. *Journal of Physical Chemistry, 116*, 9261–9268.

Kuckling, D., Harmon, M. E., & Frank, C. W. (2002). Photo-cross-linkable PNIPAAm copolymers. 1. Synthesis and characterization of constrained temperature-responsive hydrogel layers. *Macromolecules, 35*, 6377–6383.

Kwok, D. Y., Lam, C. N. C., Li, A., Leung, A., Wu, R., Mok, E., et al. (1998). Measuring and interpreting contact angles: a complex issue. *Colloids and Surfaces A, 142*, 219–235.

Kwon, O. H., Kikuchi, A., Yamato, M., & Okano, T. (2003). Accelerated cell sheet recovery by co-grafting of PEG with PIPAAm onto porous cell culture membranes. *Biomaterials, 24*, 1223–1232.

Lee, E. L., & von Recum, H. A. (2010). Cell culture platform with mechanical conditioning and nondamaging cellular detachment. *Journal of Biomedical Materials Research Part A, 93A*, 411–418.

Martins, G. V., Mano, J. F., & Alves, N. M. (2011). Dual responsive nanostructured surfaces for biomedical applications. *Langmuir, 27*, 8415–8423.

Matzelle, T. R., Ivanov, D. A., Landwehr, D., Heinrich, L. A., Herkt-Bruns, C., Reichelt, R., et al. (2002). Micromechanical properties of "smart" gels: studies by scanning force and scanning electron microscopy of PNIPAAm. *Journal of Physical Chemistry B, 106*, 2861–2866.

Nakayama, M., Yamada, N., Kumashiro, Y., Kanazawa, H., Yamato, M., & Okano, T. (2012). Thermoresponsive poly(*N*-isopropylacrylamide)-based block copolymer coating for optimizing cell sheet fabrication. *Macromolecular Bioscience, 12*, 751–760.

Nash, M. E., Carroll, W. M., Foley, P. J., Maguire, G., Connell, C. O., Gorelov, A. V., et al. (2012). Ultra-thin spin coated crosslinkable hydrogels for use in cell sheet recovery-synthesis, characterisation to application. *Soft Matter, 8*, 3889–3899.

Nitschke, M., Gramm, S., Götze, T., Valtink, M., Drichel, J., Voit, B., et al. (2007). Thermo-responsive poly(NiPAAm-co-DEGMA) substrates for gentle harvest of human corneal endothelial cell sheets. *Journal of Biomedical Materials Research A, 80*, 1003–1010.

Rochev, Y., O'Halloran, D., Gorelova, T., Gilcreest, V., Selezneva, I., Gavrilyuk, B., et al. (2004). Rationalising the design of polymeric thermoresponsive biomaterials. *Journal of Materials Science – Materials in Medicine, 15*, 513–517.

Rzaev, Z. M. O., Dincer, S., & Piskin, E. (2007). Functional copolymers of *N*-isopropylacrylamide for bioengineering applications. *Progress in Polymer Science, 32*, 534–595.

Schmaljohann, D., Beyerlein, D., Nitschke, M., & Werner, C. (2004). Thermo-reversible swelling of thin hydrogel films immobilized by low-pressure plasma. *Langmuir, 20*, 10107–10114.

Schmaljohann, D., Nitschke, M., Schulze, R., Eing, A., Werner, C., & Eichhorn, K. J. (2005). In situ study of the thermoresponsive behavior of micropatterned hydrogel films by imaging ellipsometry. *Langmuir, 21*, 2317–2322.

Schmaljohann, D., Oswald, J., Jorgensen, B., Nitschke, M., Beyerlein, D., & Werner, C. (2003). Thermo-responsive PNiPAAm-g-PEG films for controlled cell detachment. *Biomacromolecules, 4*, 1733–1739.

Teichmann, J., Valtink, M., Gramm, S., Nitschke, M., Werner, C., Funk, R., et al. (2013). Human corneal endothelial cell sheets for transplantation: thermo-responsive cell culture carriers to meet cell-specific requirements. *Acta Biomaterialia*, *9*, 5030–5038.

Tsuda, Y., Kikuchi, A., Yamato, M., Sakurai, Y., Umezu, M., & Okano, T. (2004). Control of cell adhesion and detachment using temperature and thermoresponsive copolymer grafted culture surfaces. *Journal of Biomedical Materials Research*, *69A*, 70–78.

Uhlmann, P., Skorupa, S., Werner, C., & Grundke, K. (2005). Characterization of maleic acid/anhydride copolymer films by low-rate dynamic liquid-fluid contact angle measurements using axisymmetric drop shape analysis. *Langmuir*, *21*, 6302–6307.

Xu, X. D., Zhang, X. Z., Cheng, S. X., Zhuo, R. X., & Kennedy, J. F. (2007). A strategy to introduce the pH sensitivity to temperature sensitive PNIPAAm hydrogels without weakening the thermo-sensitivity. *Carbohydrate Polymers*, *68*, 416–423.

Yang, L., Pan, F., Zhao, X. B., Yaseen, M., Padia, F., Coffey, P., et al. (2010). Thermoresponsive copolymer nanofilms for controlling cell adhesion, growth, and detachment. *Langmuir*, *26*, 17304–17314.

Zimmermann, R., Kuckling, D., Kaufmann, M., Werner, C., & Duval, J. F. L. (2010). Electrokinetics of a poly(*N*-isopropylacrylamide-co-carboxyacrylamide) soft thin film: evidence of diffuse segment distribution in the swollen state. *Langmuir*, *26*, 18169–18181.

Part Two

Biological interactions and biomedical applications of switchable surfaces

Interaction of switchable biomaterials surfaces with proteins

7

Q. Yu, H. Chen
Soochow University, Suzhou, China

7.1 Introduction

Protein–surface interactions play a well-known and important role in various biological phenomena and can determine the ultimate biofunctionality of a given material after it is in contact with a biological environment. Understanding and further modulating the interactions between proteins and material surfaces are critical (Chen, Yuan, Song, Wu, & Li, 2008; Yuan, Yu, Li, & Chen, 2011). Selective physical adsorption or covalent immobilization of proteins is essential for many applications. For example, preadsorption or immobilization of these adhesion proteins on synthetic material surfaces to promote cell adhesion and proliferation is important for regenerative medicine and tissue engineering (Wilson, Clegg, Leavesley, & Pearcy, 2005). Selective attachment of proteins with bioactivity (such as antibodies) is the key step for the fabrication of biosensors and microarrays (Lin, Weinrich, & Waldmann, 2010). On the negative side, nonspecific protein adsorption on synthetic surfaces leads to various undesired effects such as thrombosis, microbial contamination, and the reduction of sensitivity of diagnostic devices (Banerjee, Pangule, & Kane, 2011; Krishnan, Weinman, & Ober, 2008). Many studies have been conducted on surface modification that aims to either activate surfaces for selective immobilization of proteins or passivate surfaces for resistance to nonspecific protein adsorption; readers with interest in these topics are referred to other excellent reviews (Jiang & Cao, 2010; Liu, Xu, Wu, & Chen, 2013; Wang, Chen et al., 2013; Yu, Zhang, Wang, Brash, & Chen, 2011).

For applications in controlled drug delivery, separation science, and biological engineering, the modulation of protein adsorption on the surfaces (attract proteins in one condition and resist them in another condition) are of importance (Cole, Voelcker, Thissen, & Griesser, 2009; Mendes, 2008). Toward this aim, good candidates include stimuli-responsive surfaces or so-called "smart surfaces," which can undergo a dramatic change in physicochemical properties (such as wettability, charge, roughness, and adhesion) in response to specific environmental stimuli (such as temperature, pH/ionic strength, electrical fields, and light) (Stuart et al., 2010; Zhou & Huck, 2006). In this chapter, we focus on recent progress in the investigation

Switchable and Responsive Surfaces and Materials for Biomedical Applications. http://dx.doi.org/10.1016/B978-0-85709-713-2.00007-9

of interactions between protein and stimuli-responsive surfaces. Examples of using stimuli-responsive surfaces to control protein adsorption for biomedical and biotechnology applications are also included.

7.2 Protein adsorption on surfaces

Protein adsorption onto a surface is a rather complicated and dynamic process, theoretically consisting of a sequence of several, possibly interchangeable steps (Tsapikouni & Missirlis, 2008). The protein–surface interactions have been investigated by molecular simulation (Latour, 2008) and experimental techniques, quantitatively and qualitatively (Szott & Horbett, 2011). Because of the inherent amphiphilic (containing both hydrophobic and hydrophilic domains) and zwitterionic (containing both positive and negative patches) properties of proteins, hydrophobic interactions, electrostatic interactions are two main driving forces for protein adsorption on surfaces. Moreover, van der Waals attraction, hydrogen bonding, supermolecular interactions, affinity interactions, and other intermolecular interactions are also involved in protein adsorption processes in specific situations (Tsapikouni & Missirlis, 2008).

Specifically, the interactions between proteins and surfaces modified with end-grafted polymer brushes can be interpreted by three generic modes, as proposed by Halperin (Halperin, 1999; Halperin & Kröger, 2009): (1) primary adsorption to the grafting surface, (2) ternary adsorption to polymer segments within the brush layer, and (3) secondary adsorption at the outer edge of the brush (as illustrated in Figure 7.1). By theoretical calculation, Halperin analyzed the thermodynamics and kinetics of adsorption of model proteins with different sizes and the effects of chain length and graft density of grafted polymers on protein adsorption. Large proteins are prone to secondary adsorption, in which van der Waals attraction to the surface may be important. This adsorption mode may be repressed by increasing the brush thickness. In contrast, for relatively small proteins, the primary and tertiary adsorption modes would be of particular importance. These two modes may be repressed by increasing the grafting density.

Various key factors determine the extent of protein adsorption onto a surface. Briefly, these factors can be classified into three categories: (1) protein properties, such as

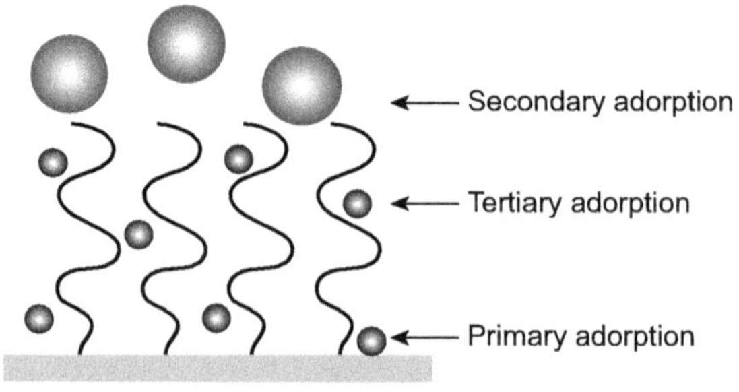

Figure 7.1 Schematic models of protein adsorption on a surface grafted with polymer brushes.

charge distribution, rigidity of conformation, hydrophilic and hydrophobic domains, and strength of intramolecular bonds; (2) surface properties, such as wettability, charge, roughness, and topography; and (3) solution properties, such as temperature, pH, and ionic strength. It is common to modulate the solution properties to induce the change of surface properties (mostly wettability and charge) and thus the interactions between proteins and surfaces.

7.3 Protein adsorption on thermo-responsive surfaces

Temperature is a widely used stimulus for controlling protein adsorption on solid surfaces. Although few reports have mentioned different protein adsorption behavior on mixed self-assembled monolayers (SAMs) containing both hydrophilic and hydrophobic segments (Balamurugan et al., 2005), most reports on the interaction between protein and thermo-responsive surfaces are based on polymers. To date, the most commonly used and best studied thermo-responsive polymer is poly(*N*-isopropylacrylamide) (PNIPAAm) (Rzaev, Dincer, & Piskin, 2007). PNIPAAm displays reversible soluble/insoluble changes in response to temperature changes across a lower critical solution temperature (LCST) at 32 °C in aqueous solution (Schild, 1992). The LCST is between room temperature and physiological temperature, making PNIPAAm an attractive material for bio-related applications. When PNIPAAm is immobilized onto a solid surface, the LCST behavior leads to temperature-dependent surface wettability (Xia & Jiang, 2008), which is expected to play an important role in protein adsorption. In this section, we focus on progress in studies of the interaction between proteins and PNIPAAm surfaces as well as some applied examples using PNIPAAm-modified surfaces to control protein adsorption.

7.3.1 Analysis and evaluation of protein adsorption on PNIPAAm-modified surfaces

Several surface characterization methods have been used to evaluate protein adsorption on PNIPAAm-modified surfaces. The time-of-flight secondary ion mass spectrometry technique is well-known to be highly sensitive to even small amounts of biomolecules adsorbed on surfaces, and for which the intensity of protein-characteristic peaks reflects the extent of protein adsorption. Using time-of-flight secondary ion mass spectrometry in combination with principal component analysis, several groups found that a PNIPAAm surface is more protein-attractive above the LCST than below the LCST from a semiquantitative point of view (Canavan, Graham, Cheng, Ratner, & Castner, 2007; Cheng, Canavan, Graham, Castner, & Ratner, 2006; Cole, Jasieniak, Thissen, Voelcker, & Griesser, 2009; Heinz et al., 2008). The fluorescence-labeling assay is another semiquantitative method for evaluating protein adsorption onto solid surfaces. After incubation in a solution of fluorescently labeled protein above and below the LCST, a remarkable contrast in the amount of fluorescence signal on micropatterned PNIPAAm surfaces was found, indicating different amounts of proteins adsorbed onto the surface (Alarcon, Pennadam, & Alexander, 2005; Yu et al., 2013). The adsorbed proteins on PNIPAAm surfaces can be eluted using aqueous sodium dodecyl sulfate, which can be further evaluated using an immunoblotting assay to analyze the type and amount. This assay is especially effective for protein adsorption in a complex medium (Yu, Zhang, Chen, Zhou et al., 2010; Yu, Zhang, Li, Xu, & Chen, 2011).

Compared with the qualitative methods discussed above, surface plasmon resonance (SPR) and quartz crystal microbalance (QCM) are two instrumentation methods for quantitative evaluation of protein adsorption that are highly sensitive and examine adsorption kinetics. For example, the adsorption of two proteins (fibrinogen and lysozyme) with different charges and sizes on a PNIPAAm film was compared using SPR, and the results indicated that the effect of temperature on adsorption (amount and kinetics) is different (Teare et al., 2005). In another report, QCM was employed to analyze the kinetics of bovine serum albumin (BSA) adsorption on P(NIPAAm-co-di(ethylene glycol) divinyl ether) cross-linked films at different temperatures (Alf, Hatton, & Gleason, 2011). Above the LCST, a simple monolayer of BSA was adsorbed on the surface, whereas below the LCST, there are two processes involved: initial protein adsorption onto the surface followed by protein diffusion into the swollen hydrogel matrix.

A radiolabeling assay also provides a quantitative method to measure the amount of adsorbed protein accurately to the ng/cm^2 scale. Compared with SPR and QCM, a radiolabeling assay can be used effectively for almost any substrate regardless of shape or surface chemistry (Sheardown, Cornelius, & Brash, 1997). More importantly, it can be used to evaluate single protein adsorption in a complex medium (e.g., plasma, serum) without interference (Tan & Brash, 2009). Chen's group used this method to evaluate adsorption of different proteins on PNIPAAm-modified surfaces including silicon (Yu, Zhang, Chen, Wu et al., 2010; Yu, Zhang, Chen, Zhou et al., 2010; Yu, Zhang, Li et al., 2011), polyurethane (Zhao et al., 2011), and silicon nanowires arrays (SiNWAs; Yu, Li, Zhang et al., 2011) in either single protein solution or plasma.

From another point of view, atomic force microscopy is focused on the intermolecular interactions between proteins (coated on the atomic force microscopy probe) and PNIPAAm surfaces at different temperatures. Using this method, Cho et al. reported several studies on intermolecular force between BSA and PNIPAAm surfaces (Cho, Kim, & Cho, 2004; Cho, Kim, & Cho, 2005; Cho, Kim, & Cho, 2008). They observed a temperature-dependent adhesion force (defined as the maximum attractive force during retraction of the surface from the tip), which is negligible at lower temperatures and suddenly increases to a maximum value as the temperature increases above a transition temperature (27–29 °C). The lower transition temperature compared with the LCST of PNIPAAm in water (32 °C) may be due to the addition of salts because the experiments were conducted in a phosphate buffer solution. The change in adhesion force is consistent with the affinity between soluble proteins and a PNIPAAm surface, which, at room temperature, is approximately an order of magnitude lower than at 37 °C (Cheng et al., 2006).

7.3.2 Mechanism of protein adsorption at different temperatures

Understanding the mechanism of protein–PNIPAAm interaction is beneficial for both fundamental studies and many related biomedical and biotechnology applications. Two factors, namely chain conformation and hydrophobicity (or the interaction with binding water), should be considered regarding the temperature-dependence of protein adsorption on PNIPAAm surfaces. When the temperature is below the LCST, PNIPAAm chains exhibit an extended and flexible coil conformation. As the protein approaches the surface, the configurational entropy loss of PNIPAAm chains leads to a steric repulsion and prevents protein adsorption (Jeon, Lee, Andrade, & Gennes, 1991). In addition, below the LCST, intermolecular hydrogen bonds are formed between C=O and N—H groups on PNIPAAm chains with water molecules (Sun et al., 2004). The bound water molecules would create a strong surface hydration "barrier" layer and prevent intimate protein contact between proteins and the surface, similar to other hydrophilic nonfouling polymers (Jiang & Cao, 2010). In contrast, when the temperature is above the LCST, PNIPAAm chains collapse to a molten globular conformation, resulting in the disappearance of osmotic–entropic repulsion, which can prevent protein adsorption. Moreover, the breakdown of hydrogen bonds between water molecules and PNIPAAm chains makes the surface more hydrophobic, favoring protein adsorption via hydrophobic interaction. The interaction between proteins and surface-tethered PNIPAAm brushes was also investigated using the self-consistent field theory by Halperin and Kröger; readers who are interested in this aspect should refer to Halperin and Kröger (2011).

Many studies have demonstrated that PNIPAAm-modified surfaces show thermo-responsive wettability (Xia, Zhu, Feng, & Jiang, 2009). Because wettability is a main factor in determining the extent of protein adsorption on solid surfaces (Sigal, Mrksich, & Whitesides, 1998), it is naturally expected that the surfaces will also exhibit thermo-responsive protein adsorption behavior. However, this hypothesis only works when the thickness of grafted PNIPAAm brushes (corresponding to the molecular weight of PNIPAAm chains) is in a certain range. Yu et al. observed thickness-dependent protein adsorption behavior on a highly dense PNIPAAm brushes-grafted silicon surface. In a certain thinner thickness range (<15 nm), PNIPAAm surfaces exhibited higher protein-resistance against human serum albumin regardless of temperature. In contrast, for thicker PNIPAAm-grafted (thickness = ~38.1 nm) surfaces, the temperature sensitivity of protein adsorption is significant (Yu, Zhang, Chen, Wu et al., 2010). It has also been found that protein adsorption is greatly prevented for PNIPAAm brushes with higher molecular weights, which may be due to the higher hydration of outermost regions even above the LCST, providing a hydration layer that prevents intimate molecular contact between proteins and the surface (Xue et al., 2011; Zhao et al., 2011). For end-grafted PNIPAAm brushes, graft density is another important factor that influences protein adsorption. PNIPAAm brushes with lower graft density allow protein penetration into the brushes and interact more extensively with the grafted chains, leading to remarkable adsorption. In contrast, densely packed brushes suppress neither the protein penetration into the layer nor the adsorption at the outer edge of brushes (Xue et al., 2011). In addition, when an adsorption assay is conducted in a more complex medium, such as blood plasma, the thermo-responsivity of protein adsorption onto PNIPAAm-modified surfaces become weaker (Yu, Zhang, Chen, Zhou et al., 2010). Regardless of adsorption temperature, the adsorption of fibrinogen from plasma at either room temperature or body temperature was less than 5 ng/cm^2. Moreover, adsorption of most plasma proteins to the PNIPAAm-modified surfaces was also significantly reduced compared with unmodified silicon surfaces (Yu, Zhang, Li et al., 2011). These results suggest that PNIPAAm, used as a "traditional" thermo-responsive polymer, is also resistant to nonspecific protein adsorption when the graft parameters are suitable, which may lead to the utilization of PNIPAAm as a nonfouling polymer for surface modification to improve blood compatibility.

It is interesting to note that the temperature-sensitive protein adsorption behavior on PNIPAAm surfaces is dependent not only on the properties of PNIPAAm chains but also protein size (Yu, Zhang, Chen, Wu et al., 2010). Compared with larger proteins (human serum albumin and fibrinogen), the smallest protein, lysozyme, exhibits the largest adsorption below the LCST, whereas as the temperature rises above the LCST, the extent of adsorption is minimized. This difference may be interpreted by Halperin's model (Halperin, 1999), which assumes that the smaller protein can penetrate into the PNIPAAm chains regardless of temperature, whereas the conformational transition of PNIPAAm chains in response to temperature change the extent of primary and tertiary adsorption of larger proteins.

7.3.4 Control protein adsorption for applications

Controlling protein adsorption on synthetic material surfaces is of great importance for protein purification and separation (Kawai, Saito, & Lee, 2003). Okano's group pioneered the development of chromatographic separations based on PNIPAAm-modified beads as the stationary phase. Using the intrinsic temperature-dependent hydrophilic/hydrophobic alteration as well as extended/collapsed conformational change, they developed several kinds of aqueous chromatographic methods for the separation of protein and other biomolecules; readers who are interested in this aspect should refer to their reviews (Kikuchi & Okano, 2002; Nagase, Kobayashi, & Okano, 2009). Stayton's group demonstrated the potential of PNIPAAm-coated particles for rapid purification, enrichment, and detection of target proteins (Malmstadt, Yager, Hoffman, & Stayton, 2003, Malmstadt, Hoffman, & Stayton, 2004). In their system, the PNIPAAm coating particles were functionalized with affinity ligands. Above the LCST, these particles adhered to the microfluidic channel wall and captured target proteins from complex mixtures under flow conditions. Subsequently, decreasing the temperature below the LCST induced the release of particles with target proteins for further analysis (the whole process in illustrated in Figure 7.2). In another approach, a microfluidic device was developed to show the capability to adsorb proteins from solution, hold them with negligible denaturation, and subsequently release them on

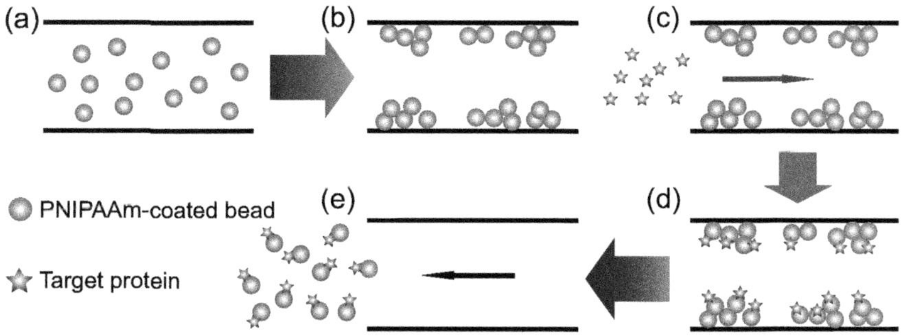

Figure 7.2 Schematic of "smart" mobile affinity matrix for microfluidic immunoassays. (a) The channel is initially filled at room temperature with a suspension of PNIPAAm-coated beads (immobilized with specific ligand). (b) The temperature in the channel is then raised to 37 °C, and the beads aggregate and adhere to the channel walls. (c) A sample containing target proteins is then introduced into the flow stream. (d) Target proteins bind the beads through affinity interaction. (e) Finally, the temperature is reduced to room temperature, leading to the breakup of the bead aggregates. Beads, bound to labeled streptavidin, elute from the channel.

demand (Huber, Manginell, Samara, Kim, & Bunker, 2003). Using a microheater to heat or cool specific sites, the device can be programmed to capture and release proteins in less than a second. The device selectively captured specific proteins depending upon fluid residence and hot line activation times.

PNIPAAm-modified surfaces with immobilized proteins or enzymes have been employed as mediators to switch the surface bioactivity. For example, laccase was immobilized on PNIPAAm gel which was firmly attached to an indium-tin oxide electrode, and the phase transition of the gel provided the possibility of controlling catalytic efficiency by changing temperature (Klis, Karbarz, Stojek, Rogalski, & Bilewicz, 2009). Shrinking the gel considerably decreases the efficiency of the catalytic reaction, whereas reswelling the gel at temperatures lower than 30 °C restores the catalytic efficiency to its original value. In another report, a dynamic scaffold based upon a thin PNIPAAm hydrogel coating was developed, which can reversibly hide and display surface patterns functionalized with enzymes through an elastic creasing instability in response to temperature changes, leading to the change of surface biocatalytic activity (Kim, Yoon, & Hayward, 2010). Combining a nanolithography technique and surface-initiated polymerization, Yu et al. developed a new platform with switchable bioactivity based on nanopatterned PNIPAAm brushes (Yu et al., 2013). At 25 °C (below the LCST), the surface was relatively smooth with weak nanopattern features resulting from the lateral spreading of PNIPAAm brushes. However, when the temperature increased to 37 °C (above the LCST) the pattern appeared. This temperature triggered expanded/collapsed conformational change was switchable and repeatable for several cycles, providing the ability to spatially regulate nonspecific and biospecific protein adsorption on the intervals between the lines of nanopatterned brushes (Figure 7.3).

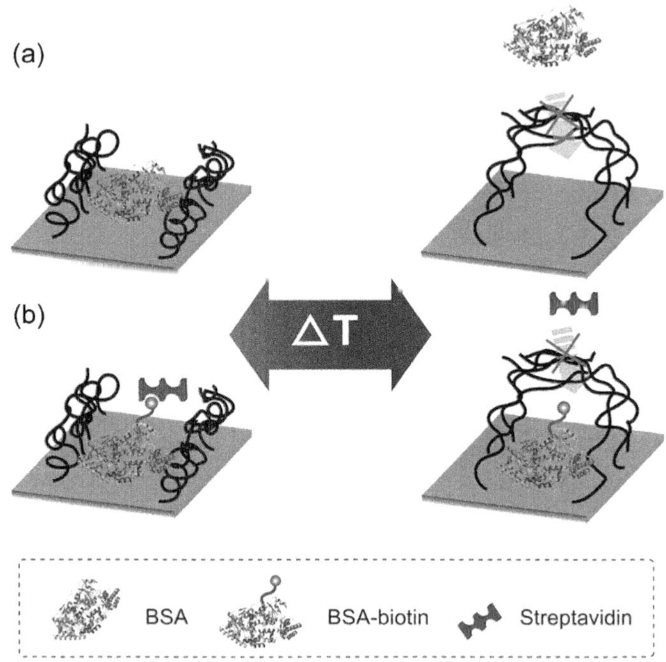

Figure 7.3 Schematic of modulation of (a) nonspecific and (b) specific protein adsorption using nanopatterned PNIPAAm brushes.

7.4 Protein adsorption on pH and/or ionic strength-responsive surfaces

Surfaces modified with polyelectrolytes represent a second commonly used responsive surfaces for controlling protein adsorption. Electrostatic interaction is the main driving force for this kind of surface to attract (with the opposite charge) or repel (with the same charge) proteins. It is well known that the pH of an aqueous solution can influence the overall charge of polyelectrolyte-modified surfaces due to the protonation/deprotonation of ionizable groups. Moreover, the ionic strength (or salt concentration) of the solution has a significant effect on surface charge. Therefore, the changes of either pH or ionic strength affect the electrostatic interaction between the surface and the proteins in solution, which can be used to control adsorption or desorption of proteins.

7.4.1 Protein adsorption on spherical polyelectrolyte brushes

PAA is a typical weak polyacid and has been widely used as a model to investigate protein–polyelectrolyte interactions. Ballauff's group systematically studied BSA adsorption on so-called "spherical polyelectrolyte brushes" (SPBs) consisting of a polystyrene (PS) core covered with PAA brushes (as shown in Figure 7.4)

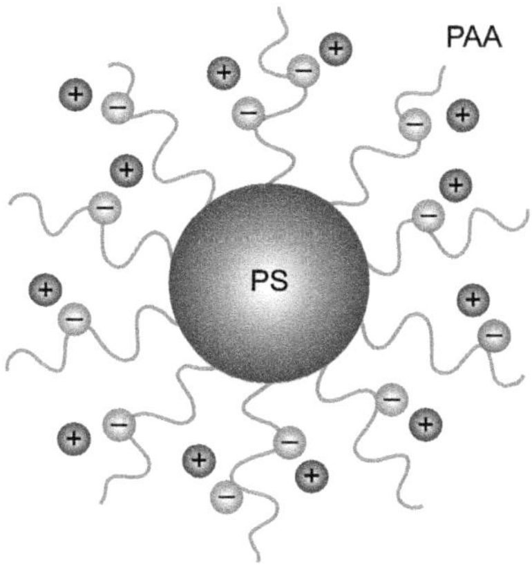

Figure 7.4 Schematic of SPB with PS core covered with PAA brushes.

(Anikin et al., 2005; Czeslik, Jackler, Hazlett et al., 2004; Leermakers, Ballauff, & Borisov, 2007; Wittemann, Haupt, & Ballauff, 2003). At pH = 6.1, they observed strong adsorption of BSA onto the SPBs although both the particles and the dissolved BSA are charged negatively, which should lead to electrostatic repulsion (Wittemann et al., 2003). Furthermore, they found that BSA adsorption is salt concentration dependent such that (1) the amount of adsorbed protein is highest at a low salt concentration and decreases drastically with increasing amounts of added salt, and (2) the adsorbed protein can be washed out again by raising the ionic strength. Based on these results, they proposed a driving force called "counterion release force," which dominates strong protein adsorption at low salt concentrations (they called it polyelectrolyte-mediated protein adsorption). In that case, the pH within the brush is lower than the solution, and the negative charge of the PAA interacts with the patches of positive charges on the protein (the protein can be considered a multivalent counterion within the brush and monovalent counterions will be released). At high salt concentrations, the pH as well as the salt concentration within the brush is the same as in the solution. Under these conditions, the driving forces are weakened. Steric repulsion between the protein and the brush layer becomes dominated and no adsorption takes place. The physical mechanism of the uptake and release of protein by SPBs was further unraveled using two-gradient self-consistent field calculations, and the calculation results are in excellent agreement with recent experimental data (Leermakers et al., 2007).

In another report, the same group used confocal fluorescence microscopy with single molecule sensitivity to characterize uptake and release of fluorescent protein molecules by individual SPB nanoparticles consisting of PS/poly(styrene sulfonate). The photophysical properties of fluorescent proteins before, during, and after adsorption were examined by fluorescence emission spectra. The results indicated that polyelectrolyte-mediated protein adsorption onto SPB particles presents a viable process for protein immobilization while preserving the structural integrity of the proteins (Anikin et al., 2005). These properties make SPBs a good candidate for enzyme immobilization. As examples, glucoamylase and β-glucosidase were adsorbed onto SPBs consisted of PS/PAA or PS/poly(styrene sulfonate) and analysis showed that both enzymes kept nearly full activity after immobilization (Haupt, Neumann, Wittemann, & Ballauff, 2005; Henzler, Haupt, & Ballauff, 2008).

7.4.2 Protein adsorption on planar polyelectrolyte brushes

Protein adsorption on planar polyelectrolyte brushes has also attracted considerable attention. Czeslik's group chose PAA and BSA as the model polyelectrolyte and protein to investigate protein–surface interactions using experimental studies (Czeslik, Jackler, Hazlett et al., 2004) and theoretical calculations (Czeslik, Jackler, Steitz, & von Grunberg, 2004). Using fluorescence correlation spectroscopy and neutron reflectometry, they studied *in situ* the degree of BSA binding to PAA brushes with changing ionic strength, but at neutral pH values both the protein and the brushes carry a negative charge. Similar to Ballauff's results, in which the adsorption of BSA at spherical PAA brush particles was measured *ex situ*, the planar PAA brushes also exhibit a salt-dependent protein affinity. At low ionic strength, PAA brushes strongly bind BSA under electrostatic repulsion, whereas increasing ionic strength makes PAA brushes become protein-resistant, although the direct repulsive electrostatic forces between BSA and PAA are screened under these conditions. These results suggested that BSA–PAA interactions cannot be explained by simple electrostatic screening effects. Rather, they proposed a simple mean-field argument to explain these experimental findings. The model predicts a large gain of free energy associated with the release of BSA counterions when transferring a BSA molecule from the solution onto the PAA brush. The free energy of this counterion "evaporation" is entropic in nature and dominates the electrostatic repulsion between the BSA molecule and the like-charged PAA brush.

Furthermore, they investigate the adsorption of two proteins with different isoelectric point (pI) on planar PAA brushes as a function of ionic strength using total internal reflection fluorescence spectroscopy (Hollmann & Czeslik, 2006). At neutral pH values, BSA and lysozyme have net negative and positive charges, respectively. However, PAA brushes strongly bind both proteins at low ionic strength. When the ionic strength of the protein solution is increased to approximately 100–200 mM, a planar PAA brush appears largely protein resistant, regardless of the protein net charge. These results further confirmed a new protein-binding mechanism concerned with PAA brushes for which a release of counterions is the most probable driving force for protein adsorption. In addition to ionic strength, they also studied the effect of temperature on protein adsorption and found the mechanism of BSA adsorption onto a PAA brush appears to be different at 20 than at 40 °C (Hollmann, Gutberlet, & Czeslik, 2006). The adsorption at 40 °C is dominated by conformational changes of the protein molecules, instead of a counterion release.

Applying attenuated total reflection Fourier transform infrared and total internal reflection fluorescence spectroscopy, the same group studied the secondary structure and aggregation properties of different proteins that are adsorbed onto planar PAA brushes (Hollmann, Steitz, & Czeslik, 2008; Reichhart & Czeslik, 2008). They found that at neutral pH lysozyme, BSA, and α-lactalbumin largely preserve their secondary structures and maintain high mobility after adsorption, suggesting a planar PAA brush is a promising materials coating for enzyme immobilization and for controlled protein delivery.

Although pH has been widely applied as a trigger to alter the properties (especially the charge) of polyelectrolytes in drug delivery (Qiu & Park, 2001; Schmaljohann, 2006), the use of pH to control protein adsorption on planar polyelectrolyte-modified surfaces is less prominent. It has been found that a polyacid, poly(methacrylic acid) (PMAA)–grafted silicon surface exhibits pH-dependent surface wettability and protein adsorption. With a decreasing pH, from 9 to 4, the modified surface became more hydrophobic and adsorption of fibrinogen increased dramatically by nearly 39 times (Li, Yu, Zhang, Wu, & Chen, 2011). In another report, using poly(2-(dimethylamino) ethyl methacrylate) (PDMAEMA), a surface with a high protein-binding capacity and charge selectivity was developed (Kusumo et al., 2007). At lower pH (~5.8), PDMAEMA carried a positive charge due to the protonation of tertiary amine groups, leading to the selective binding of net negatively charged BSA proteins but the rejection of net positively charged lysozyme proteins. Increasing pH or salt concentration in solution eliminates the electrostatic attraction, leading to the desorption of proteins. Similar pH-dependent protein adsorption behavior was also found on membranes based on polyelectrolyte membranes made of a poly(acrylonitrile) and 2-dimethylaminoethyl methacrylate copolymer (Su & Li, 2007). The pH-induced change of surface charge can be used for the generation of protein gradients (Ionov, Houbenov, Sidorenko, Stamm, & Minko, 2009). Mixed polyelectrolyte brushes containing the polyacid PAA and polybase poly(2-vinyl pyridine) (P2VP) with a composition gradient were fabricated. The pH of the solution influenced the charge of PAA and P2VP as well as the protein (e.g., BSA), leading to different protein adsorption rates on surface regions in response to pH (as shown in Figure 7.5).

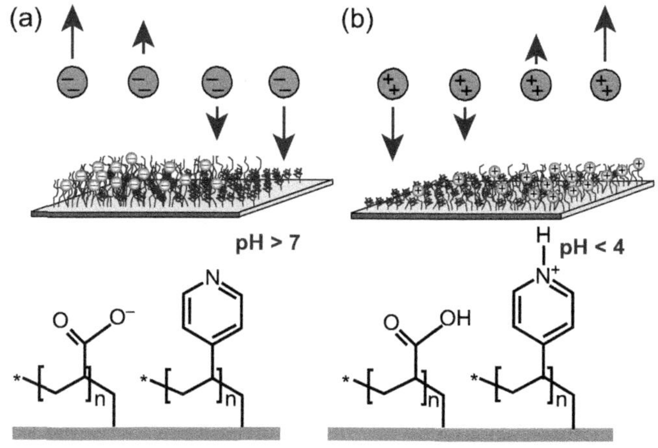

Figure 7.5 Schematic of interaction between proteins and PAA-mix-P2VP mixed-gradient brush under different pH: (a) pH > 7 and (b) pH < 4. (Note: this figure is from an open access journal, no permission is needed.)

7.5 Protein adsorption on other responsive surfaces

7.5.1 Modulation of protein adsorption and desorption using electric potential

Modulation of electric potential to control protein adsorption and release is mainly based on the electrostatic interactions between proteins and the electrode surface (Fang & Szleifer, 2006). Proteins themselves carry net positive, negative, or neutral charge according to the pI of proteins and the pH of the solution. Surfaces with a charge opposite that of interacting proteins tend to attract proteins and surfaces with the same charge as proteins tend to repel proteins. Based on this strategy, SAMs containing loosely packed carboxylic-terminated groups (with a negative charge) or amino-terminated groups (with a positive charge) were generated on gold electrodes for the regulation of protein adsorption (Liu et al., 2004). Modulation of electric potential changes the gold surface charge and thus the electrostatic interaction between surface and ionized terminal groups, which further affects the adsorption/desorption of charged proteins. These SAMs were integrated into microfluidic chips, which can reversibly and selectively adsorb and release differently charged proteins under electrical control (Mu, Liu, Cai, & Kong, 2007). Positively charged avidin and negatively charged streptavidin were used as model proteins for the test. It was found that avidin can be adsorbed onto carboxylic-functionalized chips under a negative potential, whereas it can be released by applying a positive potential. In contrast, streptavidin can be adsorbed on amino-functionalized chips under a positive potential, whereas it can be released by applying a negative potential (as illustrated in Figure 7.6). Under specific applied potentials, the chips adsorbed a certain protein from the mixture

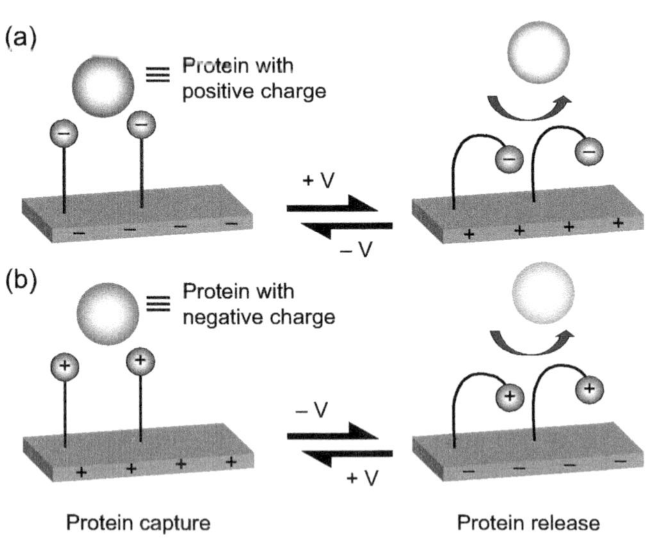

Figure 7.6 Electrically controlled adsorption and release of proteins with different charges by low-density (a) negatively charged group-terminated and (b) positively charged group-terminated monolayers on gold surfaces.

whereas the other protein was allowed to flow out, after which the adsorbed protein could be released by switching the applied potential, suggesting the potential of "on chip" protein adsorption.

7.5.2 Modulation of protein adsorption and desorption using photoinduced supermolecular interaction

Azobenzene is a typical photoresponsive molecule. The azo chromophore isomerizes by illumination with ultraviolet (UV) light (λ=300–400 nm) from the stable linear *trans* form to the bent *cis* form, whereas reverse isomerization can be triggered by irradiation with visible light (λ=425–500 nm). Photoswitched protein adsorption was achieved on electrostatic self-assembly azobenzene films containing silica nanoparticles (Zhang et al., 2012). After UV irradiation, the hybrid film exhibited a much lower capability for BSA adsorption, suggesting weaker adsorption affinity of the *cis* form of azobenzene. In another report, a photosensitive azopolymer brush–grafted gold surface was prepared by surface-initiated polymerization (Zhang, Wang, Yan, Peng, & Ma, 2011). The unique *trans–cis* photoisomerization property induced reversible surface wettability and adsorption of BSA on azopolymer brush surfaces, suggesting that protein biochips could be regenerated safely by UV irradiation rather than by being rinsed with chemical reagents.

On the basis of photocontrolled reversible host–guest interactions between photoresponsive azobenzene-containing SAM and the pH-responsive PAA polymer grafted with β-cyclodextrin moieties, a light/pH-dual-controlled surface for reversible protein adsorption was developed (Wan, Wang, Jiang, Xu, & Zhang, 2009). As shown in Figure 7.7, the photocontrolled switching can be employed to reversibly adsorb and release electroactive cytochrome c. The PAA polymer grafted with β-cyclodextrin

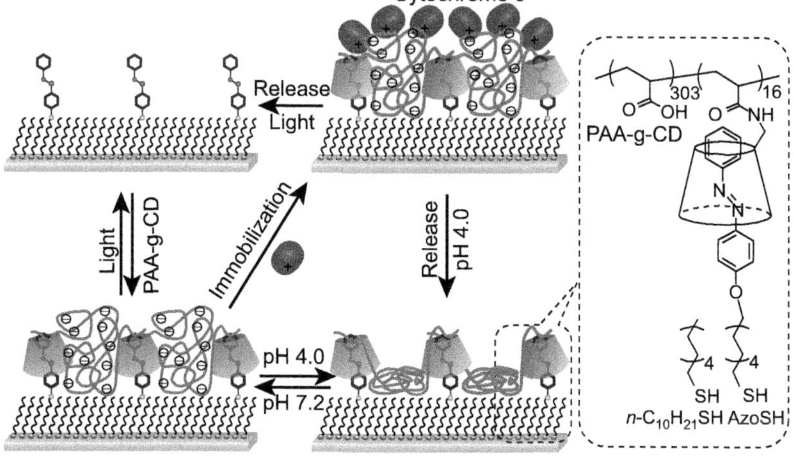

Figure 7.7 Dual-controlled reactive biointerfaces and reversible protein adsorption. (Wan, Wang, Jiang, Xu, & Zhang, 2009.)

moieties–attached azobenzene-containing SAM possesses pH-sensitivity, which can be used to reversibly immobilize electroactive cytochrome c triggered by pH.

7.6 Synergistic effect of surface chemistry and nanostructures on protein adsorption

In addition to surface chemistry, protein adsorption on material surfaces is also affected by surface topography. The influence of surface nanotopographic factors, including roughness, curvature, and geometry, on protein adsorption behavior, such as the amount of protein adsorbed and the activity and morphology of adsorbed protein has been reviewed (Song & Chen, 2007). In this section, we focus on the synergistic effect of stimuli-responsive polymers and nanostructures on protein adsorption.

It has been demonstrated that the introduction of nanostructure is an effective way to enhance the stimuli-responsiveness of surface properties such as wettability and adhesion (Liu, Liang, Zhou, & Liu, 2012). For example, in response to a change of temperature, reversible switching between superhydrophilicity (water contact angle = 0°) and superhydrophobicity (water contact angle >150°) can be achieved on PNIPAAm-grafted rough silicon surface with micro- and nano-hierarchical structure (Sun et al., 2004). Therefore, it is natural to explore whether protein adsorption can also be enhanced when nanoscale topography is introduced on stimuli-responsive surfaces (Wang, Chen et al., 2013). Recently, Chen's group employed a topographical SiNWAs for modification with various stimuli-responsive polymer brushes and systematically investigated the synergistic effect of surface chemistry and nanostructures on protein adsorption under external stimuli (Wang et al., 2011, Wang, Wang et al., 2013; Yu, Chen, Zhang, Yuan et al., 2010; Yu, Li, Zhang et al., 2011).

The first example is SiNWAs modified with the thermo-responsive polymer PNIPAAm (SiNWAs-PNIPAAm). Contact angle results demonstrated that the thermo-responsiveness of surface wettability is greatly enhanced by the introduction of a nanoscale structure. Surprisingly, the level of protein adsorption on SiNWAs-PNIPAAm above and below the LCST is similar, although wettability is believed to be one of the most important factors to determine protein adsorption on surface (Yu, Li, Zhang et al., 2011). From another point of view, SiNWAs-PNIPAAm-reduced protein adsorption by more than 99% compared with unmodified SiNWAs. This high protein resistance resulted from water molecules that were trapped in the interstices of the nanowire arrays and formed a strong hydration layer, preventing intimate molecular contact between proteins and the surface.

Interestingly, although the SiNWAs-PNIPAAm surface shows low levels of native protein adsorption, it favors the adsorption of more hydrophobic denatured proteins. The adsorption of denatured proteins onto SiNWAs-PNIPAAm is especially thermo-responsive (Figure 7.8) (Wang, Wang et al., 2013). The amount of denatured-lysozyme adsorbed at 40 °C was about twice that at 10 °C. Moreover, SiNWAs-PNIPAAm surfaces could selectively adsorb denatured proteins from the native/denatured protein mixtures of either the same or different proteins, but the amount of native proteins in

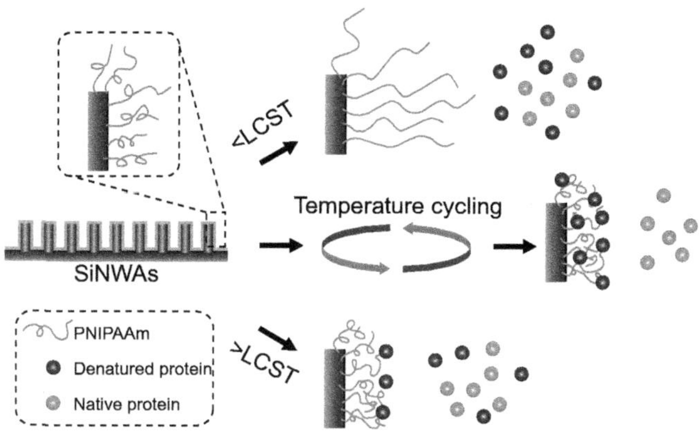

Figure 7.8 Schematic illustration of the SiNWAs-PNIPAAm and its interaction with native and denatured proteins at temperatures below the LCST of PNIPAAm (<LCST), above the LCST of PNIPAAm (>LCST), and under temperature cycling conditions. (© IOP Publishing. Reproduced by permission of IOP Publishing.)

solution remained largely unchanged. These results suggest that SiNWAs-PNIPAAm provides an effective platform for bioseparation and protein purification.

Inspired by the pH-responsive adsorption of proteins onto PMAA modified flat silicon surfaces (Li et al., 2011), and nano-enhanced effects of SiNWAs, a pH-switchable system for protein adsorption and release was developed, based upon PMAA-modified SiNWAs (SiNWAs-PMAA) (Yu, Chen, Zhang, Yuan et al., 2010). SiNWAs-PMAA possesses several unique advantages including (1) an extremely high capacity for protein (e.g., lysozyme, fibrinogen) binding at low pH, (2) high efficiency (more than 90%) of release of adsorbed proteins by increasing the pH without the loss of protein activity, and (3) the binding and release of proteins is reversible and repeatable. These properties are very important in many biomedical and biotechnology applications, including bioseparations, biocatalysis, and biosensors.

In another report, stepwise control of protein adsorption was achieved on PDMAEMA-grafted SiNWAs (SiNWAs-PDMAEMA) by varying the ionic strength (Wang et al., 2011). PDMAEMA is a weak cationic polyelectrolyte, the charge of which is affected by either the pH or ionic strength of the solution. Although no significant change was observed in protein adsorption onto SiNWAs-PDMAEMA in response to pH changes, the adsorption of lysozyme and horseradish peroxidase decreased with increasing NaCl concentration in a stepwise manner.

7.7 Aspects for future research

In the past few decades, a significant number of engineering stimuli-responsive surfaces have been developed for modulating surface interactions with proteins in response to external factors including temperature, pH, ionic strength, light, and electrical field. Switching on and off protein adsorption/desorption on surfaces enables many potential uses of these mechanisms for biomedical and biotechnological applications such as protein purification and separation, tissue engineering, electrochemical biosensors, and biocatalysis. Although considerable progress has been made, many challenges remain and further efforts may be guided by the following possibilities.

Deeply understanding the inherent mechanism of protein–surface interaction. Because of the highly diverse properties of different proteins, protein–surface interactions are complex. Although there are numerous reports to investigate protein adsorption behavior in response to different environmental triggers, the detailed mechanism is still unclear. Molecular simulation has been performed to study protein adsorption on SAMs with OEG or zwitterionic phosphorylcholine groups, providing an insight into the protein resistance of these materials (Chen, Zheng, Li, & Jiang, 2005). Therefore, it is foreseen that this technique may also provide a powerful way to explain the experimental data and give guidance to those designing surfaces for controlling protein adsorption.

Development of new stimuli-responsive surface systems. Although various stimuli-responsive surfaces have been reported to modulate protein adsorption, so far, temperature and thermo-responsive PNIPAAm-based surfaces are still the most widely used stimuli and best studied systems for controlling protein adsorption, respectively. One difficulty faced by the use of pH- or ionic strength–induced switching is that for a system to exhibit reversibility, the aqueous solution environment must be changed in order to elicit a response, making this a more complicated approach compared with "remote"-switching modalities such as temperature. For light- or electrical potential–responsive surfaces, integration of proper stimuli-responsive molecules or polymers onto surfaces is still challenging. Therefore, more efforts should be paid to develop new systems for various applications. In addition, it should be noted that all the systems listed here have limitations. There is no universal platform; hence, selection of an appropriate platform will depend on specific applications.

Development of multi-responsive and hierarchical systems. Considering the complex conditions in the biological environment (with variations in temperature, pH, and salt concentration), it is of great interest to incorporate multiple switching mechanisms into coatings to enable switching via a number of different triggers. Such multi-responsive surfaces have been developed for controlling surface wettability (Xia et al., 2007), although to date there is little research on the control of protein adsorption using multiple triggers. However, it has been demonstrated that the combination of surface chemistry and topography provide synergistic effects on protein adsorption. It is expected that further development in technologies for creating stimuli-responsive surfaces with well-defined micro/nano structures will uncover new interfacial biological phenomena and functions.

Acknowledgments

This work was supported by the National Natural Science Foundation of China (21125418, 21334004, and 21404076)" and the Natural Science Foundation of Jiangsu Province (BK20140316).

References

Alarcon, C. D. H., Pennadam, S., & Alexander, C. (2005). Stimuli responsive polymers for biomedical applications. *Chemical Society Reviews*, *34*, 276–285. http://dx.doi.org/10.1039/b406727d.

Alf, M. E., Hatton, T. A., & Gleason, K. K. (2011). Insights into thin, thermally responsive polymer layers through quartz crystal microbalance with dissipation. *Langmuir*, *27*, 10691–10698. http://dx.doi.org/10.1021/la201935r.

Anikin, K., Rocker, C., Wittemann, A., Wiedenmann, J., Ballauff, M., & Nienhaus, G. U. (2005). Polyelectrolyte-mediated protein adsorption: fluorescent protein binding to individual polyelectrolyte nanospheres. *Journal of Physical Chemistry B*, *109*, 5418–5420. http://dx.doi.org/10.1021/jp0506282.

Balamurugan, S., Ista, L. K., Yan, J., Lopez, G. P., Fick, J., Himmelhaus, M., et al. (2005). Reversible protein adsorption and bioadhesion on monolayers terminated with mixtures of oligo(ethylene glycol) and methyl groups. *Journal of the American Chemical Society*, *127*, 14548–14549. http://dx.doi.org/10.1021/ja054156g.

Banerjee, I., Pangule, R. C., & Kane, R. S. (2011). Antifouling coatings: recent developments in the design of surfaces that prevent fouling by proteins, bacteria, and marine organisms. *Advanced Materials*, *23*, 690–718. http://dx.doi.org/10.1002/adma.201001215.

Canavan, H. E., Graham, D. J., Cheng, X. H., Ratner, B. D., & Castner, D. G. (2007). Comparison of native extracellular matrix with adsorbed protein films using secondary ion mass spectrometry. *Langmuir*, *23*, 50–56. http://dx.doi.org/10.1021/la062330o.

Cheng, X., Canavan, H. E., Graham, D. J., Castner, D. G., & Ratner, B. D. (2006). Temperature dependent activity and structure of adsorbed proteins on plasma polymerized N-isopropyl acrylamide. *Biointerphases*, *1*, 61–72. http://dx.doi.org/10.1116/1.2187980.

Chen, H., Yuan, L., Song, W., Wu, Z., & Li, D. (2008). Biocompatible polymer materials: role of protein-surface interactions. *Progress in Polymer Science*, *33*, 1059–1087. http://dx.doi.org/10.1016/j.progpolymsci.2008.07.006.

Chen, S., Zheng, J., Li, L., & Jiang, S. (2005). Strong resistance of phosphorylcholine self-assembled monolayers to protein adsorption: insights into nonfouling properties of zwitterionic materials. *Journal of American Chemical Society*, *127*, 14473–14478. http://dx.doi.org/10.1021/ja054169u.

Cho, E. C., Kim, Y. D., & Cho, K. (2004). Thermally responsive poly (N-isopropylacrylamide) monolayer on gold: synthesis, surface characterization, and protein interaction/adsorption studies. *Polymer*, *45*, 3195–3204. http://dx.doi.org/10.1016/j.polymer.2004.02.052.

Cho, E. C., Kim, Y. D., & Cho, K. (2005). Temperature-dependent intermolecular force measurement of poly(N-isopropylacrylamide) grafted surface with protein. *Journal of Colloid and Interface Science*, *286*, 479–486. http://dx.doi.org/10.1016/j.jcis.2004.11.001.

Cho, E. C., Kim, D.-H., & Cho, K. (2008). Contact angles of oils on solid substrates in aqueous media: correlation with afm data on protein adhesion. *Langmuir*, *24*, 9974–9978. http://dx.doi.org/10.1021/la801444q.

Cole, M. A., Jasieniak, M., Thissen, H., Voelcker, N. H., & Griesser, H. J. (2009). Time-of-flight-secondary ion mass spectrometry study of the temperature dependence of protein adsorption onto poly(N-isopropylacrylamide) graft coatings. *Analytical Chemistry*, *81*, 6905–6912. http://dx.doi.org/10.1021/ac9009337.

Cole, M. A., Voelcker, N. H., Thissen, H., & Griesser, H. J. (2009). Stimuli-responsive interfaces and systems for the control of protein–surface and cell–surface interactions. *Biomaterials*, *30*, 1827–1850. http://dx.doi.org/10.1016/j.biomaterials.2008.12.026.

Czeslik, C., Jackler, G., Hazlett, T., Gratton, E., Steitz, R., Wittemann, A., et al. (2004). Salt-induced protein resistance of polyelectrolyte brushes studied using fluorescence correlation spectroscopy and neutron reflectometry. *Physical Chemistry Chemical Physics, 6*, 5557–5563. http://dx.doi.org/10.1039/b410805a.

Czeslik, C., Jackler, G., Steitz, R., & von Grunberg, H. H. (2004). Protein binding to like-charged polyelectrolyte brushes by counterion evaporation. *Journal of Physical Chemistry B, 108*, 13395–13402. http://dx.doi.org/10.1021/jp0488766.

Fang, F., & Szleifer, I. (2006). Controlled release of proteins from polymer-modified surfaces. *Proceedings of the National Academy of Sciences of the United States of America, 103*, 5769–5774. http://dx.doi.org/10.1073/pnas.0509688103.

Halperin, A. (1999). Polymer brushes that resist adsorption of model proteins: design parameters. *Langmuir, 15*, 2525–2533. http://dx.doi.org/10.1021/la981356f.

Halperin, A., & Kröger, M. (2009). Ternary protein adsorption onto brushes: strong versus weak. *Langmuir, 25*, 11621–11634. http://dx.doi.org/10.1021/la9008569.

Halperin, A., & Kröger, M. (2011). Collapse of thermoresponsive brushes and the tuning of protein adsorption. *Macromolecules, 44*, 6986–7005. http://dx.doi.org/10.1021/ma201006h.

Haupt, B., Neumann, T., Wittemann, A., & Ballauff, M. (2005). Activity of enzymes immobilized in colloidal spherical polyelectrolyte brushes. *Biomacromolecules, 6*, 948–955. http://dx.doi.org/10.1021/bm0493584.

Heinz, P., Bretagnol, F., Mannelli, I., Sirghi, L., Valsesia, A., Ceccone, G., et al. (2008). Poly(N-isopropylacrylamide) grafted on plasma-activated poly(ethylene oxide): thermal response and interaction with proteins. *Langmuir, 24*, 6166–6175. http://dx.doi.org/10.1021/la800575f.

Henzler, K., Haupt, B., & Ballauff, M. (2008). Enzymatic activity of immobilized enzyme determined by isothermal titration calorimetry. *Analytical Biochemistry, 378*, 184–189. http://dx.doi.org/10.1016/j.ab.2008.04.011.

Hollmann, O., & Czeslik, C. (2006). Characterization of a planar poly(acrylic acid) brush as a materials coating for controlled protein immobilization. *Langmuir, 22*, 3300–3305. http://dx.doi.org/10.1021/la053110y.

Hollmann, O., Gutberlet, T., & Czeslik, C. (2006). Structure and protein binding capacity of a planar PAA brush. *Langmuir, 23*, 1347–1353. http://dx.doi.org/10.1021/la061881b.

Hollmann, O., Steitz, R., & Czeslik, C. (2008). Structure and dynamics of alpha-lactalbumin adsorbed at a charged brush interface. *Physical Chemistry Chemical Physics, 10*, 1448–1456. http://dx.doi.org/10.1039/b716264b.

Huber, D. L., Manginell, R. P., Samara, M. A., Kim, B.-I., & Bunker, B. C. (2003). Programmed adsorption and release of proteins in a microfluidic device. *Science, 301*, 352–354. http://dx.doi.org/10.1126/science.1080759.

Ionov, L., Houbenov, N., Sidorenko, A., Stamm, M., & Minko, S. (2009). Stimuli-responsive command polymer surface for generation of protein gradients. *Biointerphases, 4*, FA45–FA49. http://dx.doi.org/10.1116/1.3119722.

Jeon, S. I., Lee, J. H., Andrade, J. D., & De Gennes, P. G. (1991). Protein–surface interactions in the presence of polyethylene oxide: I. Simplified theory. *Journal of Colloid and Interface Science, 142*, 149–158. http://dx.doi.org/10.1016/0021-9797(91)90043-8.

Jiang, S., & Cao, Z. (2010). Ultralow-fouling, functionalizable, and hydrolyzable zwitterionic materials and their derivatives for biological applications. *Advanced Materials, 22*, 920–932. http://dx.doi.org/10.1002/adma.200901407.

Kawai, T., Saito, K., & Lee, W. (2003). Protein binding to polymer brush, based on ion-exchange, hydrophobic, and affinity interactions. *Journal of Chromatography B, 790*, 131–142. http://dx.doi.org/10.1016/S1570-0232(03)00090-4.

Kikuchi, A., & Okano, T. (2002). Intelligent thermoresponsive polymeric stationary phases for aqueous chromatography of biological compounds. *Progress in Polymer Science*, *27*, 1165–1193. http://dx.doi.org/10.1016/S0079-6700(02)00013-8.

Kim, J., Yoon, J., & Hayward, R. C. (2010). Dynamic display of biomolecular patterns through an elastic creasing instability of stimuli-responsive hydrogels. *Nature Materials*, *9*, 159–164. http://dx.doi.org/10.1038/nmat2606.

Klis, M., Karbarz, M., Stojek, Z., Rogalski, J., & Bilewicz, R. (2009). Thermoresponsive poly(N-isopropylacrylamide) gel for immobilization of laccase on indium tin oxide electrodes. *Journal of Physical Chemistry B*, *113*, 6062–6067. http://dx.doi.org/10.1021/jp8094159.

Krishnan, S., Weinman, C. J., & Ober, C. K. (2008). Advances in polymers for anti-biofouling surfaces. *Journal of Materials Chemistry*, *18*, 3405. http://dx.doi.org/10.1039/b801491d.

Kusumo, A., Bombalski, L., Lin, Q., Matyjaszewski, K., Schneider, J. W., & Tilton, R. D. (2007). High capacity, charge-selective protein uptake by polyelectrolyte brushes. *Langmuir*, *23*, 4448–4454. http://dx.doi.org/10.1021/la063660b.

Latour, R. A. (2008). Molecular simulation of protein-surface interactions: benefits, problems, solutions, and future directions. *Biointerphases*, *3*, FC2–FC12. http://dx.doi.org/10.1116/1.2965132.

Leermakers, F. A. M., Ballauff, M., & Borisov, O. V. (2007). On the mechanism of uptake of globular proteins by polyelectrolyte brushes: a two-gradient self-consistent field analysis. *Langmuir*, *23*, 3937–3946. http://dx.doi.org/10.1021/la0632777.

Lin, P. C., Weinrich, D., & Waldmann, H. (2010). Protein biochips: oriented surface immobilization of proteins. *Macromolecular Chemistry and Physics*, *211*, 136–144. http://dx.doi.org/10.1002/macp.200900539.

Liu, X., Liang, Y., Zhou, F., & Liu, W. (2012). Extreme wettability and tunable adhesion: biomimicking beyond nature? *Soft Matter*, *8*, 2070. http://dx.doi.org/10.1039/c1sm07003g.

Liu, Y., Mu, L., Liu, B. H., Zhang, S., Yang, P. Y., & Kong, J. L. (2004). Controlled protein assembly on a switchable surface. *Chemical Communications*, 1194–1195. http://dx.doi.org/10.1039/b400776j.

Liu, X., Xu, Y., Wu, Z., & Chen, H. (2013). Poly(N-vinylpyrrolidone)-modified surfaces for biomedical applications. *Macromolecular Bioscience*, *13*, 147–154. http://dx.doi.org/10.1002/mabi.201200269.

Li, X., Yu, Q., Zhang, Y., Wu, Z., & Chen, H. (2011). Regulation of protein adsorption on ph-responsive surfaces. *Acta Polymerica Sinica*, 812–816. http://dx.doi.org/10.3724/SP.J.1105.2011.11075.

Malmstadt, N., Hoffman, A. S., & Stayton, P. S. (2004). "Smart" mobile affinity matrix for microfluidic immunoassays. *Lab on a Chip*, *4*, 412–415. http://dx.doi.org/10.1039/b315394k.

Malmstadt, N., Yager, P., Hoffman, A. S., & Stayton, P. S. (2003). A smart microfluidic affinity chromatography matrix composed of poly(N-isopropylacrylamide)-coated beads. *Analytical Chemistry*, *75*, 2943–2949. http://dx.doi.org/10.1021/ac034274r.

Mendes, P. M. (2008). Stimuli-responsive surfaces for bio-applications. *Chemical Society Reviews*, *37*, 2512–2529. http://dx.doi.org/10.1039/b714635n.

Mu, L., Liu, Y., Cai, S. Y., & Kong, J. L. (2007). A smart surface in a microfluidic chip for controlled protein separation. *Chemistry: A European Journal*, *13*, 5113–5120. http://dx.doi.org/10.1002/chem.200601624.

Nagase, K., Kobayashi, J., & Okano, T. (2009). Temperature-responsive intelligent interfaces for biomolecular separation and cell sheet engineering. *Journal of the Royal Society Interface*, *6*, S293–S309. http://dx.doi.org/10.1098/rsif.2008.0499.focus.

Qiu, Y., & Park, K. (2001). Environment-sensitive hydrogels for drug delivery. *Advanced Drug Delivery Reviews*, *53*, 321–339. http://dx.doi.org/10.1016/S0169-409X(01)00203-4.

Reichhart, C., & Czeslik, C. (2008). Native-like structure of proteins at a planar poly(acrylic acid) brush. *Langmuir*, *25*, 1047–1053. http://dx.doi.org/10.1021/la802905s.

Rzaev, Z. M. O., Dincer, S., & Piskin, E. (2007). Functional copolymers of N-isopropylacrylamide for bioengineering applications. *Progress in Polymer Science*, *32*, 534–595. http://dx.doi.org/10.1016/j.progpolymsci.2007.01.006.

Schild, H. G. (1992). Poly(N-Isopropylacrylamide): experiment, theory and application. *Progress in Polymer Science*, *17*, 163–249. http://dx.doi.org/10.1016/0079-6700(92)90023-R.

Schmaljohann, D. (2006). Thermo- and pH-responsive polymers in drug delivery. *Advanced Drug Delivery Reviews*, *58*, 1655–1670. http://dx.doi.org/10.1016/j.addr.2006.09.020.

Sheardown, H., Cornelius, R. M., & Brash, J. L. (1997). Measurement of protein adsorption to metals using radioiodination methods: a caveat. *Colloids and Surfaces B*, *10*, 29–33. http://dx.doi.org/10.1016/S0927-7765(97)00045-3.

Sigal, G. B., Mrksich, M., & Whitesides, G. M. (1998). Effect of surface wettability on the adsorption of proteins and detergents. *Journal of American Chemical Society*, *120*, 3464–3473.

Song, W., & Chen, H. (2007). Protein adsorption on materials surfaces with nano-topography. *Chinese Science Bulletin*, *52*, 3169–3173. http://dx.doi.org/10.1007/s11434-007-0504-6.

Stuart, M. A. C., Huck, W. T. S., Genzer, J., Muller, M., Ober, C., Stamm, M., et al. (2010). Emerging applications of stimuli-responsive polymer materials. *Nature Materials*, *9*, 101–113. http://dx.doi.org/10.1038/nmat2614.

Su, Y. L., & Li, C. (2007). The reorientation of poly(2-dimethylamino ethyl methacrylate) after environment stimuli improves hydrophilicity and resistance of protein adsorption. *Journal of Colloid and Interface Science*, *316*, 344–349. http://dx.doi.org/10.1016/j.jcis.2007.08.017.

Sun, T. L., Wang, G. J., Feng, L., Liu, B. Q., Ma, Y. M., Jiang, L., et al. (2004). Reversible switching between superhydrophilicity and superhydrophobicity. *Angewandte Chemie International Edition*, *43*, 357–360. http://dx.doi.org/10.1002/anie.200352565.

Szott, L. M., & Horbett, T. A. (2011). Protein interactions with surfaces: Computational approaches and repellency. *Current Opinion in Chemical Biology*, *15*, 683–689. http://dx.doi.org/10.1016/j.cbpa.2011.04.016.

Tan, J., & Brash, J. L. (2009). Nonfouling biomaterials based on polyethylene oxide-containing amphiphilic triblock copolymers as surface modifying additives: adsorption of proteins from human plasma to copolymer/polyurethane blends. *Journal of Biomedical Materials Research Part A*, *90A*, 196–204. http://dx.doi.org/10.1002/jbm.a.32074.

Teare, D. O. H., Barwick, D. C., Schofield, W. C. E., Garrod, R. P., Beeby, A., & Badyal, J. P. S. (2005). Functionalization of solid surfaces with thermoresponsive protein-resistant films. *Journal of Physical Chemistry B*, *109*, 22407–22412. http://dx.doi.org/10.1021/jp052869f.

Tsapikouni, T. S., & Missirlis, Y. F. (2008). Protein–material interactions: from micro-to-nano scale. *Materials Science and Engineering B*, *152*, 2–7. http://dx.doi.org/10.1016/j.mseb.2008.06.007.

Wan, P., Wang, Y., Jiang, Y., Xu, H., & Zhang, X. (2009). Fabrication of reactivated biointerface for dual-controlled reversible immobilization of Cytochrome c. WILEY-VCH Verlag GmbH & Co. KGaA, Weinheim.

Wang, D., Chen, G., Wang, H., Tang, W., Pan, W., Li, N., et al. (2013). A reusable quartz crystal microbalance biosensor for highly specific detection of single-base DNA mutation. *Biosensors and Bioelectronics*, *48*, 276–280. http://dx.doi.org/10.1016/j.bios.2013.04.035.

Wang, H., Wang, Y., Yuan, L., Wang, L., Yang, W., Wu, Z., et al. (2013). Thermally responsive silicon nanowire arrays for native/denatured-protein separation. *Nanotechnology*, *24*, 105101. http://dx.doi.org/10.1088/0957-4484/24/10/105101.

Wang, L., Wang, H., Yuan, L., Yang, W., Wu, Z., & Chen, H. (2011). Step-wise control of protein adsorption and bacterial attachment on a nanowire array surface: tuning surface wettability by salt concentration. *Journal of Materials Chemistry, 21*, 13920–13925. http://dx.doi.org/10.1039/C1JM12148K.

Wan, P. B., Wang, Y. P., Jiang, Y. G., Xu, H. P., & Zhang, X. (2009). Fabrication of reactivated biointerface for dual-controlled reversible immobilization of cytochrome C. *Advanced Materials, 21*, 4362. http://dx.doi.org/10.1002/adma.200901237.

Wilson, C. J., Clegg, R. E., Leavesley, D. I., & Pearcy, M. J. (2005). Mediation of biomaterial-cell interactions by adsorbed proteins: a review. *Tissue Engineering, 11*, 1–18. http://dx.doi.org/10.1089/ten.2005.11.1.

Wittemann, A., Haupt, B., & Ballauff, M. (2003). Adsorption of proteins on spherical polyelectrolyte brushes in aqueous solution. *Physical Chemistry Chemical Physics, 5*, 1671–1677. http://dx.doi.org/10.1039/B300607g.

Xia, F., Ge, H., Hou, Y., Sun, T. L., Chen, L., Zhang, G. Z., et al. (2007). Multiresponsive surfaces change between superhydrophilicity and superhydrophobicity. *Advanced Materials, 19*, 2520–2524. http://dx.doi.org/10.1002/adma.200602334.

Xia, F., & Jiang, L. (2008). Bio-inspired, smart, multiscale interfacial materials. *Advanced Materials, 20*, 2842–2858. http://dx.doi.org/10.1002/adma.200800836.

Xia, F., Zhu, Y., Feng, L., & Jiang, L. (2009). Smart responsive surfaces switching reversibly between super-hydrophobicity and super-hydrophilicity. *Soft Matter, 5*, 275–281. http://dx.doi.org/10.1039/b803951h.

Xue, C., Yonet-Tanyeri, N., Brouette, N., Sferrazza, M., Braun, P. V., & Leckband, D. E. (2011). Protein adsorption on poly(N-isopropylacrylamide) brushes: dependence on grafting density and chain collapse. *Langmuir, 27*, 8810–8818. http://dx.doi.org/10.1021/la2001909.

Yuan, L., Yu, Q., Li, D., & Chen, H. (2011). Surface modification to control protein-surface interaction. *Macromolecular Bioscience, 11*, 1031–1040. http://dx.doi.org/10.1002/mabi.201000464.

Yu, Q., Chen, H., Zhang, Y., Yuan, L., Zhao, T., Li, X., et al. (2010). Ph-reversible, high-capacity binding of proteins on a substrate with nanostructure. *Langmuir, 26*, 17812–17815. http://dx.doi.org/10.1021/la103647s.

Yu, Q., Li, X., Zhang, Y., Yuan, L., Zhao, T., & Chen, H. (2011). The synergistic effects of stimuli-responsive polymers with nano-structured surfaces: wettability and protein adsorption. *RSC Advances, 1*, 262–269. http://dx.doi.org/10.1039/c1ra00201e.

Yu, Q., Shivapooja, P., Johnson, L. M., Tizazu, G., Leggett, G. J., & Lopez, G. P. (2013). Nanopatterned polymer brushes as switchable bioactive interfaces. *Nanoscale, 5*, 3632–3637. http://dx.doi.org/10.1039/C3NR00312D.

Yu, Q., Zhang, Y., Chen, H., Wu, Z., Huang, H., & Cheng, C. (2010). Protein adsorption on poly(N-isopropylacrylamide)-modified silicon surfaces: effects of grafted layer thickness and protein size. *Colloids and Surfaces B, 76*, 468–474. http://dx.doi.org/10.1016/j.colsurfb.2009.12.006.

Yu, Q., Zhang, Y., Chen, H., Zhou, F., Wu, Z., Huang, H., et al. (2010). Protein adsorption and cell adhesion/detachment behavior on dual-responsive silicon surfaces modified with Poly(N-isopropylacrylamide)-block-polystyrene copolymer. *Langmuir, 26*, 8582–8588. http://dx.doi.org/10.1021/la904663m.

Yu, Q., Zhang, Y., Li, X., Xu, Y., & Chen, H. (2011). Interaction between PNIPAAm modified surfaces and plasma proteins. *Acta Polymerica Sinica, 5*, 537–542. http://dx.doi.org/10.3724/SP.J.1105.2011.10124.

Yu, Q., Zhang, Y., Wang, H., Brash, J. L., & Chen, H. (2011). Anti-fouling bioactive surfaces. *Acta Biomaterialia*, *7*, 1550–1557. http://dx.doi.org/10.1016/j.actbio.2010.12.021.

Zhang, J., Hu, T., Liu, Y., Ma, Y., Dong, J., Xu, L., et al. (2012). Photoswitched protein adsorption on electrostatically self-assembled azobenzene films. *ChemPhysChem*, *13*, 2671–2675. http://dx.doi.org/10.1002/cphc.201200231.

Zhang, Z., Wang, Y., Yan, F., Peng, D., & Ma, Z. (2011). Photosensitive azopolymer brushes via atom transfer radical polymerization for protein sensing. *Chinese Journal of Chemistry*, *29*, 153–158. http://dx.doi.org/10.1002/cjoc.201190044.

Zhao, T., Chen, H., Zheng, J., Yu, Q., Wu, Z., & Yuan, L. (2011). Inhibition of protein adsorption and cell adhesion on PNIPAAm-grafted polyurethane surface: effect of graft molecular weight. *Colloids and Surfaces B*, *85*, 26–31. http://dx.doi.org/10.1016/j.colsurfb.2010.10.047.

Zhou, F., & Huck, W. T. S. (2006). Surface grafted polymer brushes as ideal building blocks for "smart" surfaces. *Physical Chemistry Chemical Physics*, *8*, 3815–3823. http://dx.doi.org/10.1039/B606415A.

Interaction of responsive/switchable surfaces with cells

8

B.L. Banik, J.L. Brown
The Pennsylvania State University, University Park, PA, USA

8.1 Introduction

Switchable and responsive materials, also known as "smart," "designer," or "intelligent" materials, have found a niche in the biomedical engineering field as a way to create materials with defined physical, chemical, and biological properties to synthesize more physiologically designed biological systems. These surfaces are dynamic with the capability to switch between active and inactive states such as in areas of wettability, adhesiveness, bioinertness, and protein adsorption (Ulijn, 2006). Bioengineering applications extend from piezoelectric, shape memory, pH-sensitive (LeDuc & Robinson, 2007), light, enzyme-responsive (Ulijn, 2006), and thermoresponsive to output useful functions such as phase transition, shape change, biomolecule release, and activity (Chan & Mooney, 2008), wettability, and cell adhesion, migration, and phenotype progression (Mendes, 2008). Current research is beginning to breech questions regarding relationships between cell behavior and material responses for biomedical engineering advancements in the regenerative medicine and tissue engineering fields.

Cells are complex and sensitive structures adaptable to a variety of external stimuli ranging from mechanical (e.g. force, stiffness, curvature), chemical (e.g. hormones, biological grafting, surface energy), and environmental (e.g. pH, temperature, geometry) cues (LeDuc & Robinson, 2007). Cell proliferation, differentiation, and behavior are a few dynamic cellular processes of consideration when a cell is exposed to local or systemic environmental changes (Chan & Mooney, 2008). The cell's cytoskeleton, primarily composed of microtubules and intermediate and actin filaments, plays a primary role in the determination of cell-material properties and interactions when exposed to switchable and reversible surfaces (LeDuc & Robinson, 2007). Although mimicking the robust and efficient nature of cells is a difficult undertaking, switchable and reversible materials provide a platform for materials science and engineering to escalate the intricate processes found in cellular systems. As a tissue develops, cells will respond to the various local and systemic environmental changes, such as pH or mechanical stimulation, which can be stimulated by switchable and reversible materials (Chan & Mooney, 2008). The scope of switchable and reversible materials extends across various disciplines. With regard to research into cellular responses, mesenchymal stem cells (MSCs) are of particular interest, specifically in the biomedical field of tissue engineering, because of their expansion and differentiation potential and responsiveness to materials (Curran, Chen, & Hunt, 2006).

Switchable and Responsive Surfaces and Materials for Biomedical Applications. http://dx.doi.org/10.1016/B978-0-85709-713-2.00008-0

This chapter discusses the interaction between switchable properties of materials and cells to explore the material property effects on cellular behavior. The scope of this chapter encompasses interfacial material properties (i.e. surface energy, wettability, topography, and ligand presentation) and the physical behavior (i.e. adhesion, migration, and differentiation) of cells as a response. Within these sections, the chapter is organized to address areas such as elasticity, geometry, and wettability. In conclusion, future trends are presented to encourage forward thinking in biomedical engineering research to utilize switchable and responsive materials and ultimately improve biomedical applications, designs, and products

8.2 Controlling stem cell behavior

The material design of switchable and responsive materials defines the stem cell microenvironment and can significantly affect stem cell fate and phenotypic expression. A cell interacts with its environment and adjacent cells through subcellular microstructures, which mediate "ruffled membranes," focal adhesions, adhesion plaques, and focal contacts (Borm et al., 2005; Boyan et al., 1996). These supporting structures aid in migration, adhesion, transmission of regulatory signals and force, organization, orientation, and structural support. The stimuli-responsive materials may be utilized as self-assembled monolayers, polymer films (Mendes, 2008), or polymer brushes (Anselme, Ploux, & Ponche, 2010) such that the biospecific binding material properties can be tailored specifically for applications.

8.2.1 Adhesion and adsorption

Adhesion is connected to many cellular functions—spreading, migration, proliferation, and differentiation (Anselme et al., 2010)—and, therefore, it is imperative to understand the cascade of events that can both positively and negatively impact it. Noteworthy to point out is that when a cell comes in contact with a surface, the substrate is not bare but rather covered with proteins and water that have adsorbed on the surface upon introduction to the system. Thus, when describing the "phases of adhesion," often times the first phase will describe an initial integrin-ligand binding event. Anselme et al. (2010) provide an organized table with adhesion phase profiles from four different authors for various cell types (Anselme et al., 2010). Figure 8.1 summarizes criteria to consider for ruffling formation, which ultimately affects migration efficiency and adhesiveness. These are noteworthy considerations when investigating cell adhesion and adsorption processes.

8.2.2 Spreading and migration

During attachment, cells become various morphologies as they adapt to the topography and geometry of the surface, which in turn affects the spreading, orientation, and migration of the cells. Differentiation of stem cells is impacted by the substrate's material chemistry as demonstrated by reports on lineage commitment of stem cells in relation to spreading and matrix stiffness (Phadke, Chang, & Varghese, 2010).

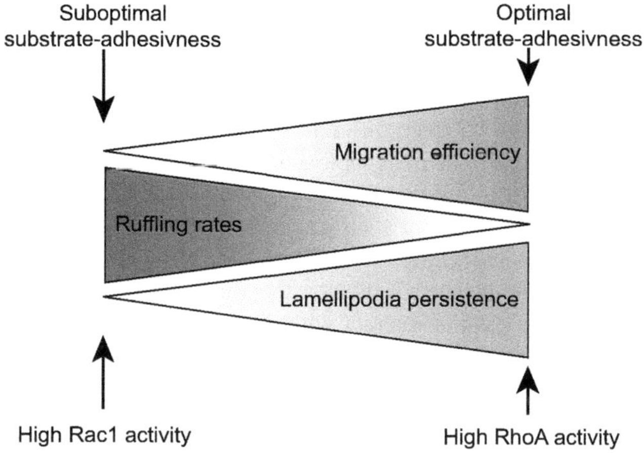

Figure 8.1 Summary of relationships between adhesion-dependent lamellipodia persistence, migration efficiency, and ruffling rates.

Reprinted with permission from *Experimental Cell Research*: 302, 83–95. Copyright 2005. Science Direct. (Borm, Requardt, Herzog & Kirfel 2005)

McBeath et al. (2004) investigated the effects of cell shape on human MSC commitment. Interestingly, the group found that MSCs that had the opportunity to adhere, flatten, and spread trended toward an osteogenic lineage, whereas MSCs that took a cell shape more unspread and rounded suggested an adipogenic phenotype (McBeath et al., 2004). The cellular mechanism behind the shape dependency for phenotype progression was proposed to be modulated by Rho-associated protein kinase (ROCK)-mediated cytoskeletal tension, centering on RhoA activity (McBeath et al., 2004).

8.3 Interfacial properties

Excitement has risen for reversible stimuli-responsive materials because the substrates now provide the potential to control intricate cellular processes that are often a challenge to mimic. For example, with systems that can fluctuate between interacting and noninteracting surfaces, bio–interfacial interactions such as biomolecular adsorption and desorption states can be created allowing for an on/off switch of proteins or bacteria adhesion (Cole et al., 2009). The adsorbed protein layer significantly dictates the cell–material interface reactions. Therefore, it is important to identify the effects of stimuli-responsive materials—surface energy, hydrophobicity, stiffness, curvature, and ligands—on cell behavior and response. Switching interfacial interactions between cells and surfaces is based on two principles: either a stimulus is applied to the biomolecules, which alters the binding affinity to the surface, or a stimulus is applied to the surface, which in turn affects the material chemistry (Cole et al., 2009). The focus of this section will be on the latter.

8.3.1 Surface energy

Biomaterial surface energy is characterized by general charge density and net polarity of charge and can profoundly alter cell attachment and proliferation because of cells capabilities to discriminate between different surface chemistries (Boyan et al., 1996). Cell binding can be tailored via switchable and responsive materials based on surface presentation either via incorporation of ions, which may change the orientations of binding proteins (Boyan et al., 1996), or a surface coating that could result in the switching between different functional groups following an external stimulus. The charge and structural arrangement on the surface dictates the ability for certain ions, proteins, minerals, biomolecules, and cells to interact with the biomaterial and in turn participates in defining the biological response. Surface chemistry effects on cellular processes trickle down through a series of steps: alterations of binding orientations, changes in cell attachment through cell adhesion molecules, transformations in cell shape, and ultimately variations in cell behavior (Boyan et al., 1996; Phadke et al., 2010).

Through experiments with silane-modified surfaces, Curran et al. demonstrated the ability to use surface energy as a way to control MSC adhesion and differentiation pathways through cell adhesion, real-time polymerase chain reaction, and protein expression (Curran et al., 2006). Modified surfaces included the following functional groups: methyl (—CH_3) and clean glass (TAAB) control that maintain the MSC phenotype, amino (—NH_2) and silane (—SH) that promoted osteogenesis, and hydroxyl (—OH) and carboxyl (—COOH) that promoted chondrogenesis. Table 8.1 provides the corresponding surface energy and contact angle for each modified surface (Curran et al., 2006).

8.3.2 Wettability

The hydrophobicity or hydrophilicity of a surface can result from a change in molecular chemical conformation or polarity (Xia et al., 2009) and thus modify the surface chemistry and wettability. This in turn affects protein adsorption and cell–material interactions (Alves et al., 2010; Phadke et al., 2010). Surface wettability has a determining role in cell adhesion (Arima & Iwata, 2007). Arima and Iwata (2007) prepared a series of alkanethiol self-assembled monolayers and compared the cell adhesion

Table 8.1 Surface energy and dynamic contact angle measurements for specified surfaces

Material	Surface energy (mJ/m^2)	Contact angle
TAAB	53.39	62.16 ± 2.33
—OH	52.89	63.06 ± 2.01
—COOH	49.51	68.89 ± 4
—NH_2	43.52	78.71 ± 5.69
—SH	39.6	84.96 ± 4.11
—CH_3	35.53	91.36 ± 1.12

and protein adsorption for human umbilical vein endothelial cells and human cervical carcinome HeLa cells based on the wettability of the materials. A moderate surface wettability of 40°–60° was found to be suitable for cell adhesion for both cell types (Arima & Iwata, 2007).

8.3.3 Topography—elasticity and curvature

Because of the diversity of tissue microenvironments, the question as to whether the topography of a substrate is a factor in MSC lineage commitment has been investigated. Topographical features such as geometry, size, stiffness, and curvature have been shown to be critical in cell polarization, migration, and signaling (Alves et al., 2010). Engler et al., (2006) sought to grasp the concept of signal transduction from various matrix elasticities to morphological transformations elicited by cells. The mechanotransduction process can be described in two mechanisms—the cell's ability to apply force against the matrix and the signal generation dependent upon the force required to deform the matrix (Engler et al., 2006). Using a polyacrylamide gel system elastically tunable via bis-acrylamide concentration for various crosslinking ranges, Engler et al., (2006) specified phenotypic progression toward neurons, myoblasts, and osteoblasts based strictly on the matrix elasticity. Pelham and Wang (1997) suggested the ability of cells to detect environmental elasticity and to adjust adhesion and motility appropriately. The molecular involvement in elasticity is suggested to be tied to tyrosine phosphorylation as the cell pushes and pulls the integrin receptors through the cytoskeleton. Myosin is also interpreted as being a key player as the cell probes for substrate properties (Pelham & Wang, 1997).

Curvature of a surface deforms the actin cytoskeleton of cells as the cells contour in or around the region of interest. Cell receptors would be either stretched or compressed based on the cell's conformation on the surface topography (Anselme et al., 2010). The reorganization of the cytoskeleton to adapt to the change in morphology can affect the cell behavior and response.

8.3.4 Ligands

An understanding of extracellular membrane (ECM) components can also be used to control stem cell differentiation. Integrins are fundamental receptors of the ECM that aid in cell adhesion and attachment as well as assist in responding to the ECM for cell locomotion and anchoring sites (Borm et al., 2005). By harnessing ECM–cell interactions of exposured integrin binding sites and activation of integrins by ligand binding (Phadke et al., 2010), approaches have explored the use of cell adhesive peptides to direct stem cell fate: RGD, GFOGER, and YIGSR (Stabenfeldt, Brown, & Barker, 2010). Ligand binding is also under investigation to understand stem cell responses to pathways initiated by activation of GTPases in the Rho family such as RhoA, Rac1, and Cdc42 (Borm et al., 2005). The goal is to tailor cell adhesive peptides and improve integrin specificity to more directly control cell direction. Figure 8.2 illustrates how ligand presentation (e.g. surface functional groups, ligand density) is involved in the design of synthetic stem cell scaffolding.

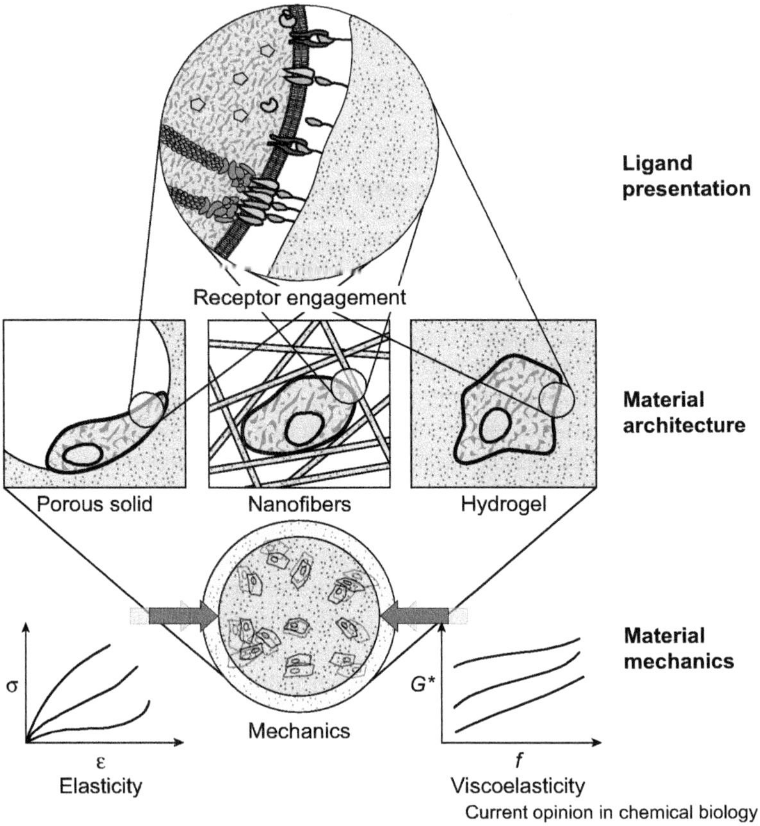

Figure 8.2 Design parameters for stem cell scaffolding.
Reprinted with permission from *Current Opinion in Chemical Biology*: 11, 381–87. Copyright 2007. Science Direct. (Saha, Pollock, Schaffer & Healy 2007)

8.4 Physical behavior

The response of a cell to a surface, switchable or otherwise, can be divided into two basic mechanisms. The first, integrin signaling, involves the formation of aggregates of proteins involved in linking the extracellular matrix to the cytoskeleton. These protein aggregates are referred to as focal adhesions and, in addition to linking the cytoskeleton and extracellular matrix, also demonstrate a surprising amount of intracellular biochemical signaling. The second mechanism a cell can respond to a surface is through altered population dynamics regulating cell:cell signaling. A single cell on an adhesive island has no mechanism to attach to surrounding cells and as such demonstrates no cell:cell signaling, which regulates numerous signaling cascades such as cadherin:beta-catenin signaling. Likewise, numerous surfaces, either through dramatic elevation changes or limiting the ability of a monolayer of cells to form, decrease the amount of cell:cell signaling. Figure 8.3 provides a simple overview of how surfaces can modulate both focal adhesion and cell:cell signaling.

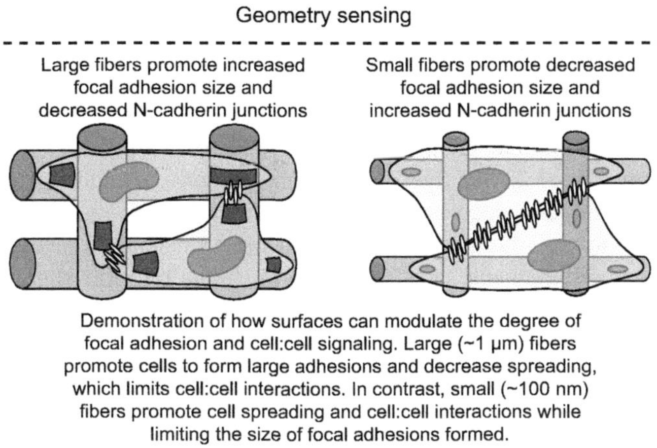

Figure 8.3 Illustration demonstrating how cell:matrix and cell:cell signaling can be regulated by a parameter as simple as the characteristic dimension of the underlying substrate.

8.4.1 Adhesion

The physical behavior of a cell in response to a surface depends on multiple factors, the first of which is whether the cell is capable of adhesion to the surface. Section 8.2.1 discussed adhesion as a phenotype; this section will explore adhesion as the beginning of an array of cascades ultimately leading to cell fate determination. The simplest fate determination through adhesion is whether adhesion is possible. Most stem and somatic cells, with the exception of hematopoietic stem cells and blood cells, are anchorage dependent. These anchorage dependent stem and somatic cells will undergo apoptosis if adhesion is not possible. In anchorage dependent cells, integrin association with extracellular matrix blocks caspase 8 and 9 activation through Src, ERK 1/2, and Akt signaling preventing apoptosis (Grigoriou et al., 2005; Ozaki et al., 2006). Working as an inhibitor is far from the only mechanism associated with integrin attachment to the ECM. A thorough discussion of how a cell forms adhesion to the surface will make clear how a cell also responds to the unique properties of the surface.

Once an integrin is capable of binding, physical changes in the cytoskeleton as well as chemical changes through soluble signaling begin to occur. Initially, the single integrin binding the ECM reduces the energy required for adjacent integrins to bind the underlying ECM, which results in the formation of small integrin clusters (Irvine et al., 2002). Next, a series of proteins begin to associate with the integrin clusters eventually tethering the integrins to cytoskeleton of the cell. The integrins and associated intermediate proteins between the integrins and the cytoskeleton are classified as the adhesome. The adhesome currently contains 156 proteins with close to 700 interactions among them. Adhesome proteins involved with adhesion to an ECM vary with respect to time, surface chemistry, surface geometry, stiffness, and available ligands present in the underlying ECM. With respect to time, the early integrin clusters are referred to as a focal complex; as the focal complex increases in size and links to the cytoskeleton it is referred to as the focal adhesion. There are variants of focal adhesions dependent on the integrins involved and activation state of adhesome proteins; however, for simplicity in this chapter they are all referred to as focal adhesions since they all relate to the linkage of the ECM and intracellular cytoskeleton. Examining the proteins within the adhesome, 58 are enzymes (receptors, kinases, phosphatases, GTPases, GTPase activating proteins, and guanine nucleotide exchange factors) capable of biochemical signal propagation. The ability of these components to participate in signal transduction in response to a surface (mechanotransduction) in addition to the physical linkage of the cytoskeleton to both the nucleus and the substrate (mechanotransmission) provide focal adhesions with the necessary potency to modulate far more than just whether a cell can adhere (Kritikou, 2007; Zaidel-Bar & Geiger, 2010).

8.4.2 Migration

The next typical phenotype demonstrated by a cell after attaching to a surface is migration. Migration rates can be affected by surface chemistry, surface ligand concentration, surface stiffness, and surface geometry. Similar to adhesion, surface chemistry can affect migration through binding affinity affecting matrix proteins capable of binding integrins. Surfaces with very high energies prevent matrix protein accumulation and cell adhesion; likewise, those with very low energies either bind and distort protein conformation to a point where integrin binding is inhibited or do not form an interface capable of permitting protein absorption. However, moderate surface energies that permit matrix accumulation can affect migration rates. Typically with these moderate surface energies, the lower the surface energy, the lower the migration rate because of a decrease in the rate of unbinding integrins at the trailing edge of a cell caused by an increased overall adhesion strength (Hallab et al., 2001; Keselowsky, Collard, & García, 2003). Similar to surface energy, ligand concentration has a bimodal correlation with migration rate. Generally, higher ligand concentrations capable of binding integrins result in higher rates of migration. However, once ligand concentration surpasses the optimal level, the rate of migration decreases (Doyle et al., 2009; Shi & Boettiger, 2003). Additionally, the stiffness of a substrate demonstrates a bimodal correlation with migration velocity. Surfaces that are very stiff reduce the rate of focal adhesion turnover and instead strengthen the existing focal adhesions limiting forward motion. On the other end, surfaces that have a low stiffness promote rapid focal adhesion turnover where the stressed adhesions slide inward toward the center of the cell while also disassembling. The increased focal adhesion sliding and turnover on low stiffness surfaces reduces the rate of migration (Angelini et al., 2010; Jiang et al., 2006; Peyton & Putnam, 2005). Ligand density can be combined with stiffness to modulate where the peak rate of migration will occur. Increasing the ligand density decreases the stiffness required to achieve maximal migration velocity (Peyton & Putnam, 2005). Geometry can affect the migration of cells in several ways. First, geometries that present a characteristic dimension an order of magnitude larger than the cell, but present continuous features (e.g. a raised edge, fiber), will promote contact guidance of cells. This serves to direct the cells along the length of the surface feature (Doyle et al., 2009; Tan & Saltzman, 2004). Geometry can affect migration through the focal adhesion dynamics as well if the geometry presents a characteristic dimension on the order of proteins and protein complexes/organelles, which would be 10 nm to 1 μm. This occurs through mechanisms similar to ligand density for uniformly rough surfaces where the rough islands present a limited space for assembly of focal adhesions. As such, roughness on the order of 1 μm can increase the rate of migration through promoting small focal adhesions and increased focal adhesion turnover (Bettinger, Langer, & Borenstein, 2009; Ventre, Causa, & Netti, 2012; Von der Mark et al. 2010). Although not well understood, researchers are beginning to find that the focal adhesion can serve as a geometry sensor. In these cases, it appears that curvature can alter the mechanics of focal adhesion turnover. Once understood this would be another mechanism by which geometry can control cell migration. Preliminary results from our laboratory indicate this is through controlling the interaction between focal adhesion kinase and Src.

8.4.3 Differentiation

Finally, the phenotype or differentiation of the cell can be affected by geometry. The notion that a substrate can dictate the differentiation of a stem cell is of great interest to tissue engineers and has been investigated previously in the context of surface chemistry or stiffness. One of the most striking examples was the work of Engler et al., (2006) demonstrating that stiffness of a surface directly controlled the differentiation of stem cells. Substrates presenting a stiffness matching brain ECM resulted in neural differentiation, those presenting the stiffness of unmineralized osteoid ECM resulted in osteogenic differentiation and those in between the two were myogenic (Engler et al., 2006). However, even surfaces presenting a similar degree of stiffness but altering the presentation down to a single variable in that presentation can alter the differentiation of stem cells. One of the more compelling evaluations of how altering a single parameter within a geometry can alter differentiation was carried out by Christopherson, Song, and Mao (2009). They observed altered differential differentiation of neural stem cells on 283 nm fibers and 749 nm fibers as compared with control. The smaller fibers promoted oligodendrocyte differentiation whereas the larger fibers promoted neuronal differentiation and the control tissue culture polystyrene promoted maintenance of the progenitor population (Christopherson et al. 2009). This research is substantial in the context of regenerative medicine because it demonstrates the diameter dependence over the differentiation of an adult stem cell population. We recently reported on the development of a novel technique where nanofiber diameter was controlled post-spinning through an isotropic wet etching technique. Surprisingly there was an enormous amount of variation in the steady state activation pattern of ERK1/2 and p38 mitogen-activated protein kinase after 24 h post-seeding within a relatively narrow range of diameters, with significant shifts occurring with diameter changes of less than 200 nm (Jaiswal & Brown, 2012). Not surprisingly, ERK1/2 and p38 are upstream regulators of phenotype and can control ges (Bost et al., 2002; McMahon, Prendergast, & Campbell, 2008; Morooka & Nishida, 1998; Tuan, Boland, & Tuli, 2003). Examining the regulation of lineage commitment in response to either the altered stiffness of an underlying substrate or a change in the diameter of a nanofiber suggests that focal adhesion signaling is nexus regulating phenotype. However, other possibilities exist. Cell polarity can be induced through substrates demonstrating either aligned fibers or aligned parallel ridges. Both have demonstrated an ability to promote lineages such as muscle, tendon, and ligament, which all contain elongated spindle shaped cells *in vivo* (Pijnappels et al. 2008). Furthermore, numerous studies have demonstrated that forcing a stem cell to conform to a shape, either through large or small patterned area or patterned areas with unique aspect ratios or other features and a uniform shape will alter the lineage commitment of stem cells. Typically, the patterning follows the results observed on surfaces demonstrating different stiffnesses. Patterns that promote increased intracellular tension, e.g. a square, rectangle, or star shape, promote osteogenic differentiation; whereas, those surfaces that promote decreased intracellular tension (e.g. a circle, or flower shape, promote adipogenic differentiation). In both cases, the intracellular tension governed by RhoA/ROCK signaling demonstrated control over the eventual phenotype (Kilian et al., 2010; McBeath et al., 2004). Extrapolating further, RhoA/ROCK signaling is increased on surfaces that promote large stable focal adhesions, which could be stiff surfaces or fibers with diameters near 1 μm, whereas surfaces that promote rapid focal adhesion turnover are also associated with decreased RhoA/ROCK signaling (Ozdemir et al., 2013). Although interesting, the patterned surfaces provide engineering challenges to incorporate into a scaffold; however, they may have value in directing stem cell differentiation prior to implantation, and they certainly provide clues as to which intracellular signaling processes play a role in geometry mediated differentiation.

8.5 Future trends

Looking forward, the key to scaffold design in the future will use switchable and responsive surfaces in an instructive and multimodal way. Already, there are culture dishes on the market that prevent cell adhesion at room temperature, allowing for enzyme-free passaging of cells. Future researchers should look to expand on this idea and generate surfaces that respond to the needs of the cell population over time. Early after seeding or implantation, a surface needs to promote migration and proliferation of the target cells. This could be accomplished through a surface demonstrating a low initial surface energy or a high stiffness to promote early migration of the cell population, which leads to a uniform distribution of cells across the surface. After the cell population reaches confluence on the surface, the expression of cytokines and the forces generated by the cell population are altered. Future materials should respond to these cell generated cues to provide instruction on what do next. For example, the compression of the surface provided by the cell population could expose previously hidden ligands capable of initiating differentiation that rely on the compression of the material to return to a conformation that promotes binding the appropriate receptor. Alternatively, the material could respond to cytokines expressed by the cell population to trigger the release of aptamer bound growth factors. The possibilities are vast; however, the message is simple: future materials must maintain bi-directional dynamic communication with the cell population.

References

Alves, N. M., Pashkuleva, I., Reis, R. L., & Mano, J. F. (2010). Controlling cell behavior through the designofpolymersurfaces.*Small*,*6*(20),2208–2220.http://dx.doi.org/10.1002/smll.201000233.

Angelini, T. E., Hannezo, E., Trepat, X., Fredberg, J. J., & Weitz, D. A. (2010). Cell migration driven by cooperative substrate deformation patterns. *Physical Review Letters*, *104*(16), 168104. http://dx.doi.org/10.1103/PhysRevLett.104.168104.

Anselme, K., Ploux, L., & Ponche, A. (2010). Cell/material interfaces: influence of surface chemistry and surface topography on cell adhesion. *Journal of Adhesion Science and Technology*, *24*(5), 831–852. http://dx.doi.org/10.1163/016942409X12598231568186.

Arima, Y., & Iwata, H. (2007). Effect of wettability and surface functional groups on protein adsorption and cell adhesion using well-defined mixed self-assembled monolayers. *Biomaterials*, *28*(20), 3074–3082. http://dx.doi.org/10.1016/j.biomaterials.2007.03.013.

Bettinger, C. J., Langer, R., & Borenstein, J. T. (2009). Engineering substrate topography at the micro- and nanoscale to control cell function. *Angewandte Chemie International Edition in English*, *48*(30), 5406–5415. http://dx.doi.org/10.1002/anie.200805179.

Borm, B., Requardt, R. P., Herzog, V., & Kirfel, G. (2005). Membrane ruffles in cell migration: indicators of inefficient lamellipodia adhesion and compartments of actin filament reorganization. *Experimental Cell Research*, *302*(1), 83–95. http://dx.doi.org/10.1016/j.yexcr.2004.08.034.

Bost, F., Caron, L., Marchetti, I., Dani, C., Le Marchand-Brustel, Y., & Binetruy, B. (2002). Retinoic acid activation of the ERK pathway is required for embryonic stem cell commitment into the adipocyte lineage. *The Biochemical Journal*, *361*(Pt 3), 621–627.

Boyan, B. D., Hummert, T. W., Dean, D. D., & Schwartz. Z. (1996). Role of material surfaces in regulating bone and cartilage cell response. *Biomaterials*, *17*(2), 137–146.

Chan, G., & Mooney, D. J. (2008). New materials for tissue engineering: towards greater control over the biological response. *Trends in Biotechnology*, *26*(7), 382–392. http://dx.doi.org/10.1016/j.tibtech.2008.03.011.

Christopherson, G. T., Song, H., & Mao, H.-Q. (2009). The influence of fiber diameter of electrospun substrates on neural stem cell differentiation and proliferation. *Biomaterials*, *30*(4), 556–564. http://dx.doi.org/10.1016/j.biomaterials.2008.10.004.

Cole, M. A., Voelcker, N. H., Thissen, H., & Griesser, H. J. (2009). Stimuli-responsive interfaces and systems for the control of protein-surface and cell-surface interactions. *Biomaterials*, *30*(9), 1827–1850. http://dx.doi.org/10.1016/j.biomaterials.2008.12.026.

Curran, J. M., Chen, R., & Hunt, J. A. (2006). The guidance of human mesenchymal stem cell differentiation in vitro by controlled modifications to the cell substrate. *Biomaterials*, *27*(27), 4783–4793. http://dx.doi.org/10.1016/j.biomaterials.2006.05.001.

Doyle, A. D., Wang, F. W., Matsumoto, K., & Yamada, K. M. (2009). One-dimensional topography underlies three-dimensional fibrillar cell migration. *The Journal of Cell Biology*, *184*(4), 481–490. http://dx.doi.org/10.1083/jcb.200810041.

Engler, A. J., Sen, S., Sweeney, H. L., & Discher, D. E. (2006). Matrix elasticity directs stem cell lineage specification. *Cell*, *126*(4), 677–689. http://dx.doi.org/10.1016/j.cell.2006.06.044.

Grigoriou, V., Shapiro, I. M., Cavalcanti-Adam, E. A., Composto, R. J., Ducheyne, P., & Adams, C. S. (2005). Apoptosis and survival of osteoblast-like cells are regulated by surface attachment. *The Journal of Biological Chemistry*, *280*(3), 1733–1739. http://dx.doi.org/10.1074/jbc.M402550200.

Hallab, N. J., Bundy, K. J., O'Connor, K., Moses, R. L., & Jacobs, J. J. (2001). Evaluation of metallic and polymeric biomaterial surface energy and surface roughness characteristics for directed cell adhesion. *Tissue Engineering*, *7*(1), 55–71. http://dx.doi.org/10.1089/107632700300003297.

Irvine, D. J., Hue, K. -A., Mayes, A. M., & Griffith, L. G. (2002). Simulations of cell-surface integrin binding to nanoscale-clustered adhesion ligands. *Biophysical Journal*, *82*(1 Pt 1), 120–132. http://dx.doi.org/10.1016/S0006-3495(02)75379-4.

Jaiswal, D., & Brown, J. L. (2012). Nanofiber diameter-dependent MAPK activity in osteoblasts. *Journal of Biomedical Materials Research Part A*, *100*(11), 2921–2928. http://dx.doi.org/10.1002/jbm.a.34234.

Jiang, G., Huang, A. H., Cai, Y., Tanase, M., & Sheetz, M. P. (2006). Rigidity sensing at the leading edge through alphavbeta3 integrins and RPTPalpha. *Biophysical Journal*, *90*(5), 1804–1809. http://dx.doi.org/10.1529/biophysj.105.072462.

Keselowsky, B. G., Collard, D. M., & García, A. J. (2003). Surface chemistry modulates fibronectin conformation and directs integrin binding and specificity to control cell adhesion. *Journal of Biomedical Materials Research Part A*, *66*(2), 247–259. http://dx.doi.org/10.1002/jbm.a.10537.

Kilian, K. A., Bugarija, B., Lahn, B. T., & Mrksich, M. (2010). Geometric cues for directing the differentiation of mesenchymal stem cells. *Proceedings of the National Academy of Sciences of the United States of America*, *107*(11), 4872–4877. http://dx.doi.org/10.1073/pnas.0903269107.

Kritikou, E. (2007). Networks: the complexity of adhesion. *Nature Reviews Molecular Cell Biology*, *8*(9), 674–675. http://dx.doi.org/10.1038/nrm2237.

LeDuc, P. R., & Robinson, D. N. (2007). Using lessons from cellular and molecular structures for future materials. *Advanced Materials*, *19*(22), 3761–3770. http://dx.doi.org/10.1002/adma.200701286.

McBeath, R., Pirone, D. M., Nelson, C. M., Bhadriraju, K., & Chen, C. S. (2004). Cell shape, cytoskeletal tension, and RhoA regulate stem cell lineage commitment. *Developmental Cell*, *6*(4), 483–495.

McMahon, L. A., Prendergast, P. J., & Campbell, V. A. (2008). A comparison of the involvement of p38, ERK1/2 and PI3K in growth factor-induced chondrogenic differentiation of mesenchymal stem cells. *Biochemical and biophysical research communications*, *368*(4), 990–995. http://dx.doi.org/10.1016/j.bbrc.2008.01.160.

Mendes, P. M. (2008). Stimuli-responsive surfaces for bio-applications. *Chemical Society Reviews*, *37*(11), 2512–2529. http://dx.doi.org/10.1039/b714635n.

Morooka, T., & Nishida, E. (1998). Requirement of p38 mitogen-activated protein kinase for neuronal differentiation in PC12 cells. *The Journal of Biological Chemistry*, *273*(38), 24285–24288.

Ozaki, K., Minoda, A., Kishikawa, F., & Kohno, M. (2006). Blockade of the ERK pathway markedly sensitizes tumor cells to HDAC inhibitor-induced cell death. *Biochemical and Biophysical Research Communications*, *339*(4), 1171–1177. http://dx.doi.org/10.1016/j.bbrc.2005.11.131.

Ozdemir, T., Xu, L., Siedlecki, C., & Brown, J. (2013). Substrate curvature sensing through Myosin IIa upregulates early osteogenesis. *Integrative Biology: Quantitative Biosciences from Nano to Macro*, *5*(11), 1407–1416. http://dx.doi.org/10.1039/c3ib40068a.

Pelham, R., & Wang, Y.-L. (1997). Cell locomotion and focal adhesions are regulated by. *Proceedings of the National Academy of Sciences, 94*(December), 13661–13665 .

Peyton, S. R., & Putnam, A. J. (2005). Extracellular matrix rigidity governs smooth muscle cell motility in a biphasic fashion. *Journal of Cellular Physiology*, *204*(1), 198–209. http://dx.doi.org/10.1002/jcp.20274.

Phadke, A., Chang, C., & Varghese, S. (2010). Functional biomaterials for controlling stem cell differentiation. In K. Roy (Ed.), *Biomaterials as stem cell niche* (pp. 19–44). Springer. http://dx.doi.org/10.1007/8415.

Pijnappels, D. A., Schalij, M. J., Ramkisoensing, A. A., van Tuyn, J., de Vries, A. A. F., & Ypey, D. L., et al. (2008). Forced alignment of mesenchymal stem cells undergoing cardiomyogenic differentiation affects functional integration with cardiomyocyte cultures. *Circulation Research*, *103*(2), 167–176. http://dx.doi.org/10.1161/CIRCRESAHA.108.176131.

Saha, K., Pollock, J. F., Schaffer, D. V., & Healy, K. E. (2007). *Designing synthetic materials to control stem cell phenotype*, 381–387.

Shi, Q., & Boettiger, D. (2003). A novel mode for integrin-mediated signaling: tethering is required for phosphorylation of FAK Y397. *Molecular Biology of the Cell, 14*(October) 4306–4315. http://dx.doi.org/10.1091/mbc.E03 .

Stabenfeldt, S. E., Brown, A. C., & Barker, T. H. (2010). Engineering ECM complexity into biomaterials for directing cell fate. In K. Roy (Ed.), *Biomaterials as stem cell niche* (pp. 1–18). Springer. http://dx.doi.org/10.1007/8415.

Tan, J., & Saltzman, W. M. (2004). Biomaterials with hierarchically defined micro- and nanoscale structure. *Biomaterials*, *25*(17), 3593–3601. http://dx.doi.org/10.1016/j.biomaterials.2003.10.034.

Tuan, R. S., Boland, G., & Tuli, R. (2003). Adult mesenchymal stem cells and cell-based tissue engineering. *Arthritis Research and Therapy*, *5*(1), 32–45. http://dx.doi.org/10.1186/ar614.

Ulijn, R. V. (2006). Enzyme-responsive materials: a new class of smart biomaterials. *Journal of Materials Chemistry*, *16*(23), 2217. http://dx.doi.org/10.1039/b601776m.

Ventre, M., Causa, F., & Netti, P. A. (2012). Determinants of cell-material crosstalk at the interface: towards engineering of cell instructive materials. *Journal of the Royal Society, Interface/The Royal Society*, *9*(74), 2017–2032. http://dx.doi.org/10.1098/rsif.2012.0308.

Von der Mark, K., Park, J., Bauer, S., & Schmuki, P. (2010). Nanoscale engineering of biomimetic surfaces: cues from the extracellular matrix. *Cell and tissue research*, *339*(1), 131–153. http://dx.doi.org/10.1007/s00441-009-0896-5.

Xia, F., Zhu, Y., Feng, L., & Jiang, L. (2009). Smart responsive surfaces switching reversibly between super-hydrophobicity and super-hydrophilicity. *Soft Matter*, *5*(2), 275. http://dx.doi.org/10.1039/b803951h.

Zaidel-Bar, R., & Geiger, B. (2010). The switchable integrin adhesome. *Journal of Cell Science*, *123*(Pt 9), 1385–1388. http://dx.doi.org/10.1242/jcs.066183.

Further reading

Lutolf, M., & Hubbell, J. (2005). Synthetic biomaterials as instructive extracellular microenvironments for morphogenesis in tissue engineering. *Nature Biotechnology*, *23*(1), 47–55.

Mei, Y., et al. (2010). Combinatorial development of biomaterials for clonal growth of human pluripotent stem cells. *Nature Materials*, *9*, 768–778.

Santos, E., et al. (2012). Novel advances in the design of three-dimensional bio-scaffolds to control cell fate: translation from 2D to 3D. *Trends in Biotechnology*, *30*(6), 331–341.

Temperature-responsive polymers for cell culture and tissue engineering applications

9

Y. Akiyama, T. Okano
Tokyo Women's Medical University, Tokyo, Japan

9.1 Introduction

Temperature-responsive polymer, poly(*N*-isopropylacrylamide) (PIPAAm), has been widely used in biomedical field because of its unique property, which is a reversible hydration and dehydration accompanying with dynamic conformational changes in the polymer chains by temperature change (Heskins & Guillet, 1968), across 32 °C, the lower critical solution temperature (LCST) of PIPAAm. Because of the reversible hydration and dehydration, PIPAAm-grafted surfaces exhibit a hydrophilic and hydrophobic alternation in response to temperature change (Figure 9.1) (Takei et al., 1994; Yakushiji et al., 1998). The intensive investigation of PIPAAm-grafted surfaces has revealed that the grafted polymer chain conformation affects the wettability of resultant polymer-grafted surface (Kikuchi & Okano, 2005; Takei et al., 1994; Yakushiji et al., 1998). The degree of hydrophilic and hydrophobic alternation of a PIPAAm-grafted surface also depends on the alternation of mobility of the graft polymer chains. In 1990, the application of intelligent surfaces to a cell culture system newly establishes temperature-responsive cell culture surface (TRCS) (Yamada et al., 1990), of which cell attachment and detachment properties of TRCS are modulated by a mild temperature change. The cell attachment and detachment process involves cellular metabolism (Okano, Yamada, Okuhara, Sakai, & Sakurai, 1995; Okano, Yamada, Sakai, & Sakurai, 1993). The unique properties of TRCS have further produced cell sheet engineering. Since then, the vigorous investigation of TRCS has been performed and been found to give new insights into the mechanism of temperature-induced cell attachment and detachment, the characteristics of recovered cells as a cell sheet, and the properties of TRCS (Kushida et al., 1999; Yamato et al., 1999, 2001). TRCSs were initially invented by an electron beam (EB) irradiation method, and the various types of TRCSs have been intensively developed by various methods for applying these unique and attractive properties for biomedical field.

In this chapter, the characteristics of these TRCSs are systematically described in terms of their preparation methods, and some of the recent clinical applications of TRCS-based cell sheet engineering are discussed. In the final part of this chapter, other TRCS applications for the medical field are also explained and discussed.

Switchable and Responsive Surfaces and Materials for Biomedical Applications. http://dx.doi.org/10.1016/B978-0-85709-713-2.00009-2

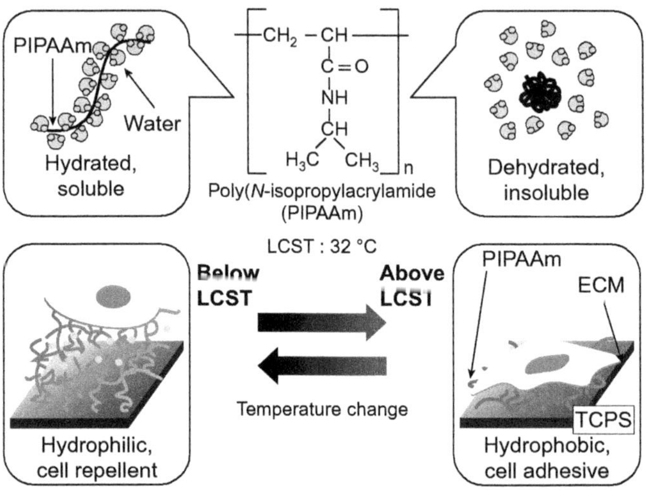

Figure 9.1 Characteristics of PIPAAm and conventional TRCS, which is PIPAAm-grafted TCPS prepared by the electron beam irradiation (EB) irradiation method. TRCS shows cell adhesion and detachment (deadhesion) because of the reversible temperature-induced hydrophilic and hydrophobic alternation. Detached cells contain an ECM component at the basal site of the cell. Cell adhesion and detachment process involves in metabolic adenosine triphosphate (ATP) consumption in the cell.

9.2 Methods for preparing temperature-responsive cell culture surfaces (TRCSs) and their characteristics

Temperature-responsive polymers are grafted on the surface of various kinds of materials for expressing a TRCS character by using various methods as listed in Table 9.1. Polymer-grafting procedures for preparing TRCS are mainly divided into three types: covalently bonding, physically, and electrostatically surface coating modifications. The first is methodologically further categorized into the following three different approaches: (1) "grafting by irradiation," (2) "grafting-to base material," and (3) "grafting-from base material (by a surface-initiated living radical polymerization)" (Table 9.1) (Barbey et al., 2009; Genzer & Bhat, 2008; Nagase, Kobayashi, & Okano, 2009). The second method is performed by spin coating or drop casting methods (Table 9.2) (Morikawa & Matsuda, 2002; Rollason, Davies, & Sefton, 1993; Takezawa, Mori, & Yoshizato, 1990). Recently, a semi-interpenetrating polymer network–based PIPAAm modification, which may be a kind of physical coating method, was also applied to the preparation of TRCS (Reddy, Kano, Maruyama, Hadano, & Takahara, 2008, 2009; Wang, Chow, Phan, Lim, & Yang, 2006; Yang, Yang, Lin, & Chen, 2007). The third method is the construction of a polyelectrolyte multilayer by sequentially and alternatively coating thermoresponsive polymer that has cationic or anionic units (Table 9.3) (Liao, Moussallem, Kim, Schlenoff, & Ma, 2010).

As substrates for the deposition of temperature-responsive polymers, commercially available tissue culture polystyrene (TCPS) (Yamada et al., 1990), polystyrene (PSt) (Della

Table 9.1 **Covalently polymer bonding surface modifications**

Polymer deposition procedure	Subcategory of the procedure	Preparations and characteristics of temperature-responsive cell culture surfaces
Grafting by irradiation	Mutual radiation	Polymerization and polymer grafting to substrate surface is initiated by electron beam, plasma, gamma ray, and ultraviolet irradiations in the presence of monomer (and photoreactive compounds).
	Activation of the substrate surface by preirradiation Peroxide formation on the substrate surface	After activating substrate surface or forming peroxide units on the surface by plasma, gamma ray, or laser irradiation, polymerization and polymer grafting to the surfaces is initiated with heating or ultraviolet irradiation in the presence of monomer
Grafting-to base material		Presynthesized polymer chains as intended is grafted via one terminal or multiple points of the polymer.
Grafting-from base material	Atom transfer radical polymerization Reversible addition-fragmentation chain transfer radical polymerization	Grafting polymer brush by living radical polymerization. Length and density of resultant graft polymer brush are extremely uniform and dense.

Volpe, Cassinelli, & Morra, 1998; Morra & Cassinelli, 1997), polyethyrene terephthalate (Cheng, Wang, Hanein, Bohringer, & Ratner, 2004; Kikuchi, Okuhara, Karikusa, Sakurai, & Okano, 1998), glass (Fukumori et al., 2010, 2009; Nagase, Watanabe, Kikuchi, Yamato, & Okano, 2011; Takahashi, Nakayama, Yamato, & Okano, 2010), silicon (Li, Zhu, Li, & Gao, 2008; Sui et al., 2011; Xu, Zhong, Yung, Kang, & Neoh, 2004), gold (Kessel et al., 2010; Wischerhoff et al., 2008), polydimethylsiloxane (PDMS) (Lee & Von Recum, 2010; Ma, Chen, Shi, Li, & Wang, 2009; Shi et al., 2010), and polyurethane derivatives (Reddy et al., 2008, 2009; Wang et al., 2006; Yang et al., 2007) have been frequently used. This section discusses the characteristics of these TRCSs as well as their preparation methods.

Table 9.2 **Polymer-coated surface modifications**

Approach for deposition procedure	Subcategory of the procedure	Preparations and characteristics of temperature-responsive cell culture surfaces
Spin coating or drop casting	Coating polymer (homo-polymer or copolymer) Coating mixture of polymer (homo-polymer or copolymer) and extracellular matrix	The polymer or mixture of polymer and extracellular matrix solutions is physically coated on substrate surfaces. Coated polymer is eluted from the surfaces during cell detachment.

Table 9.3 **Electrostatically polymer-coated surface**

Approach for procedure	Subcategory of the procedure	Preparations and characteristics of temperature-responsive cell culture surfaces
Drop casting	An alternate layer-by-layer assembling method	Polyelectrolyte multilayer is constructed by alternating positive and negative charged thermoresponsive polymers. Negatively charged thermoresponsive micro-gels are electrostatically adsorbed on the substrate surface with positive charge.

9.2.1 Preparation of TRCSs by covalent polymer bonding surface modification

9.2.1.1 Grafting polymer by irradiation procedure

Grafting polymer by irradiation procedure has been used for the modification of biomaterial surfaces (Stannett, 1990). As irradiation sources, EB (Yamada et al., 1990), plasma (Pan, Wesley, Luginbuhl, Denton, & Ratner, 2001), gamma ray (Biazar et al., 2010), and ultraviolet (UV) (Biazar, Khorasani, & Daliri, 2011; Morra & Cassinelli, 1997) are often used for preparing TRCS. These irradiation procedures can be subcategorized into (1) mutual radiation (Pan et al., 2001; Yamada et al., 1990), (2) preirradiation (Biazar et al., 2010, 2011; Della Volpe et al., 1998), and (3) peroxide (Chen, Ito, & Imanishi, 1997; Rayatpisheh, Li, & Chan-Park, 2012) formation approaches. Mutual radiation includes grafting polymer by EB, plasma, and UV irradiation. In the case of UV irradiation, photoinitiator or photocrosslinker moieties are required for initiating polymerization and grafting. In the preirradiation and peroxide approaches, polymerization and grafting are initiated by the immersion of polymeric-substrate surfaces, which was chemically activated by plasma or UV irradiation, into the monomer solution. Characteristic of TRCS by those subcategorized approaches are discussed in the following sections.

9.2.1.2 TRCSs prepared by the electron beam and plasma irradiation methods

As mutual grafting method has developed, EB irradiation or plasma-induced grafting and polymerization are often used for preparing TRCSs. Most initial descriptions of TRCS are found in a patent filed in 1989 by Okano et al. The patent shows the preparation of TRCS, PIPAAm-grafted TCPS surfaces by using EB irradiation, and the temperature-dependent cell attachment and detachment character of TRCS surface (Okano et al.). TRCS is prepared as follows: N-isopropylacrylamide (IPAAm) monomers dissolved in 2-propanol solution are uniformly spread on TCPS surfaces, and the dish is immediately subjected to EB irradiation. During EB irradiation, polymerization, grafting, and cross-linking among polymer molecules occur simultaneously. PIPAAm-grafted TCPS is washed with cold water to remove the unreacted monomer and ungrafted polymers. After being dried, PIPAAm-grated TCPS is obtained as TRCS. The authors' first scientific report after the patent shows that (1) the rate of rat primary hepatocyte adhesion on TRCS is quite similar to that on TCPS at 37 °C and (2) the adhered cells are detached and efficiently recovered from the TRCS surface by lowering temperatures without trypsin treatment or chelating agents such as EDTA, which are conventionally used for cell recovery from TCPS (Yamada et al., 1990). The optimal temperature is dependent on the cultured cell species (Okano et al., 1993, 1995). For example, the optimal temperature for recovering the adhered hepatocytes and endothelial cells were found to be 10 °C and 20 °C, respectively. Confluently cultured cells are allowed to detach themselves from the TRCS surface as a contiguous monolayer cell sheet, which keeps its intact cell–cell junctions and extracellular matrix (ECM) deposited at the basal site of the cell sheet by only lowering the temperature (Kushida et al., 1999, 2000; Kushida, Yamato, Kikuchi, & Okano, 2001). Temperature-induced cell and cell-sheet recovery processes are noninvasive for cultured cells and cell sheets, whereas cell recovery process with trypsin and chelating agents destroys the cell–cell junction and ECM components.

Their further investigations have revealed a cell detachment mechanism from TRCS by using various inhibitors (Okano et al., 1995; Yamato et al., 1999). The cell adhesion process consists of two steps. The first step is termed passive adhesion, which represents cells physicochemically attaching to the substrate. The adhesion process moves to active adhesion as the second step. Cell attachment could trigger the cells to spread on the substrate, consuming ATP molecules for cellular metabolism. By lowering the temperature, TRCS hydrates to become hydrophilic and initiates a cell detachment process. ATP molecules are also required for this process, because they are consumed for intercellular signal transduction and the reorganization of their cytoskeletons to change their shapes.

Unique properties found in TRCS prepared by EB irradiation are grafted PIPAAm thickness and density, which are observed to affect the temperature-induced cell attachment and detachment character of TRCS (Akiyama, Kikuchi, Yamato, & Okano, 2004; Fukumori et al., 2009, 2010; Sakai, Okano, Yamada, & Sakurai, 1996). This polymer thickness dependency is illustrated in Figure 9.2. At the interface of basal hydrophobic TCPS surface, grafted PIPAAm chains are dynamically restricted, highly aggregated, and dehydrated because of the hydrophobic TCPS substrate. When PIPAAm layer is below 20 nm in thickness, the highly aggregated and dehydrated PIPAAm chains are promote dehydratoin PIPAAm chains at the outermost surfaces, which allows PIPAAm surfaces to be cell adhesive and lose its temperature-induced cell detachment character. In the surfaces with more than a 20-nm thicknesses, the aggregation and dehydration of the basal PIPAAm chains are unable to effectively reach and affect PIPAAm chains in the outermost surfaces region because of the thicker PIPAAm layer. The resulting TRCS shows a cell-repellent character. For TRCS with an optimal PIPAAm thickness (around 20 nm), such aggregation and dehydration adequately dehydrate PIPAAm chains at the outermost surface, which favorably express a temperature-induced cell attachment and detachment ability as TRCS. Other than basal TCPS, a polymer thickness dependency is also observed in PIPAAm-grafted coverslips (Fukumori et al., 2009, 2010). The PIPAAm-grafted coverslips with an 8.8-nm PIPAAm layer at a dry state shows a cell-repellent character and a larger temperature induced polymer thickness alteration under aqueous condition in comparison with that with a 3.3-nm polymer layer. This phenomena thickness dependency on cell adhesion character is strongly supported. Swelling of the thinner PIPAAm layer is strongly suppressed. Other than PIPAAm, poly(vinyl methyl ether) is also used for preparing TRCS by grafting on PSt surfaces with EB irradiation (Gramm et al., 2011).

As a next-generation TRCS prepared by EB irradiation, TRCS with rapid cell detachment character (Hyeong Kwon, Kikuchi, Yamato, & Okano, 2003; Kwon, Kikuchi, Yamato, Sakurai, & Okano, 2000; Tang, Akiyama, Yamato, & Okano, 2010)

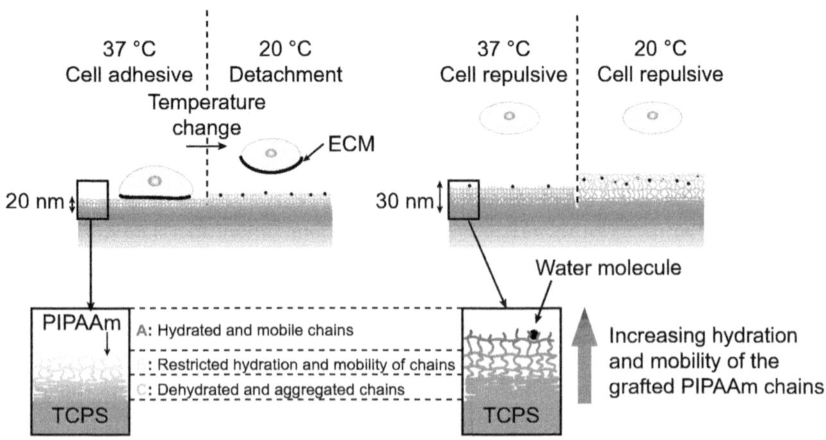

Figure 9.2 Schematic drawing of the graft polymer thickness dependency of hydration and mobility of the graft polymer chains and cell adhesive character of TRCS.

and TRCS enabling cell culture in serum-free conditions (Ebara et al., 2004a,b) are newly developed. Incorporating the hydrophilic unit, polyethyleneglycol (PEG), 2-carboxyisopropylacrylamide, or PIPAAm macromonomer into a PIPAAm layer of TRCS is used for acceleration of the hydration of grafted PIPAAm chains and subsequent detachment of a cell sheet. Human umbilical vein endothelial cells spread and are cultured on TRCS modified with RDGS cell adhesion without serum.

Ratner et al. have used a plasma polymerization method as an alternative approach to prepare TRCS (Canavan, Cheng, Graham, Ratner, & Castner, 2005a,b, 2007; Cheng et al., 2004, 2005; Galperin, Long, & Ratner, 2010) and has intensively characterized the TRCS before and after the cell sheet recovery process. For the plasma polymerization method, in a low vacuum and vaporized IPAAm monomer atmosphere, high plasma power is employed to form an adhesion-promoting layer. The layer is necessary to graft subsequent functional polymer deposition, which is carried out at successively reduced plasma power (Pan et al., 2001). Their TRCSs show a temperature-induced bovine aortic endothelial cell (BAEC) sheet recovery, which is similar to that of TRCS prepared by the EB irradiation method (Canavan et al., 2005a; Cheng et al., 2004). The principal component analysis of their TRCS surfaces after spontaneous cell sheet recovery using time-of-flight secondary ion mass spectrometry showed that laminin and bovine serum albumin components still remain on the TRCS surfaces (Canavan et al., 2005a,b, 2007). Immunoassay of the surfaces show that collagen component also remains on TRCS, although the immunoassay reveal that the fibronectin component is associated with the recovered cell sheet and hardly remained on TRCS (Canavan et al., 2005a,b, 2007). The characteristic of fibronectin recovery is also in consistent with a cell sheet recovered from EB-irradiated TRCS (Canavan et al., 2005a,b, 2007). However, TRCS prepared by a plasma polymerization barely shows a polymer thickness dependency on the following cell adhesion event as observed in EB-irradiated TRCS (Cheng et al., 2005). This is probably due to the polymer thickness variation of the deposited adhesion-promoting layer by the initial plasma polymerization as described previously. Their plasma polymerization method further attains the successful modification of polyethyreneterephtarate (Cheng et al., 2004), PDMS (Tunc, Humayun, Cheng, & Ratner, 2008), and poly (monochloro-*p*-xyllene) (Tunc, Cheng, Ratner, Meng, & Humayun, 2007; Tunc et al., 2009) substrates with PIPAAm as TRCS.

9.2.1.3 TRCSs prepared by a photo-initiator or photo-cross-linker–induced polymer grafting method

PIPAAm-grafted TCPS prepared by EB irradiation is now commercially available, which has been released with the product name UpCell®, from Thermo Fisher Science Technology, Waltham, MA, USA. However, as some researchers pointed out (Nash et al., 2012; Von Recum et al., 1999; Park et al., 2010; Moran, Carroll, Gorelov, & Rochev, 2007a; Nash et al., 2011; Mukundan, Nirmal, Thomas, Sajeev, & Nair, 2011; Reed et al., 2010; Patel, Cavicchia, Zhang, & Zhang Newby, 2012), the convenient preparation of TRCS has been a goal of researchers in the average laboratory without expensive EB and plasma equipments. As an alternative approach, UV irradiation is preferred as a convenient preparation method for TRCSs, because it is inexpensive and easily assembled from commercial available components (e.g., a UV lamp, filter).

Morra et al. first used a UV irradiation method (Morra & Cassinelli, 1997; Von Recum et al., 1999) to graft PIPAAm on PSt substrate. IPAAm monomer and benzophenone (BP), a photoinitiator, is dissolved in 2-propanol solution and subjected to UV irradiation to prepare PIPAAm-based TRCS. PIPAAm-grafted PSt surfaces reveal TRCS by an L-929 mouse fibroblast cell attachment and detachment assay.

PIPAAm-based copolymers containing photoreactive units such as cinnamoyl, azidophenyl, and benzophenone are used to fabricate TRCSs by UV irradiation methods. Recum et al. have prepared P(IPAAm-*co*-4-(*N*-cinnamoylcarbamide)methylstyrene containing cynnamoyl groups as a photo-activating cross-linker (Von Recum et al., 1998). To our knowledge, their report is the first one pointing out the importance of preparing TRCS without expensive EB equipment. The copolymers dissolved in solvent are added to a polystyrene dish, which becomes soft with the solvent, partially entrapping the copolymers. The entrapped copolymers are cross-linked to form the copolymer hydrogel partially incorporated in the polystyrene surfaces by UV irradiation. They have showed a temperature-induced BAEC attachment and detachment behavior on the prepared TRCS.

Ito et al. have used a thermoresponsive polymer containing azidophenyl groups as a photocrosslinker unit for preparing TRCS (Chen, Imanishi, & Ito, 1998; Ito, Chen, Guan, & Imanishi, 1997). The copolymer-coated polystyrene surfaces is irradiated by UV through a patterned photomask. The resultant surface has a patterned thermoresponsive polymer-grafted surface, because the copolymers are covalently attached to only the UV-irradiated region. They have demonstrated that the copolymer-grafted surfaces are TRCSs. Mouse fibroblast STO cells successfully adhere to the copolymer grafted region at 37 °C, and the cells can detach themselves under a lower temperature. In a similar way, Nash et al. report that TCPS modified with PIPAAm, which is prepared by a photo-crosslinkable PIPAAm-containing benzophenone unit, shows the characteristic of TRCS (Nash et al., 2012). The TRCS demonstrates a successful 3T3 cell sheet harvest and a graft-polymer-thickness dependency as described previously.

PDMS is a polymeric material widely used in chemical, biological, and medical fields because of its advantageous characters: low cost, biocompatibility, availability, oxygen permeability, and convenient processibility (Zhou, Ellis, & Voelcker, 2010; Zhou, Khadakov, Ellis, & Voelcker 2012). Highly hydrophobic PDMS surface is difficult to alter to be hydrophilic because of the migration and rearrangement of uncured PDMS oligomer (Zhou et al., 2010), although a hydrophilically modified PDMS surface is desired for biomedical applications. Since Hu et al. developed a direct photo-polymerization procedure from BP preadsorbed PDMS surface (Hu et al., 2004), their direct polymer grafting method has been preferably used for preparing PDMS-substrate based TRCS (Lin et al., 2012; Ma et al., 2009), because their method is convenient for modifying PDMS surface with PIPAAm without a special chemistry. Vascular smooth muscle cells and osteoblast cells are reported to be recovered as a contiguous cell sheet from PDMS-substrate based TRCS. Ma et al. have found that optimal BP preadsorption treatment is required after UV-initiated polymer grafting, because the diffusion of the adsorbed BP molecules into the inside of PDMS substrate depends on the thickness of PDMS and affects resulting properties of the grafted PIPAAm surfaces. Lin et al. have showed that the PIPAAm thickness of their PDMS-based TRCS is higher than that of PIPAAm-grafted TCPS by EB irradiation (Lin et al., 2012). They have further demonstrated an aligned vascular smooth-muscle cell sheet form microtextured PDMS-based TRCS.

Recently, thermoresponsive star polymers have been reported to be applied to TRCS preparation by using a UV-irradiated polymer grafting method (Park et al., 2010). The star polymer consists of thermoresponsive units, 2-(2-methoxyethoxy) ethyl methacrylate and oligo(ethylene oxide)methacrylate segments, and benzophenone photoreactive unit, at the internal and peripheral location of the star polymer, respectively. After the star polymers were physically deposited on TCPS surface, the star polymers are mutually cross-linked and covalently grafted on TCPS surfaces by UV irradiation. When the star polymer deposited on TRCS is 31 nm in thickness, cell attachment and detachment is satisfactory. However, TRCS having a thicker polymer layer with a thickness of 39–53 nm shows a poor cell attachment character.

9.2.1.4 TRCSs prepared by using preactivated surface methods

Gas plasma, gamma ray, and laser treatments allow polymeric surfaces to form radical and peroxide moieties, which can graft polymer on the surface in the presence of monomer (Ikada, 1994; Ratner, 1995). Few cases include both initiator and monomer for initiating polymerization and graft the polymer on the activated surfaces. Morra et al. have obtained TRCS by grafting PIPAAm on PSt surfaces activated with air plasma, where alcoholic hydroxyl groups formed, by using ammonium cerium nitrate induced polymerization chemistry (Della Volpe et al., 1998). Prepared TRCS shows a temperature-induced L-929cell attachment and detachment character. Similarly, Biazar et al. have prepared TRCSs by grafting PIPAAm onto PSt substrate, which is preactivated by gamma rays (Biazar et al., 2010). Fibroblast and epithelial cells are recovered as a cell sheet from the TRCS surfaces by lowering temperature. Rayatpisheh et al. have fabricated micropatterned PDMS-based TRCS by an argon plasma–activated polymerization technique (Rayatpisheh et al., 2012). They have described that the plasma-activated polymerization is less expensive and has more accessible approaches than EB irradiation. After immersing the activated surfaces into IPAAm monomer solution, PIPAAm is readily grafted on the PDMS surfaces by UV irradiation or heating treatment. Human aortic smooth muscle cell attachment and detachment behavior is observed on the TRCS. In addition, cells in a human aortic smooth muscle cell sheet recovered from the TRCS are aligned.

9.2.2 Grafting polymer by "grafting-to" methods

Grafting-to methods have been widely used for grafting various polymers on material surfaces. In general, polymers having a functional group at the terminals of the polymer chains are tethered to substrate surfaces at one end of the polymer chains through the reaction of the complementary functional groups of the substrate, whereas polymers with the plural number of functional groups are immobilized on the substrate at the multipoint of the polymers chains.

9.2.2.1 TRCSs prepared by "grafting-to" methods

To our knowledge, there are few reports describing the preparation of temperature-responsive cell-culture surfaces by using polymer grafting-to methods. Lee et al. have prepared TRCS by conjugating a copolymer, P(IPAAm-*co*-PAAc) activated with *N*, *N′*-dicyclohexylcarbodiimide, to an amine-bonded elastic silicone membrane, and found that 3T3 cells adhere and detach themselves from the polymer-conjugated surfaces at a stretched and nonstretched states (Lee & Von Recum, 2010). By constantly applying a mechanical stretching force to the polymer-conjugated and aminated membranes during cell culture, the cells align with the stretching direction.

Other types of temperature-responsive polymers are also used for preparing TRCS by a grafting-to method. Higuchi et al. have prepared Pluronic®-based TRCS, individually grafting Pluronic F68 and F127, which are terminally modified with carbonyldiimidazole groups, on a poly-L-lysine–coated tissue culture flask (Higuchi et al., 2006). Pluronic F68 and F127 are commercially available nonionic triblock copolymers consisting of a central hydrophobic polypropylene oxide segment, which is flanked by two hydrophilic polyethylene oxide segments, exhibiting a thermoresponsive property. Their uses in food and pharmaceutical areas are approved by the Food and Drug Administration. The polymers reversibly alter their molecular forms from a soluble form to a liquid crystal form in aqueous solution around 30 °C without showing phase separation above LCTS like PIPAAm in an aqueous solution. Temperature-induced L929 cell attachment and detachment is observed on Pluronic-modified surfaces by optimizing the amount of the graft polymers. However, the cells are unable to be recovered as a cell sheet as observed in conventional PIPAAm-grafted TCPS because of the weak cell–cell interaction.

9.2.3 Grafting polymer by "grafting-from" methods

The advantage of the grafting-to method is experimentally quite simple, and the polymers with molecular weight intended are immobilized, whereas the disadvantage of this method is difficult to achieve the fabrication of surfaces having a high and dense polymer graft density, because the preimmobilized and physically adsorbed polymers sterically interfere the further reaction between the functional groups of the polymers and the complementary functional groups on the substrate.

By contrast, in the case of conventional grafting-from method, azo-initiator–immobilized or vinylized surfaces are used for allowing polymers to grow from the surfaces with a conventional radical polymerization for producing functional polymer brush. However, it is difficult to regulate the graft polymer chain length.

To overcome these problems, a surface-initiated living radical polymerization technique is used for preparing polymer brush surfaces, where precisely controlled polymer chains are densely tethered. The technique also can be used to construct block copolymer on the surfaces by sequentially grafting another polymer via the active polymer-end groups of a first-grafted polymer brush. As a typical surface-initiated living radical polymerization, atom transfer radical polymerization (ATRP) and reversible addition-fragmentation chain transfer radical (RAFT) polymerization are reported to be used for preparing TRCS. Their polymerization processes as well as the unique character of the polymer brush surfaces by the surface initiated polymerization methods are discussed in excellent review articles (Barbey et al., 2009; Edmondson, Osborne, & Huck, 2004; Genzer & Bhat, 2008).

9.2.3.1 TRCSs prepared by "grafting-from" methods

Several groups have reported the fabrication of polymer brush–based TRCS by using the ATRP or RAFT polymerization method. Xu et al. have prepared PIPAAm brush–based TRCS on silicon by using the ATRP method and observed 3T3 Swiss albino cells attachment/detachment behavior by temperature change (Xu et al., 2004, 2006). The graft polymer layer is 31 nm in thicknesses, and a thicker PIPAAm layer increases the number of adhering cells. This polymer thickness–dependent cell attachment and detachment character is different from that of TRCS prepared by EB irradiation. The authors further evaluate temperature-modulated cell attachment/detachment character on two types of diblock copolymer brush surfaces comprising of poly (ethylene glycol monomethacrylate) (PEGMA) and PIPAAm composition. One is a PEGMA-*b*-PIPAAm brush (PIPAAm is sequentially grafted on the end of the first-grafted PEGMA brush) and another is PIPAAm-*b*-PEGMA brush surface. PEGMA-*b*-PIPAAm brush-grafted surface successfully shows a temperature-induced cell attachment/detachment character. However, cells failed to adhere to PIPAAm-*b*-PEGMA brush-grafted surfaces, where the PEGMA component is exposed to the outermost side, even at 37 °C. The results demonstrate that the active polymer-end groups can be used for further polymer grafting.

Li et al. have fabricated a PIPAAm brush surface with a gradient in polymer thickness on silicon by the ATRP method and investigated the influence of the PIPAAm brush thickness on the cell attachment and detachment event (Li et al., 2008). For preparing a gradient in polymer brush length, a silicon wafer immobilized with ATPR initiator is gradually and vertically immersed into a reaction vessel, whereas the reaction solution is continuously added to precede the reaction. As a result, PIPAAm brush surfaces with a various polymer thickness from 5 to 80 nm are fabricated on the silicon surfaces. When the polymer thickness is between 20 and 45 nm, satisfactory HepG2 cell attachment and detachment character is observed. However, adhered cells are unable to detach themselves sufficiently from a surface having less than 20 nm polymer thickness, and the cells fail to adhere to a surface having more than 45 nm polymer thicknesses.

Mizutani et al. and Nagase et al. have investigated the influence of graft polymer thickness and polymer graft density on the cell attachment/detachment character of their polymer brush based TRCS by ATRP method, in terms of molecular dynamics of the graft polymer chains (Mizutani, Kikuchi, Yamato, Kanazawa, & Okano, 2008; Nagase et al., 2011). A high densely grafted polymer brush surface with short-chain length shows a temperature-induced cell attachment/detachment character, because the mobility of polymer chains is relatively restricted and dehydrated. Decreasing the density of short polymer brush to the middle density, resulting polymer brush surface is much more cell adhesive. However, the TRCS fails to show cell detachment character even below LCST, because probably basal surfaces character is exposed to the outer surfaces. In contrast, by increasing the length of polymer brush chain at a middle and higher polymer graft density from the short polymer chain length at those densities, the cell adhesive character of resulting polymer brush surfaces is altered to be poor. Eventually, the TRCS shows a cell repellent character, probably because the outermost polymer chains are induced to be more movable and more hydrated. In addition, longer PIPAAm chains may also cancel the exposed hydrophobic properties of basal surface.

Takahashi et al. have prepared glass surfaces modified with PIPAAm brush by using a surface-initiated RAFT polymerization method and successfully obtain polymer brush–based TRCS by optimizing both the graft PIPAAm density and its chain length (Table 9.4) (Takahashi et al., 2010). PIPAAm brushed TRCS having longer graft polymer chains at a middle polymer graft density (0.03 chains/nm^2) and having middle length of graft polymer chains at a high density of 0.04 chains/nm^2 show a sufficient cell proliferation. BAEC sheet was successfully recovered from those two types of TRCSs. TRCS having longer graft polymer chains at a high density shows a poor cell adhesive character, whereas that having shorter graft polymer chains at a low density is unable to show a cell recovery character even by lowering temperature. In addition, they have showed that the conversion of the terminal dithiobenzoate groups of TRCS to hydrophobic or hydrophilic groups also affected its cell detachment property (Takahashi et al., 2012). The conversion technique is useful to recover cell sheets that are hardly to be detached. In addition, the technique might also be applicable to functionalize TRCS having bioactive molecules for enhancing the proliferation or culture of cells without serum.

After these three reports, the detail correlation among PIPAAm brush density, its polymer brush thickness, and the resulting cell attachment and detachment character has been investigated (Table 9.4) (Choi, Choi, Xue, & Leckband, 2013; Sui et al., 2011; Xue, Choi, Choi, Braun, & Leckband, 2012).

Sui et al. have prepared polymer brush on grafted silicon or glass surfaces with various polymer graft densities by using a surface-initiated ATRP method, and demonstrated that a polymer brush surfaces with 0.03 PIPAAm chains/nm^2 can give a sufficient cell adhesive and detachable character as TRCS (Sui et al., 2011). MC-3T3 cell sheet is successfully fabricated on the TRCS. Grafted polymer thickness of the surfaces at a dried state is determined to be 9 nm by ellipsometric measurement. Polymer brush surfaces with higher polymer graft densities—0.11 and 0.30 chains/nm^2—show poor cell adhesive properties. Their polymer thicknesses are estimated to be 100 nm for 0.11 chains/nm^2 and 220 nm for 0.30 chains/nm^2.

Xue et al. have investigated the effect of graft PIPAAm brush density on the resulting cell adhesion character in terms of the number and occupied area of adhered cells, and the circularity of the cells, and they find the correlation of the polymer graft density with cell attachment and detachment character (Choi et al., 2012; Xue et al., 2012). They prepare TRCS on a self-assembled monolayer formed silicon surface, where various polymer densities are grafted by using surfaces initiated ATRP method. On TRCS with a dense polymer brush (0.11 chains/nm^2), a lower circularity is found in attached cells having an extremely poor cell adhesive character. When the polymer

Table 9.4 Correlation of character of PIPAAm brush-grafted surfaces with the molecular weight and density of the graft polymer chains

Graft polymer chain density (chains/nm^2)	Molecular weight of the graft polymer chains (Mn)	Thickness of the grafted PIPAAm layer at dried state (nm)[a]	Character of the PIPAAm-grafted surface	References
0.04[b]	58000[c]	ND	TRCS[b]	Takahashi et al. (2010)
	49000[c]	ND	TRCS[b]	
	23000[c]	ND	TRCS[b]	
0.03[b]	58000[c]	ND	TRCS[b]	
	49000[c]	ND	Cell-adhesive TRCS[b]	
	23000[c]	ND	Cell-adhesive TRCS[b]	
0.02[b]	58000[c]	ND	Cell-adhesive TRCS[b]	
	49000[c]	ND	Cell-adhesive surface[b]	
	23000[c]	ND	Cell-adhesive surface[b]	
0.30[d,e]	ND	220[d]	TRCS[a]	Sui et al. (2011)
0.11[d,e]	ND	100[d]	Poor cell-adhesive TRCS[a]	
0.03[d,e]	ND	9.0[d]	Poor cell-adhesive TRCS[a]	
0.21[f,g,i]	38500[h]	14.1[f]	Extremely poor cell-adhesive TRCS[f]	Xue et al. (2012)
0.11[f,g,i]	44000[h]	8.5[f]	Extremely poor cell-adhesive TRCS[f]	
0.09[f,g,i]	48000[h]	7.5[f]	Poor cell adhesive TRCS[f]	
0.04[f,g,i]	44000[h]	3.1[f]	TRCS[f]	
0.0105[f,g,i]	27000[h]	5.0[f]	Cell-adhesive surface[f]	

ND, not determined; PIPAAm, poly(*N*-isopropylacrylamide); TRCS, temperature-responsive cell culture surfaces.
[a]Determined by ellipsometric measurement.
[b]Glass modified with a carboxylated azo-initiator V-501 and hexyltriethoxysilane was used as basal substrate.
[c]Ungrafted polymers produced in the RAFT process.
[d]Silicon modified with 2-bromo-2-methylpropionyl bromide and 1-bromocarbonyl-1-methyl acetate was used as basal substrate.
[e]Estimated from the Alexander de Gennes model.
[f]Silicon modified with 11-(2-bromo-2-methyl)-propionyl undecylate trichlorosilane and 2-[methoxy(polyethyleneoxy)propyl]-trichlorosilane was used as a basal substrate.
[g]Density was calculated from the amount of immobilized initiator units.
[h]Calculated from XPS data, packing density of the initiator, and a chain initiation efficiency.
[i]Estimated from polymer thickness, chain density, density of dry IPAAm, and the Avogador's number.

brush density decreased to be 0.09 chains/nm^2, cell adhesive property is improved as suggested by a result that the area and circularity of attached cells are larger and lower, respectively. The molecular weights of these graft polymer chains are relatively similar. On TRCSs with diluted PIPAAm brush densities (0.04 and 0.0015 chains/nm^2), the adherent cell population becomes significantly larger with maintaining a lower circularity. In other words, cells can adhere densely in a compact area. By decreasing temperature below LCST, an increase in the cell circularity value as well as a decrease in the density of the cell attachment area are observed, indicating cell detachment from the TRCSs. They also demonstrate that the cell adhesion character is positively correlated with fibronectin and bovine serum albumin adsorption on the polymer graft chain. This result supports a model that the adsorption of protein by penetrating inside the polymer brush layer contributes to reversible cell attachment and detachment, when the graft polymer density of TRCS is from relatively lower to middle (0.0015–0.09 chains/nm^2) (Choi et al., 2012). Such correlation is similarly observed in polystyrene microbeads modified with a PIPAAm brush, which is applied to cell suspension culture carrier (Tamura, Kobayashi, Yamato, & Okano, 2012b).

From these reports as described previously, PIPAAm polymer brush surfaces with a density of around 0.04 PIPAAm chains/nm^2 and relatively thinner PIPAAm layer seem to be optimal for expressing a temperature-induced cell attachment and detachment character. The optimizations of graft polymer density and graft polymer chain length are expected to be useful to produce TRCS for culturing various cell species.

Polyethylene glycol–based temperature-responsive polymer were also applicable to the preparation of polymer brush–based TRCS, as reported by Kessel et al. (2010) and Wischerhoff et al. (2008).

9.2.4 Polymer coating method for preparing TRCS

The polymer coating method is categorized into physical and electrostatic polymer coating approaches (Tables 9.2 and 9.3). The former is further divided into two different polymer coating approaches, and they are (1) directly coating temperature-responsive polymer component on base substrate and (2) coating a mixture of temperature-responsive polymer and ECM components on base substrate (Table 9.2). The drop-casting method (polymer solution is dropped and spread on the substrate surfaces, and then the surfaces is dried at atmospheric or under vacuum conditions) or spin-coating method is used in the first two polymer coating approaches. Advantages of coating method are similar to that of UV light irradiation method as described previously. In other words, the method is easy to perform without expensive equipments such as EB and plasma irradiation for preparing TRCS. Furthermore, thermoresponsive polymers are able to be deposited as TRCS on various substrates, such as metal oxide, and inorganic and organic surfaces, by this polymer coating method. In contrast, an electrostatic polymer coating approach is also used for preparing TRCS (Table 9.3). Physical and electrostatic polymer coating approaches and characteristic of TRCS by these approaches are described in the following section.

9.2.4.1 TRCS prepared by physical polymer coating approach

At the early stage of the preparation of polymer coating–based TRCS, a mixture solution of PIPAAm and ECM is used. Takezawa et al. have coated polystyrene surfaces with a mixed solution at 4 °C, and the coated surface is dried at 10 °C for several hours (Takezawa et al., 1990). Human dermal fibroblasts sufficiently adhere to the polymer and collagen-coated surfaces, and proliferate to confluency. However, the cells are unable to adhere to pure PIPAAm surfaces, and cells adhering to collagen-coated surfaces are unable to detach themselves below LCST. This different cell adherent character indicates that the collagen component works for cell adhesion and proliferation.

Matsuda et al. have prepared TRCSs by coating a mixture of PIPAAm graft–copolymerized gelatin and PIPAAm solutions onto polystyrene surfaces (Morikawa & Matsuda, 2002). The coating and drying procedures are similar to the method described previously. Although only PIPAAm-coated surface shows a cell repellent character, which is similar to the character discussed previously, coating the optimal amounts of PIPAAm-grafted gelatin and PIPAAm on PSt surfaces express the characteristic of TRCS. They further attempt to prepare TRCS on the inner surface of a glass capillary tube with the mixed solution. BAECs adhere and proliferate itself to form a monolayer on the inner surface of glass capillary. The cell monolayer detaches itself as a contiguous tubular cell sheet, which is structurally similar to a vascular tissue, by lowering temperature (Takamizawa, Shoda, & Matsuda, 2002). The basal part of the tubular cell sheet has ECM with a fibrous mesh structure, which is made by the cultured cells. The results suggest that the polymer coating method is useful for preparing TRCS having a complicate shape and fabricating cell sheets having a complicate shape.

Moran et al. have also prepared TRCS with PIPAAm copolymer–containing hydrophobic *N*-tert-butylacrylamide units and ECM (Moran et al., 2007a; Moran, Carroll, Selezneva, Gorelov, & Rochev, 2007b). They sequentially coat the copolymer and ECM (collagen, laminin, or fibronectin) or poly-L-lysine as a cell adhesion promoter after initial coating with the copolymer on TCPS or Thermonox coverslips. Double polymeric layer–coated surfaces exhibit a TRCS character, which allows human umbilical vein endothelial cells and 3T3 cells to be harvested as contiguous cell sheets by lowering temperature. However, the cell detachment mechanism is different from that proposed by Okano et al. because the coated copolymer hydrates to dissolve into cell culture medium below LCST of the copolymer. Namely, it can be called the dissolution-induced cell detachment.

TRCS with the dissolution-induced cell detachment character is reported by Rollason et al. (Rollason et al., 1993) after Okano's and Takezawa's reports (Takezawa et al., 1990; Yamada et al., 1990). Recently, a few reports point out more convenient fabrication of cell sheet from the coated polymer dissolution-based TRCS (Mukundan et al., 2011; Nash et al., 2011; Varghese, Raj, Sreenivasan, & Kumary, 2010). Nash et al. have showed that various cell lines are able to be harvested from PIPAAm homopolymer-coated Thermonox disk surfaces. Cell adhesive character on a PIPAAm homopolymer–coated disk surface is independent from coated polymer layer thickness. Cell detachment mechanism described in these reports is speculated to be due to the dissolution of coated polymers, as mentioned previously, by Rollason et al. (1993).

Ito et al. have a used triblock copolymer, which contains hydrophobic poly[(R)-3-hydroxybutyrate] (PHB) sandwiched between two hydrophilic PIPAAm units (PIPAAm-PHB-PIPAAm), and preparing TRCS by a coating method for expecting that hydrophobic

PHB unit in the copolymer serves to enhance the copolymer adsorption because of the hydrophobic property (Loh, Cheong, Li, & Ito, 2009a,b). Molecular dynamics simulation suggests that the copolymer molecules dissolved in organic solvent are self-assemble to form micelle. When the micelle deposits on hydrophobic surface, the hydrophobic cores of the copolymers in the micelle aggregate and anchor to the hydrophobic surfaces through hydrophobic interaction. Hydrophilic PIPAAm segments can be entangled without the attachment of the hydrophobic surface. The PIPAAm chains entirely cover the hydrophobic surface. The shape of the deposited micelle is similar to a shuttlecock used in the game of badminton. Molecular dynamic simulation also suggests that the hydrophobic core and hydrophilic PIPAAm segments of the deposited micelle corresponds to a base and feather parts of the shuttlecock, respectively. The interaction between the substrate surfaces and the polymer solution is an important factor, which dominates the homogeneity of resulting polymer-coated surfaces (Loh et al., 2009a). Washing the polymer-coated surface with water revealed that more than 90% of the coated polymer still remained on the surfaces, whereas, in the case of PIPAAm homopolymer–coated surface, more than 80% of the PIPAAm is eluted. The triblock copolymer–coated surfaces are much more stable than PIPAAm-coated surfaces, and the hydrophobic component contributes to the stability of the coated polymer. Cell attachment character is also influenced by the amount of the coated polymer. Optimal amount and the thickness of the coated copolymer are 0.566 μg/cm^2 and 4–8 nm, respectively. They demonstrate temperature-induced MC3T3-E1 osteoblast and human mesenchymal stem cells detachments from the optimized copolymer-coated TRCS. They also further apply the copolymer and gelatin–coated surfaces to culture and recover the mouse embryonic stem cells maintaining their pluripotency (Loh et al., 2009b).

Although the polymer coating–based TRCS is inexpensive and convenient method as described above, there are some issues related to contamination by (1) the desorption of the coated polymer during cell culture and/or by lowering temperature as mentioned above and (2) difficulty in fabricating cell sheet without a defect. The issues may be problems in cell sheet-based tissue engineering.

Recently, Nakayama et al. have prepared TRCS by using molecularly designed diblock copolymer comprising of PIPAAm and hydrophobic poly(butyl methacrylate) (PBMA) segments (PBMA-*b*-PIPAAm) (Nakayama et al., 2012). The block-copolymer prepared by RAFT polymerization is coated on commercially available TCPS by a spin-coating method (Figure 9.3). Under an aqueous condition, the desorption of coated block polymer is not observed after decreasing the temperature below LCTS, whereas the significant decrease of PIPAAm form the PIPAAm-coated TCPS is observed. The PBMA, a hydrophobic segment, works as a stable anchor connecting the block-copolymer to hydrophobic TCPS surfaces and prevents desorption of the coated polymer. The block-copolymer–coated TCPS surface worked as TRCS, which can exhibit a temperature-induced cell attachment and detachment character, when the graft polymer density and thickness are 1.44 μg/cm^2 and 15.4 nm, respectively. As observed in TRCS prepared by the EB irradiation method, the characteristics of the block-copolymer–coated TRCS is dependent on both the polymer graft density and thickness. BAECs cultured and proliferated to confluency on the block-copolymer–coated TRCS are harvested as a contiguous cell sheet without defects by lowering temperature.

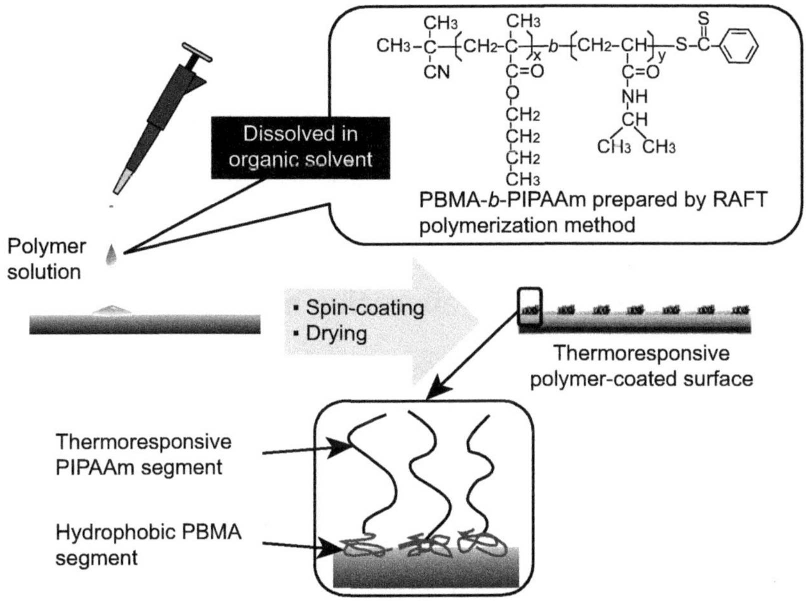

Figure 9.3 Schematic drawing of preparation of PBMA-*b*-PIPAAm–coated (TRCS).

9.2.4.2 TRCS prepared by electrostatically polymer coating approach

Liao et al. have successfully prepared TRCS by multiple and alternate layer-by-layer assembling with two different PIPAAm copolymers; one contains cationic units and another contains anionic units (Liao et al., 2010). Allylamine hydrochloride and styrene sulfonic acid are used as cationic and anionic components in cationic and anionic PIPAAm copolymers, respectively. TRCS is prepared by alternately coating cationic and anionic PIPAAm copolymers on glass coverslip surface. Bone marrow–derived human mesenchymal stem cells (hMSCs) adhere more efficiently to and proliferate on the polymeric multilayered TCRS with terminally cationic copolymer–coated than simple fetal bovine serum (FBS)-coated glass coverslip surfaces, because the cationic moiety probably promotes fibronectin adsorption as well as cell adhesion in comparison with the anionic moiety. After reaching confluency, hMSCs are harvested as a contiguous cell sheet by lowering temperature. The harvested hMSC sheet shows a higher colony-forming unit than that of hMSC harvested from an FBS-coated surface by a conventional trypsin digestion method, suggesting that their intact ECM possibly contribute to the preservation of the stemness in the primitive stage and the maintenance of their multilinage potential. TRCS comprising thermoresponsive polyelectrolyte layers may be useful for the expansion of hMSC *in vitro*.

Schmidt et al. have successfully developed TRCS by electrostatically coating PIPAAm micro-gel on poly(ethyleneimine)-coated glass coverslip surfaces (Schmidt et al., 2010). The micro-gel has an anionic character that is derived from potassium persulfate used an initiator agent. The PIPAAm micro-gel is electrostatically adsorbed on a cationic glass surface. Detailed AFM observation and the investigation of adsorbed micro-gels show that, at 37 °C, poor water content and smaller elastic modulus of the adsorbed micro-gel allow cells to adhere and to spread through a stable focal adhesion. At 25 °C, adsorbed micro-gels increase water content and remarkably decrease the modulus, leading to cell detachment. Micro-gel–adsorbed glass coverslip surface shows an L929 mouse fibroblast cells attachment and detachment behavior by temperature change.

9.3 Cell sheet-based regenerative medicine

Cell sheets harvested from TRCSs are actively applied to the fabrication of artificial tissue and organ *in vitro* and *in vitro*, and the regenerations of the damaged tissue and organ by cell-sheet transplantation procedures. Only TRCSs by EB irradiation are now commercially available and are widely used in the cell sheet-based tissue-engineering field, although various types of TRCSs can be constructed by using various methods, as described previously. As shown in Figure 9.4, TRCSs provide many successful examples for fabricating cell sheets from various types of cells, such as epithelium keratinocyte (Yamato et al., 2001), corneal epithelial cell sheets (Nishida et al., 2004a,b), cardiomyocyte cell sheet (Shimizu et al., 2002a,b; Shimizu, Yamato, Kikuchi, & Okano, 2003), oral mucosal epithelial cell sheet (Murakami et al., 2006a,b; Nishida et al., 2004b; Ohki et al., 2006; Takagi et al., 2010; Watanabe, Yamato, Shiroyanagi, Tanabe, & Okano, 2011), cells sheet from periodontal ligament (Akizuki et al., 2005; Hasegawa, Yamato, Kikuchi, Okano, & Ishikawa, 2005; Iwata et al., 2009, 2010), and hepatocyte cell sheets (Harimoto et al., 2002; Hirose et al., 2000; Ohashi et al., 2007). The transplantations of these cell sheets are carried out without the need for sutures. In this section, examples regarding cell sheet engineering toward clinical application are briefly described.

Corneal epithelial cell sheets are transplanted for reconstructing the damaged corneal surfaces of patients who suffered severe traumas (thermal or chemical burns) or eye diseases (Steve-Johnson syndrome and ocular pemphigoid), which induce limbal stem cell deficiency with the loss of vision. In the case of unilateral limbal stem cell deficiency, corneal cell sheets derived from autologous epithelial cells taken from the contralateral eye are used for the reconstruction (Nishida et al., 2004a). However, in severe clinical cases, patients lost complete bilateral corneal cells. Autologous oral mucosal cells, which are similar to native corneal epithelium cells, are used for preparing cell sheets, and the cell sheets are applied to the treatment of damaged corneas (Burillon et al., 2012; Nishida et al., 2004b). Very recently, detailed clinical trial results suggested that cultured autologous oral mucosal epithelial cell sheet is well tolerated and recognized as a safe tissue-engineered product (Burillon et al., 2012).

Oral mucosal cell sheets are also clinically applied to the treatment of esophageal ulcerations using an endoscopic transplantation method (Kanai, Yamato, Ohki, Yamamoto, & Okano, 2012; Ohki et al., 2006, 2012). Removal of large esophageal cancer tissue section by endoscopic submucosal dissection causes postoperative inflammation and stenosis from a lack of an epithelial layer. After the transplantation

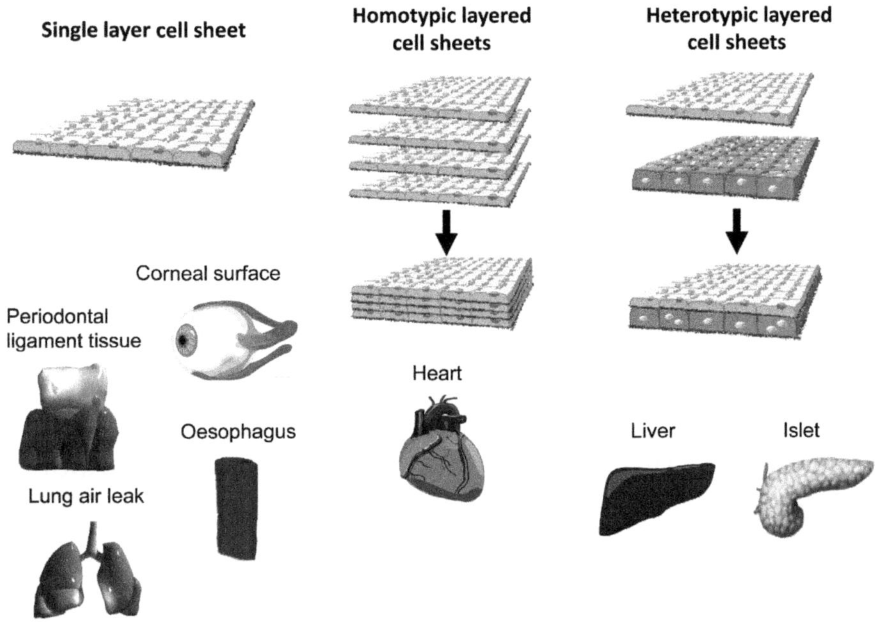

Figure 9.4 Illustration of cell sheet-based tissue engineering. (Left) Damaged cornea, esophagus, periodontal ligament tissues, and air-leaked lung are treated by the transplantation of single layer cell sheet; (middle) three-dimensional thick tissues such as myocardium are constructed by layering homotypic cell sheets; and (right) tissues and organs with a complex structure and physiological function such as liver and islet are reconstructed by layering heterotypic cell sheets.

of oral mucosal cell sheets to the ulcer site of the patients, reepithelization is successfully induced and the wound healing is promoted. The esophageal luminal surface of patients is reconstructed, preventing inflammation and stenosis (Ohki et al., 2012).

Periodontal ligament tissue plays important roles in surrounding and supporting the tooth. Periodontitis is one of the most prevalent infectious diseases, characterized by the destruction of tissues, such as alveolar bone, cementum, and the periodontal ligament. Conventional treatment is applied to regenerate periodontal tissue. However, the rapid migration and proliferation of gingival epithelial cells results in the formation of a long junctional epithelium at the destroyed tissue, preventing the regeneration of periodontal tissues. As a novel method for regeneration, PLD cell sheets are employed (Akizuki et al., 2005; Hasegawa et al., 2005; Iwata et al., 2009). The cell sheets, fabricated on temperature-responsive cell culture surfaces, are transplanted to periodontal defects, and the cell sheet induces periodontal ligament tissue healing with bone, periodontal ligament, and cementum formation in a rat xenogenic model and a dog autogenic model. In a preclinical study of cell sheet-based PLD regeneration, cell sheets from human PLD cells, which are produced in a cell processing center, are safe and have the potential to regenerate human periodontal ligaments (Washio et al., 2010).

Lung air leak followed by pulmonary resection was attempted to be sealed by skin fibroblast sheets in an animal model (Kanzaki et al., 2007, 2008, 2012). In clinical use, fibrin glue and synthetic sealants are intraoperatively used as an air leak sealant. Fibrin glue is unable to achieve the immediate closure of air leak position, and the delayed closure caused postoperative recurrence because of its poor tissue bonding strength. By contrast, other biological adhesives or synthetic sealants have a poor adhesive stability or a poor biocompatibility. Suturing and stapling techniques are also used to close the air leak. However, problems such as damage to the normal lung and failure to close the air leakage still remained. By sharp contrast, in an animal model, tissue-engineered skin fibroblast sheets are easily transplanted to the wounded lungs of pigs, and the air leakage is immediately sealed by the cell sheets without suturing and stapling procedures. In addition, cell sheets are flexible in their response to contraction and relaxation during respiration. This novel sealant is expected to be an ideal cell-based sealant that can reduce surgical complication during pulmonary resections.

Neonatal rat cardiomyocyte sheets, comprising cardiomyocytes and endothelial cells harvested from temperature-responsive cell culture surfaces, are layered to construct three-dimensional (3D) myocardial tissue (Shimizu et al., 2002b, 2003). Layered cell sheets are pulsatile and cell-dense like a native myocardium and are electrically synchronized within 45 min with a functional gap junction (Haraguchi, Shimizu, Yamato, Kikuchi, & Okano, 2006). After transplantation of the layered cell sheets to subcutaneous tissue of a nude rat, neovascularization in the transplanted sheet occurred rapidly because of the presence of endothelial cells (Sekine et al., 2008; Shimizu et al., 2006a). Three weeks after the transplantation, surface electrograms originating from the sheet and host heart were detected individually (Shimizu et al., 2002a). Actually, cardiac tissue of 1-mm thickness are regenerated in subcutaneous tissue by repeating the transplantations of triple-layer cell sheets (Shimizu et al., 2006b). The cell sheets are applicable to the treatment of severe heart failure. Cardiomyocyte sheet transplantation improves cardiac performance of ischemic myocardium (Sekine et al., 2008, 2011). Similarly, layered skeletal myoblast sheet transplantation induces reduction in myocardial fibrosis and the reorganization of the cytoskeletal proteins in dilated cardiomyopathy, improving the cardiac performance (Memon et al., 2005). Very recently, Sekine et al. have engineered cardiac tissues having perfusable blood vessels *in vitro* by using resected tissue with an artery and a vein as connectable vascular bed for the cardiac tissues (Sekine et al., 2013). Their technique may promote *in vitro* tissue fabrication in the tissue engineering field.

Cell-sheet based tissue engineering has been further applied to the treatment of complexes and functional tissues and organs such as a bladder (Shiroyanagi, Yamato, Yamazaki, Toma, & Okano, 2003, 2004), liver (Ohashi et al., 2007) and islets (Shimizu et al., 2009).

Mice hepatic-layered cell sheets, which express human α1-antitrypsin, are transplanted into mice subcutaneous locations with neovascular networks (Ohashi et al., 2007). After transplantation, liver-specific functions such as albumin production and drug metabolism are improved and kept for longer than 200 days. In addition, the transplanted cell sheet has an ability to proliferate and grow in response to a regenerative stimulus, which is induced by the resection of two-thirds of the host liver (Ohashi et al., 2007).

Recently, double-layered 3D cocultured cell sheets comprising human umbilical vein endothelial cells and rat hepatocytes cells at the upper and bottom sides, respectively, are prepared, and the cell sheets are similar to the native liver structure and functionality (Kim, Ohashi, Utoh, Kano, & Okano, 2012). The double-layered cell sheets form a hepatic-plate like microstructure, which has a hepatic perisinusoidal space-like structure between endothelial cell and rat hepatocytes cell sheets, and bile canaliculi expressing claudin 3 proteins, which are observed at the outline of the canaliculi in the normal liver. Hepatocyte-specific functions such as albumin and urea production and bile secretion are more actively preserved in contrast to a hepatocyte monolayer cell sheet. The cell sheets based tissue may be a new strategy for creating liver-targeted regenerative medicine as well as a miniaturized liver-based chip.

Rat islet cell sheets harvested from laminin-5 coated conventional PIPAAm-TCPS have secretory granules, which can secrete glucagon (Shimizu et al., 2009). In addition, the islet cell sheets show a glucose-dependent insulin secretion. The granules and the production of glucagon and insulin are the characteristics of native islet cells. After transplantation of the double-layered islets cell sheets to the subcutaneous space of severe immunodeficiency mice with streptozotocin-induced diabetic combined with the nonfasting blood glucose levels of the mice, a state of normoglycemia decreases within 1 week, and the normal state is maintained for 110 days (Saito et al., 2011). Histological and immunohistochemical analyses of the transplanted cell sheets demonstrate an intense vascular network formation within and around the islet cell sheets. C-peptide produced by the cleavage of proinsulin is also detected in the serum of only islet cell sheet-transplanted mice, indicating that insulin is secreted from the transplanted islet cell sheets. The islets cell sheet may be a new approach for treating diabetes mellitus.

9.4 Future application of temperature-responsive cell culture surface to support and promote regenerative medicine field

Besides fabricating a cell sheet, TRCSs have recently been applied to a cell separation and sorting system, which is an important technology for the regenerative medicine field. In this section, those applications, which are becoming an important technology in the regenerative medicine field in the future, will be discussed.

The scale-up or expansion of cell production is one of the important issues in the regenerative medicine field, because a huge amount of cells is required for the cell-based treatment of damaged human tissues and organs. However, a conventional cell anchorage T-flask approach is ineffective for the cell expansion and requires high labor intensity. Instead of the inefficient anchorage method for expanding target cells easily, a microcarrier method has been developed (Glacken, Fleischaker, & Sinskey, 1983; Van Wezel, 1967). To date, microcarriers are widely used as a suspension culture support for expanding various types of anchorage-dependent cells because of its lager surface area-to-volume ratio in contrast to a conventional two-dimensional cell culture system (Martin, Eldardiri, Lawrence-Watt, & Sharpe, 2011). Microcarrier-based suspension culture methods have been industrially applied to the production of valuable bioengineered products such as vaccine, enzyme, hormones, and cytokines. In contrast, the method is also expected to give the sufficient number of therapeutically useful cells for regenerative medicine, such as bone marrow–derived mesenchymal stem cells, cells differentiated from embryonic stem cells, and induced pluripotential stem cells. As described previously, however, a conventional trypsin-mediated cell recovery method for passage culture or scale-up is a drawback of the method. Some researchers have attempted to apply the characteristic of TRCS, which can harvest cells with extremely less damage to the microcarrier system.

Miller et al. have grafted PIPAAm on a commercially available microplate support (MicorHex made of PSt) for suspension culture by EB irradiation (Kenda-Ropson, Lenglois, & Miller, 2002). CHO-K1 cells are demonstrated to adhere to PIPAAm-modified palate, and the cells detached themselves by lowering temperature in a large-scale bioreactor (1 L culture volume). The string speed of the bioreactor affects cell detachment.

Yang et al. have grafted PIPAAm on commercially available denatured collagen covalently bonded cross-linked dextran particles (Cytodex-3) by the grafting-to method (Yang, Jeon, Bhang, Lee, & Kim, 2010). Human bone marrow–derived mesenchymal stem cells successfully adhere, spread, and grow on PIPAAm-grafted Cytodex-3. Although trypsin treatment induces apoptosis and the death of recovered cells, the temperature-induced cell recovery method induces practically no apoptosis and no cell death.

Tamura et al. investigated PIPAAm brush-grafted chloromethylated polystyrene microbeads as microcarriers, which are prepared by the ATRP method (Tamura, Kobayashi, Yamato, & Okano, 2012a–c) (Figure 9.5). Their extensive investigation optimizes the size of the bead, the graft polymer chain length, and the polymer graft density. Moreover, not only the dispersibility of the PIPAAm graft microcarrier, but also the cell proliferation are improved during suspension culture by incorporating anionic moieties to the grafted PIPAAm brush. Microcarriers modified with PIPAAm copolymer brush having anionic moieties are expected to be useful for an efficient and scalable suspension cell culture.

Cell separation or cell sorting from a mixture of cells is also one of important issues in tissue engineering. Monoclonal antibodies have been conventionally used for efficient cell separation and sorting. However, the methods are unable to be applicable for cells of which their surface-specific markers are unidentified.

Matsuda et al. have reported that TRCS comprising a mixture of PIPAAm and PIPAAm-grafted gelatin is used as a new cell sorting surface that can recognize differences in the adhesion character of BAECs and smooth muscle cells (Matsuda, Saito, & Shoda, 2007). In the case of TRCS with lower content-PIPAAm-grafted gelatin, the endothelial cells preferably adhere to the TRCS. However, TRCS shows an extremely poor adhesive character for smooth muscle cells. The endothelial cells are successfully sorted and purified from a mixture of endothelial and smooth muscle cells by repeating the sequential procedure of cell detachment at room temperature and replanting cells at 37 °C resulting from a difference in the cell-adherent character between the two cells. At the third cell culture procedure, only endothelial cells were found on the TRCS, indicating that the smooth muscle cells were almost completely removed from a mixture of cells through the repeating procedure. Differences in the cell adhesive character between endothelial and smooth muscle cells may be due to differences in the number of cell-adhesion receptors on the cell membrane between

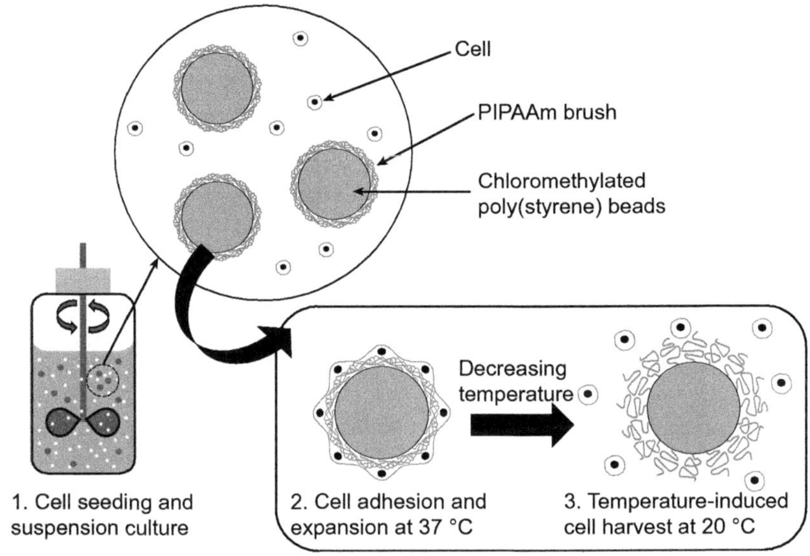

Figure 9.5 Illustration of PIPAAm brush-grafted microcarrier beads used in cell suspension in a culture system. The beads show temperature-induced cell adhesion and detachment, which is the characteristic of TRCS.

the two cells. By optimizing the mixing ratio of PIPAAm to PIPAAm-grafted gelatin, the cell adhesive property of gelatin of TRCS can be manipulated. Any kinds of cell can be sorted or purified regardless of their species by this TRCS without labeling cells with an antibody.

References

Akiyama, Y., Kikuchi, A., Yamato, M., & Okano, T. (2004). Ultrathin poly(*N*-isopropylacrylamide) grafted layer on polystyrene surfaces for cell adhesion/detachment control. *Langmuir*, *20*, 5506–5511.

Akizuki, T., Oda, S., Komaki, M., Tsuchioka, H., Kawakatsu, N., Kikuchi, A., et al. (2005). Application of periodontal ligament cell sheet for periodontal regeneration: a pilot study in beagle dogs. *Journal of Periodontal Research*, *40*, 245–251.

Barbey, R., Lavanant, L., Paripovic, D., Schuwer, N., Sugnaux, C., Tugulu, S., et al. (2009). Polymer brushes via surface-initiated controlled radical polymerization: synthesis, characterization, properties, and applications. *Chemical Reviews*, *109*, 5437–5527.

Biazar, E., Khorasani, M., & Daliri, M. (2011). Cell sheet engineering: solvent effect on nanometric grafting of poly-*N*-isopropylacrylamide onto polystyrene substrate under ultraviolet radiation. *International Journal of Nanomedicine*, *6*, 295–302.

Biazar, E., Zeinali, R., Montazeri, N., Pourshamsian, K., Behrouz, M. J., Asefnejad, A., et al. (2010). Cell engineering: nanometric grafting of poly-*N*-isopropylacrylamide onto polystyrene film by different doses of gamma radiation. *International Journal of Nanomedicine*, *5*, 549–556.

Burillon, C., Huot, L., Justin, V., Nataf, S., Chapuis, F., Decullier, E., et al. (2012). Cultured autologous oral mucosal epithelial cell sheet (CAOMECS) transplantation for the treatment of corneal limbal epithelial stem cell deficiency. *Investigative Ophthalmology and Visual Science*, *53*, 1325–1331.

Canavan, H. E., Cheng, X., Graham, D. J., Ratner, B. D., & Castner, D. G. (2005a). Cell sheet detachment affects the extracellular matrix: a surface science study comparing thermal liftoff, enzymatic, and mechanical methods. *Journal of Biomedical Materials Research Part A*, *75*, 1–13.

Canavan, H. E., Cheng, X., Graham, D. J., Ratner, B. D., & Castner, D. G. (2005b). Surface characterization of the extracellular matrix remaining after cell detachment from a thermo-responsive polymer. *Langmuir*, *21*, 1949–1955.

Canavan, H. E., Graham, D. J., Cheng, X., Ratner, B. D., & Castner, D. G. (2007). Comparison of native extracellular matrix with adsorbed protein films using secondary ion mass spectrometry. *Langmuir*, *23*, 50–56.

Cheng, X., Canavan, H. E., Stein, M. J., Hull, J. R., Kweskin, S. J., Wagner, M. S., et al. (2005). Surface chemical and mechanical properties of plasma-polymerized *N*-isopropylacrylamide. *Langmuir*, *21*, 7833–7841.

Cheng, X., Wang, Y., Hanein, Y., Bohringer, K. F., & Ratner, B. D. (2004). Novel cell patterning using microheater-controlled thermoresponsive plasma films. *Journal of Biomedical Materials Research Part A*, *70*, 159–168.

Chen, G., Imanishi, Y., & Ito, Y. (1998). Effect of protein and cell behavior on pattern-grafted thermoresponsive polymer. *Journal of Biomedical Materials Research*, *42*, 38–44.

Chen, G., Ito, Y., & Imanishi, Y. (1997). Regulation of growth and adhesion of cultured cells by insulin conjugated with thermoresponsive polymers. *Biotechnology and Bioengineering*, *53*, 339–344.

Choi, S., Choi, B. C., Xue, C., & Leckband, D. E. (2013). Protein adsorption mechanisms determine the efficiency of thermally controlled cell adhesion on poly (*N*-isopropylacrylamide) brushes. *Biomacromolecules*, *14*, 92–100.

Della Volpe, C., Cassinelli, C., & Morra, M. (1998). Wilhelmy plate measurements on poly(*N*-isopropylacrylamide)-grafted surfaces. *Langmuir*, *14*, 4650–4656.

Ebara, M., Yamato, M., Aoyagi, T., Kikuchi, A., Sakai, K., & Okano, T. (2004a). Immobilization of cell-adhesive peptides to temperature-responsive surfaces facilitates both serum-free cell adhesion and noninvasive cell harvest. *Tissue Engineering*, *10*, 1125–1135.

Ebara, M., Yamato, M., Aoyagi, T., Kikuchi, A., Sakai, K., & Okano, T. (2004b). Temperature-responsive cell culture surfaces enable "on-off" affinity control between cell integrins and RGDS ligands. *Biomacromolecules*, *5*, 505–510.

Edmondson, S., Osborne, V. L., & Huck, W. T. (2004). Polymer brushes via surface-initiated polymerizations. *Chemical Society Reviews*, *33*, 14–22.

Fukumori, K., Akiyama, Y., Kumashiro, Y., Kobayashi, J., Yamato, M., Sakai, K., et al. (2010). Characterization of ultra-thin temperature-responsive polymer layer and its polymer thickness dependency on cell attachment/detachment properties. *Macromolecular Bioscience*, *10*, 1117–1129.

Fukumori, K., Akiyama, Y., Yamato, M., Kobayashi, J., Sakai, K., & Okano, T. (2009). Temperature-responsive glass coverslips with an ultrathin poly(N-isopropylacrylamide) layer. *Acta Biomaterialia*, *5*, 470–476.

Galperin, A., Long, T. J., & Ratner, B. D. (2010). Degradable, thermo-sensitive poly(*N*-isopropyl acrylamide)-based scaffolds with controlled porosity for tissue engineering applications. *Biomacromolecules*, *11*, 2583–2592.

Genzer, J., & Bhat, R. R. (2008). Surface-bound soft matter gradients. *Langmuir, 24,* 2294–2317.
Glacken, M. W., Fleischaker, R. J., & Sinskey, A. J. (1983). Large-scale production of mammalian cells and their products: engineering principles and barriers to scale-up. *Annuals of the New York Academy of Sciences, 413,* 355–372.
Gramm, S., Teichman, J., Nitschlke, M., Goh, U., Eichhorn, K. J., & Werner, C. (2011). Electron beam immobilization of functionalized poly(vinyl methyl ether) thin films on polymer surfaces. Towards stimuli responsive coatings for biomedical purposes. *Express Polymer Letters, 5,* 970.
Haraguchi, Y., Shimizu, T., Yamato, M., Kikuchi, A., & Okano, T. (2006). Electrical coupling of cardiomyocyte sheets occurs rapidly via functional gap junction formation. *Biomaterials, 27,* 4765–4774.
Harimoto, M., Yamato, M., Hirose, M., Takahashi, C., Isoi, Y., Kikuchi, A., et al. (2002). Novel approach for achieving double-layered cell sheets co-culture: overlaying endothelial cell sheets onto monolayer hepatocytes utilizing temperature-responsive culture dishes. *Journal of Biomedical Materials Research, 62,* 464–470.
Hasegawa, M., Yamato, M., Kikuchi, A., Okano, T., & Ishikawa, I. (2005). Human periodontal ligament cell sheets can regenerate periodontal ligament tissue in an athymic rat model. *Tissue Engineering, 11,* 469–478.
Heskins, M., & Guillet, J. E. (1968). Solution properties of poly(*N*-isopropylacrylamide). *Journal of Macromolecular Science Part A – Chemistry, 2,* 1441–1455.
Higuchi, A., Aoki, N., Yamamoto, T., Miyazaki, T., Fukushima, H., Tak, T. M., et al. (2006). Temperature-induced cell detachment on immobilized pluronic surface. *Journal of Biomedical Materials Research Part A, 79,* 380–392.
Hirose, M., Yamato, M., Kwon, O. H., Harimoto, M., Kushida, A., Shimizu, T., et al. (2000). Temperature-Responsive surface for novel co-culture systems of hepatocytes with endothelial cells: 2-D patterned and double layered co-cultures. *Yonsei Medical Journal, 41,* 803–813.
Hu, S., Ren, X., Bachman, M., Sims, C. E., Li, G. P., & Allbritton, N. L. (2004). Surface-directed, graft polymerization within microfluidic channels. *Analytical Chemistry, 76,* 1865–1870.
Hyeong Kwon, O., Kikuchi, A., Yamato, M., & Okano, T. (2003). Accelerated cell sheet recovery by co-grafting of PEG with PIPAAm onto porous cell culture membranes. *Biomaterials, 24,* 1223–1232.
Ikada, Y. (1994). Surface modification of polymers for medical applications. *Biomaterials, 15,* 725–736.
Ito, Y., Chen, G., Guan, Y., & Imanishi, Y. (1997). Patterned immobilization of thermoresponsive polymer. *Langmuir, 13,* 2756–2759.
Iwata, T., Yamato, M., Tsuchioka, H., Takagi, R., Mukobata, S., Washio, K., et al. (2009). Periodontal regeneration with multi-layered periodontal ligament-derived cell sheets in a canine model. *Biomaterials, 30,* 2716–2723.
Iwata, T., Yamato, M., Zhang, Z., Mukobata, S., Washio, K., Ando, T., et al. (2010). Validation of human periodontal ligament-derived cells as a reliable source for cytotherapeutic use. *Journal of Clinical Periodontology, 37,* 1088–1099.
Kanai, N., Yamato, M., Ohki, T., Yamamoto, M., & Okano, T. (2012). Fabricated autologous epidermal cell sheets for the prevention of esophageal stricture after circumferential ESD in a porcine model. *Gastrointestinal Endoscopy, 76,* 873–881.
Kanzaki, M., Yamato, M., Takagi, R., Kikkawa, T., Isaka, T., Okano, T., et al. (2012). Controlled collagen crosslinking process in tissue-engineered fibroblast sheets for preventing scar contracture on the surface of lungs. *Journal of Tissue Engineering and Regenerative Medicine, 5,* 383–391.

Kanzaki, M., Yamato, M., Yang, J., Sekine, H., Kohno, C., Takagi, R., et al. (2007). Dynamic sealing of lung air leaks by the transplantation of tissue engineered cell sheets. *Biomaterials*, *28*, 4294–4302.

Kanzaki, M., Yamato, M., Yang, J., Sekine, H., Takagi, R., Isaka, T., et al. (2008). Functional closure of visceral pleural defects by autologous tissue engineered cell sheets. *European Journal of Cardiothoracic Surgery*, *34*, 864–869.

Kenda-Ropson, N., Lenglois, S., & Miller, A. O. (2002). Microsupport with two-dimensional geometry (2D-MS) 4. Temperature-induced detachment of anchorage-dependent CHO-K1 cells from cryoresponsive MicroHex((R)) (CryoHex). *Cytotechnology*, *39*, 163–170.

Kessel, S., Schmidt, S., Muller, R., Wischerhoff, E., Laschewsky, A., Lutz, J. F., et al. (2010). Thermoresponsive PEG-based polymer layers: surface characterization with AFM force measurements. *Langmuir*, *26*, 3462–3467.

Kikuchi, A., & Okano, T. (2005). Nanostructured designs of biomedical materials: applications of cell sheet engineering to functional regenerative tissues and organs. *The Journal of Controlled Release*, *101*, 69–84.

Kikuchi, A., Okuhara, M., Karikusa, F., Sakurai, Y., & Okano, T. (1998). Two-dimensional manipulation of confluently cultured vascular endothelial cells using temperature-responsive poly(*N*-isopropylacrylamide)-grafted surfaces. *Journal of Biomaterials Science Polymer Edition*, *9*, 1331–1348.

Kim, K., Ohashi, K., Utoh, R., Kano, K., & Okano, T. (2012). Preserved liver-specific functions of hepatocytes in 3D co-culture with endothelial cell sheets. *Biomaterials*, *33*, 1406–1413.

Kushida, A., Yamato, M., Kikuchi, A., & Okano, T. (2001). Two-dimensional manipulation of differentiated Madin-Darby canine kidney (MDCK) cell sheets: the noninvasive harvest from temperature-responsive culture dishes and transfer to other surfaces. *Journal of Biomedical Materials Research*, *54*, 37–46.

Kushida, A., Yamato, M., Konno, C., Kikuchi, A., Sakurai, Y., & Okano, T. (1999). Decrease in culture temperature releases monolayer endothelial cell sheets together with deposited fibronectin matrix from temperature-responsive culture surfaces. *Journal of Biomedical Materials Research*, *45*, 355–362.

Kushida, A., Yamato, M., Konno, C., Kikuchi, A., Sakurai, Y., & Okano, T. (2000). Temperature-responsive culture dishes allow nonenzymatic harvest of differentiated Madin-Darby canine kidney (MDCK) cell sheets. *Journal of Biomedical Materials Research*, *51*, 216–223.

Kwon, O. H., Kikuchi, A., Yamato, M., Sakurai, Y., & Okano, T. (2000). Rapid cell sheet detachment from poly(*N*-isopropylacrylamide)-grafted porous cell culture membranes. *Journal of Biomedical Materials Research*, *50*, 82–89.

Lee, E. L., & Von Recum, H. A. (2010). Cell culture platform with mechanical conditioning and nondamaging cellular detachment. *Journal of Biomedical Materials Research Part A*, *93*, 411–418.

Liao, T., Moussallem, M. D., Kim, J., Schlenoff, J. B., & Ma, T. (2010). *N*-isopropylacrylamide-based thermoresponsive polyelectrolyte multilayer films for human mesenchymal stem cell expansion. *Biotechnology Progress*, *26*, 1705–1713.

Lin, J. B., Isenberg, B. C., Shen, Y., Schorsch, K., Sazonova, O. V., & Wong, J. Y. (2012). Thermo-responsive poly(*N*-isopropylacrylamide) grafted onto microtextured poly(dimethylsiloxane) for aligned cell sheet engineering. *Colloids and Surfaces B: Biointerfaces*, *99*, 108–115.

Li, L., Zhu, Y., Li, B., & Gao, C. (2008). Fabrication of thermoresponsive polymer gradients for study of cell adhesion and detachment. *Langmuir*, *24*, 13632–13639.

Loh, X. J., Cheong, W. C. D., Li, J., & Ito, Y. (2009a). Novel poly(*N*-isopropylacrylamide)-poly[(R)-3-hydroxybutyrate]-poly(*N*-isopropylacrylamide) triblock copolymer surface as a culture substrate for human mesenchymal stem cells. *Soft Matter*, *5*, 2937–2946.

Loh, X. J., Gong, J., Sakuragi, M., Kitajima, T., Liu, M., Li, J., et al. (2009b). Surface coating with a thermoresponsive copolymer for the culture and non-enzymatic recovery of mouse embryonic stem cells. *Macromolecular Bioscience*, *9*, 1069–1079.

Ma, D., Chen, H., Shi, D., Li, Z., & Wang, J. (2009). Preparation and characterization of thermo-responsive PDMS surfaces grafted with poly(*N*-isopropylacrylamidc) by benzophenone-initiated photopolymerization. *Journal of Colloid and Interface Science*, *332*, 85–90.

Martin, Y., Eldardiri, M., Lawrence-Watt, D. J., & Sharpe, J. R. (2011). Microcarriers and their potential in tissue regeneration. *Tissue Engineering Part B Reviews*, *17*, 71–80.

Matsuda, T., Saito, Y., & Shoda, K. (2007). Cell sorting technique based on thermoresponsive differential cell adhesiveness. *Biomacromolecules*, *8*, 2345–2349.

Memon, I. A., Sawa, Y., Fukushima, N., Matsumiya, G., Miyagawa, S., Taketani, S., et al. (2005). Repair of impaired myocardium by means of implantation of engineered autologous myoblast sheets. *The Journal of Thoracic and Cardiovascular Surgery*, *130*, 1333–1341.

Mizutani, A., Kikuchi, A., Yamato, M., Kanazawa, H., & Okano, T. (2008). Preparation of thermoresponsive polymer brush surfaces and their interaction with cells. *Biomaterials*, *29*, 2073–2081.

Moran, M. T., Carroll, W. M., Gorelov, A., & Rochev, Y. (2007a). Intact endothelial cell sheet harvesting from thermoresponsive surfaces coated with cell adhesion promoters. *Journal of the Royal Society Interface*, *4*, 1151–1157.

Moran, M. T., Carroll, W. M., Selezneva, I., Gorelov, A., & Rochev, Y. (2007b). Cell growth and detachment from protein-coated PNIPAAm-based copolymers. *Journal of Biomedical Materials Research Part A*, *81*, 870–876.

Morikawa, N., & Matsuda, T. (2002). Thermoresponsive artificial extracellular matrix: *N*-isopropylacrylamide-graft-copolymerized gelatin. *Journal of Biomaterial Science Polymer Edition*, *13*, 167–183.

Morra, M., & Cassinelli, C. (1997). *Thermal modification of cell cultured on poly(N-isopropylacrylamide) surface grafated polystyrene dished*. New York: Springer.

Mukundan, L. M., Nirmal, R., Thomas, L. V., Sajeev, U. S., & Nair, P. D. (2011). Retrieval of rat aortic smooth muscle cells as intact cell sheet for regenerative medicine: a cost effective approach using photo polymerization. *Biotechnology Letters*, *33*, 2083–2089.

Murakami, D., Yamato, M., Nishida, K., Ohki, T., Takagi, R., Yang, J., et al. (2006a). The effect of micropores in the surface of temperature-responsive culture inserts on the fabrication of transplantable canine oral mucosal epithelial cell sheets. *Biomaterials*, *27*, 5518–5523.

Murakami, D., Yamato, M., Nishida, K., Ohki, T., Takagi, R., Yang, J., et al. (2006b). Fabrication of transplantable human oral mucosal epithelial cell sheets using temperature-responsive culture inserts without feeder layer cells. *Journal of Artificial Organs*, *9*, 185–191.

Nagase, K., Kobayashi, J., & Okano, T. (2009). Temperature-responsive intelligent interfaces for biomolecular separation and cell sheet engineering. *Journal of the Royal Society Interface*, *6*(Suppl 3), S293–S309.

Nagase, K., Watanabe, M., Kikuchi, A., Yamato, M., & Okano, T. (2011). Thermo-responsive polymer brushes as intelligent biointerfaces: preparation via ATRP and characterization. *Macromolecular Bioscience*, *11*, 400–409.

Nakayama, M., Yamada, N., Kumashiro, Y., Kanazawa, H., Yamato, M., & Okano, T. (2012). Thermoresponsive poly(*N*-isopropylacrylamide)-based block copolymer coating for optimizing cell sheet fabrication. *Macromolecular Bioscience*, *12*, 751–760.

Nash, M. E., Carroll, W. M., Foley, P. J., Maguire, G., Connell, C. O., Gorelov, A. V., et al. (2012). Ultra-thin spin coated crosslinkable hydrogels for use in cell sheet recovery-synthesis, characterisation to application. *Soft Matter*, *8*, 3889–3899.

Nash, M. E., Carroll, W. M., Nikoloskya, N., Yang, R., O'connell, C., Gorelov, A. V., et al. (2011). Straightforward, one-step fabrication of ultrathin thermoresponsive films from commercially available pNIPAm for cell culture and recovery. *ACS Applied Materials and Interfaces*, *3*, 1980–1990.

Nishida, K., Yamato, M., Hayashida, Y., Watanabe, K., Maeda, N., Watanabe, H., et al. (2004a). Functional bioengineered corneal epithelial sheet grafts from corneal stem cells expanded ex vivo on a temperature-responsive cell culture surface. *Transplantation*, *77*, 379–385.

Nishida, K., Yamato, M., Hayashida, Y., Watanabe, K., Yamamoto, K., Adachi, E., et al. (2004b). Corneal reconstruction with tissue-engineered cell sheets composed of autologous oral mucosal epithelium. *New England Journal of Medicine*, *351*, 1187–1196.

Ohashi, K., Yokoyama, T., Yamato, M., Kuge, H., Kanehiro, H., Tsutsumi, M., et al. (2007). Engineering functional two- and three-dimensional liver systems in vivo using hepatic tissue sheets. *Nature Medicine*, *13*, 880–885.

Ohki, T., Yamato, M., Murakami, D., Takagi, R., Yang, J., Namiki, H., et al. (2006). Treatment of oesophageal ulcerations using endoscopic transplantation of tissue-engineered autologous oral mucosal epithelial cell sheets in a canine model. *Gut*, *55*, 1704–1710.

Ohki, T., Yamato, M., Ota, M., Takagi, R., Murakami, D., Kondo, M., et al. (2012). Prevention of esophageal stricture after endoscopic submucosal dissection using tissue-engineered cell sheets. *Gastroenterology*, *143*, 582–588 e1–2.

Okano, T., Kataoka, K., Yamada, N., Sakurai, Y., Amiya, T., & Mamada, A. (1989). Bed material for cell culture. Japane Patent Application JP, TOKUGANHEI, 1–31844.

Okano, T., Yamada, N., Okuhara, M., Sakai, H., & Sakurai, Y. (1995). Mechanism of cell detachment from temperature-modulated, hydrophilic-hydrophobic polymer surfaces. *Biomaterials*, *16*, 297–303.

Okano, T., Yamada, N., Sakai, H., & Sakurai, Y. (1993). A novel recovery system for cultured cells using plasma-treated polystyrene dishes grafted with poly(*N*-isopropylacrylamide). *Journal of Biomedical Materials Research*, *27*, 1243–1251.

Pan, Y. V., Wesley, R. A., Luginbuhl, R., Denton, D. D., & Ratner, B. D. (2001). Plasma polymerized *N*-isopropylacrylamide: synthesis and characterization of a smart thermally responsive coating. *Biomacromolecules*, *2*, 32–36.

Park, S., Cho, H. Y., Yoon, J. A., Kwak, Y., Srinivasan, A., Hollinger, J. O., et al. (2010). Photo-cross-linkable thermoresponsive star polymers designed for control of cell-surface interactions. *Biomacromolecules*, *11*, 2647–2652.

Patel, N. G., Cavicchia, J. P., Zhang, G., & Zhang Newby, B. M. (2012). Rapid cell sheet detachment using spin-coated pNIPAAm films retained on surfaces by an aminopropyltriethoxysilane network. *Acta Biomaterials*, *8*, 2559–2567.

Ratner, B. D. (1995). Surface modification of polymers: chemical, biological and surface analytical challenges. *Biosensors and Bioelectronics*, *10*, 797–804.

Rayatpisheh, S., Li, P., & Chan-Park, M. B. (2012). Argon-plasma-induced ultrathin thermal grafting of thermoresponsive pNIPAm coating for contractile patterned human SMC sheet engineering. *Macromolecular Bioscience*, *12*, 937–945.

Reddy, T. T., Kano, A., Maruyama, A., Hadano, M., & Takahara, A. (2008). Thermosensitive transparent semi-interpenetrating polymer networks for wound dressing and cell adhesion control. *Biomacromolecules*, *9*, 1313–1321.

Reddy, T. T., Kano, A., Maruyama, A., Hadano, M., & Takahara, A. (2009). Synthesis and characterization of semi-interpenetrating polymer networks based on polyurethane and *N*-isopropylacrylamide for wound dressing. *Journal of Biomedical Materials Research Part B: Applied Biomaterials*, *88*, 32–40.

Reed, J. A., Lucero, A. E., Hu, S., Ista, L. K., Bore, M. T., Lopez, G. P., et al. (2010). A low-cost, rapid deposition method for "smart" films: applications in mammalian cell release. *ACS Applied Materials & Interfaces*, *2*, 1048–1051.

Rollason, G., Davies, J. E., & Sefton, M. V. (1993). Preliminary report on cell culture on a thermally reversible copolymer. *Biomaterials*, *14*, 153–155.

Saito, T., Ohashi, K., Utoh, R., Shimizu, H., Ise, K., Suzuki, H., et al. (2011). Reversal of diabetes by the creation of neo-islet tissues into a subcutaneous site using islet cell sheets. *Transplantation*, *92*, 1231–1236.

Sakai, H., Okano, T., Yamada, N., & Sakurai, Y. (1996). Thermoresponsive polymer surface for cell culture: analysis of the surface and control of the cell attachment/detachment. In N. Ogata, S. Kim, J. Feijen, & T. Okano (Eds.), *Advanced biomaterials in biomedical engineering and drug delivery systems*. Tokyo: Springer.

Schmidt, S., Zeiser, M., Hellweg, T., Duschl, C., Fery, A., Hwald, M., et al. (2010). Adhesion and mechanical properties of PNIPAM microgel films and their potential use as switchable cell culture substrates. *Advanced Functional Materials*, *20*, 3235–3243.

Sekine, H., Shimizu, T., Dobashi, I., Matsuura, K., Hagiwara, N., Takahashi, M., et al. (2011). Cardiac cell sheet transplantation improves damaged heart function via superior cell survival in comparison with dissociated cell injection. *Tissue Engineering Part A*, *17*, 2973–2980.

Sekine, H., Shimizu, T., Hobo, K., Sekiya, S., Yang, J., Yamato, M., et al. (2008). Endothelial cell coculture within tissue-engineered cardiomyocyte sheets enhances neovascularization and improves cardiac function of ischemic hearts. *Circulation*, *118*, S145–S152.

Sekine, H., Shimizu, T., Sakaguchi, K., Dobashi, I., Wada, M., Yamato, M., et al. (2013). In vitro fabrication of functional three-dimensional tissues with perfusable blood vessels. *Nature Communications*, *4*, 1399.

Shi, D., Ma, D., Dong, F., Zong, C., Liu, L., Shen, D., et al. (2010). Proliferation and multi-differentiation potentials of human mesenchymal stem cells on thermoresponsive PDMS surfaces grafted with PNIPAAm. *Bioscience Reports*, *30*, 149–158.

Shimizu, H., Ohashi, K., Utoh, R., Ise, K., Gotoh, M., Yamato, M., et al. (2009). Bioengineering of a functional sheet of islet cells for the treatment of diabetes mellitus. *Biomaterials*, *30*, 5943–5949.

Shimizu, T., Sekine, H., Isoi, Y., Yamato, M., Kikuchi, A., & Okano, T. (2006a). Long-term survival and growth of pulsatile myocardial tissue grafts engineered by the layering of cardiomyocyte sheets. *Tissue Engineering*, *12*, 499–507.

Shimizu, T., Sekine, H., Yang, J., Isoi, Y., Yamato, M., Kikuchi, A., et al. (2006b). Polysurgery of cell sheet grafts overcomes diffusion limits to produce thick, vascularized myocardial tissues. *The FASEB Journal*, *6*, 708–710.

Shimizu, T., Yamato, M., Akutsu, T., Shibata, T., Isoi, Y., Kikuchi, A., et al. (2002a). Electrically communicating three-dimensional cardiac tissue mimic fabricated by layered cultured cardiomyocyte sheets. *Journal of Biomedical Materials Research*, *60*, 110–117.

Shimizu, T., Yamato, M., Isoi, Y., Akutsu, T., Setomaru, T., Abe, K., et al. (2002b). Fabrication of pulsatile cardiac tissue grafts using a novel 3-dimensional cell sheet manipulation technique and temperature-responsive cell culture surfaces. *Circulation Research*, *90*, e40.

Shimizu, T., Yamato, M., Kikuchi, A., & Okano, T. (2003). Cell sheet engineering for myocardial tissue reconstruction. *Biomaterials*, *24*, 2309–2316.

Shiroyanagi, Y., Yamato, M., Yamazaki, Y., Toma, H., & Okano, T. (2003). Transplantable urothelial cell sheets harvested noninvasively from temperature-responsive culture surfaces by reducing temperature. *Tissue Engineering*, *9*, 1005–1012.

Shiroyanagi, Y., Yamato, M., Yamazaki, Y., Toma, H., & Okano, T. (2004). Urothelium regeneration using viable cultured urothelial cell sheets grafted on demucosalized gastric flaps. *BJU International*, *93*, 1069–1075.

Stannett, V. T. (1990). Radiation grafting — state-of-the-art. *International Journal of Radiation Applications and Instrumentation. Part C. Radiation Physics and Chemistry, 35*, 82–87.

Sui, X., Di Luca, A., Gunnewiek, M. K., Kooij, E. S., Van Blitterswijk, C. A., Moroni, L., et al. (2011). Stability and cell adhesion properties of poly(*N*-isopropylacrylamide) brushes with variable grafting densities. *Australian Journal of Chemistry, 64*, 1261–1268.

Takagi, R., Murakami, D., Kondo, M., Ohki, T., Sasaki, R., Mizutani, M., et al. (2010). Fabrication of human oral mucosal epithelial cell sheets for treatment of esophageal ulceration by endoscopic submucosal dissection. *Gastrointestinal Endoscopy, 72*, 1253–1259.

Takahashi, H., Matsuzaka, N., Nakayama, M., Kikuchi, A., Yamato, M., & Okano, T. (2012). Terminally functionalized thermoresponsive polymer brushes for simultaneously promoting cell adhesion and cell sheet harvest. *Biomacromolecules, 13*, 253–260.

Takahashi, H., Nakayama, M., Yamato, M., & Okano, T. (2010). Controlled chain length and graft density of thermoresponsive polymer brushes for optimizing cell sheet harvest. *Biomacromolecules, 11*, 1991–1999.

Takamizawa, K., Shoda, K., & Matsuda, T. (2002). Pull-out mechanical measurement of tissue-substrate adhesive strength: endothelial cell monolayer sheet formed on a thermoresponsive gelatin layer. *Journal of Biomaterials Science Polymer Edition, 13*, 81–94.

Takei, Y. G., Aoki, T., Sanui, K., Ogata, N., Sakurai, Y., & Okano, T. (1994). Dynamic contact angle measurement of temperature-responsive surface properties for poly(*N*-isopropylacrylamide) grafted surfaces. *Macromolecules, 27*, 6163–6166.

Takezawa, T., Mori, Y., & Yoshizato, K. (1990). Cell culture on a thermo-responsive polymer surface. *Nature Biotechnology, 8*, 854–856.

Tamura, A., Kobayashi, J., Yamato, M., & Okano, T. (2012a). Temperature-responsive poly(*N*-isopropylacrylamide)-grafted microcarriers for large-scale non-invasive harvest of anchorage-dependent cells. *Biomaterials, 33*, 3803–3812.

Tamura, A., Kobayashi, J., Yamato, M., & Okano, T. (2012b). Thermally responsive microcarriers with optimal poly(*N*-isopropylacrylamide) grafted density for facilitating cell adhesion/detachment in suspension culture. *Acta Biomaterilia, 8*, 3904–3913.

Tamura, A., Nishi, M., Kobayashi, J., Nagase, K., Yajima, H., Yamato, M., et al. (2012c). Simultaneous enhancement of cell proliferation and thermally induced harvest efficiency based on temperature-responsive cationic copolymer-grafted microcarriers. *Biomacromolecules, 13*, 1765–1773.

Tang, Z., Akiyama, Y., Yamato, M., & Okano, T. (2010). Comb-type grafted poly(*N*-isopropylacrylamide) gel modified surfaces for rapid detachment of cell sheet. *Biomaterials, 31*, 7435–7443.

Tunc, M., Cheng, X., Ratner, B. D., Meng, E., & Humayun, M. (2007). Reversible thermosensitive glue for retinal implants. *Retina, 27*, 938–942.

Tunc, M., Humayun, M., Cheng, X., & Ratner, B. D. (2008). A reversible thermosensitive adhesive for retinal implants: in vivo experience with plasma-deposited poly(*N*-isopropyl acrylamide). *Retina, 28*, 1338–1343.

Tunc, M., Yildirim, U., Yuksel, H., Cheng, X., Humayun, M., & Ratner, B. (2009). Conjunctival impression cytology by using a thermosensitive adhesive: polymerized *N*-isopropyl acrylamide. *Cornea, 28*, 770–773. http://dx.doi.org/10.1097/ICO.0b013e318196706d.

Van Wezel, A. L. (1967). Growth of cell-strains and primary cells on micro-carriers in homogeneous culture. *Nature, 216*, 64–65.

Varghese, V. M., Raj, V., Sreenivasan, K., & Kumary, T. V. (2010). In vitro cytocompatibility evaluation of a thermoresponsive NIPAAm-MMA copolymeric surface using L929 cells. *Journal of Materials Science: Materials in Medicine, 21*, 1631–1639.

Von Recum, H., Kikuchi, A., Yamato, M., Sakurai, Y., Okano, T., & Kim, S. W. (1999). Growth factor and matrix molecules preserve cell function on thermally responsive culture surfaces. *Tissue Engineering*, *5*, 251–265.

Von Recum, H. A., Kim, S. W., Kikuchi, A., Okuhara, M., Sakurai, Y., & Okano, T. (1998). Novel thermally reversible hydrogel as detachable cell culture substrate. *Journal of Biomedical Materials Research*, *40*, 631–639.

Wang, L. S., Chow, P. Y., Phan, T. T., Lim, I. J., & Yang, Y. Y. (2006). Fabrication and characterization of nanostructured and thermosensitive polymer membranes for wound healing and cell grafting. *Advanced Functional Materials*, *16*, 1171–1178.

Washio, K., Iwata, T., Mizutani, M., Ando, T., Yamato, M., Okano, T., et al. (2010). Assessment of cell sheets derived from human periodontal ligament cells: a pre-clinical study. *Cell and Tissue Research*, *341*, 397–404.

Watanabe, E., Yamato, M., Shiroyanagi, Y., Tanabe, K., & Okano, T. (2011). Bladder augmentation using tissue-engineered autologous oral mucosal epithelial cell sheets grafted on demucosalized gastric flaps. *Transplantation*, *91*, 700–706.

Wischerhoff, E., Uhlig, K., Lankenau, A., Borner, H. G., Laschewsky, A., Duschl, C., et al. (2008). Controlled cell adhesion on PEG-based switchable surfaces. *Angewandte Chemie International Edition in English*, *47*, 5666–5668.

Xue, C., Choi, B.-C., Choi, S., Braun, P. V., & Leckband, D. E. (2012). Protein adsorption modes determine reversible cell attachment on poly(*N*-isopropyl acrylamide) brushes. *Advanced Functional Materials*, *22*, 2394–2401.

Xu, F. J., Zhong, S. P., Yung, L. Y., Kang, E. T., & Neoh, K. G. (2004). Surface-active and stimuli-responsive polymer–Si(100) hybrids from surface-initiated atom transfer radical polymerization for control of cell adhesion. *Biomacromolecules*, *5*, 2392–2403.

Xu, F. J., Zhong, S. P., Yung, L. Y., Tong, Y. W., Kang, E. T., & Neoh, K. G. (2006). Thermoresponsive comb-shaped copolymer-Si(100) hybrids for accelerated temperature-dependent cell detachment. *Biomaterials*, *27*, 1236–1245.

Yakushiji, T., Sakai, K., Kikuchi, A., Aoyagi, T., Sakurai, Y., & Okano, T. (1998). Graft architectural effects on thermoresponsive wettability changes of poly(*N*-isopropylacrylamide)-modified surfaces. *Langmuir*, *14*, 4657–4662.

Yamada, N., Okano, T., Sakai, H., Karikusa, F., Sawasaki, Y., & Sakurai, Y. (1990). Thermo-responsive polymeric surfaces: control of attachment and detachment of cultured cells. *Die Makromolekulare Chemie Rapid Communications*, *11*, 571–576.

Yamato, M., Okuhara, M., Karikusa, F., Kikuchi, A., Sakurai, Y., & Okano, T. (1999). Signal transduction and cytoskeletal reorganization are required for cell detachment from cell culture surfaces grafted with a temperature-responsive polymer. *Journal of Biomedical Materials Research*, *44*, 44–52.

Yamato, M., Utsumi, M., Kushida, A., Konno, C., Kikuchi, A., & Okano, T. (2001). Thermo-responsive culture dishes allow the intact harvest of multilayered keratinocyte sheets without dispase by reducing temperature. *Tissue Engineering*, *7*, 473–480.

Yang, H. S., Jeon, O., Bhang, S. H., Lee, S. H., & Kim, B. S. (2010). Suspension culture of mammalian cells using thermosensitive microcarrier that allows cell detachment without proteolytic enzyme treatment. *Cell Transplantation*, *19*, 1123–1132.

Yang, J. M., Yang, S. J., Lin, H. T., & Chen, J. K. (2007). Modification of HTPB-based polyurethane with temperature-sensitive poly(*N*-isopropyl acrylamide) for biomaterial usage. *Journal of Biomedical Materials Research Part B Applied Biomaterials*, *80*, 43–51.

Zhou, J., Ellis, A. V., & Voelcker, N. H. (2010). Recent developments in PDMS surface modification for microfluidic devices. *Electrophoresis*, *31*, 2–16.

Zhou, J., Khodakov, D. A., Ellis, A. V., & Voelcker, N. H. (2012). Surface modification for PDMS-based microfluidic devices. *Electrophoresis*, *33*, 89–104.

Stimuli-sensitive polymers for drug delivery and diagnostic systems interacting with biosurfaces

10

T. Yoshida
Astellas Pharma Inc., Tsukuba-shi, Japan

10.1 Introduction

Drug delivery systems and pharmaceutical formulations are very important for effective, safe, and convenient administration of drugs. Many formulations have contributed to high efficacy of drugs in the body by improvement of water solubility of the drugs (Alani et al., 2010; Yamaguchi & Tominaga, 2007; Yamashita et al., 2003; Yoshida et al., 2012), site-specific drug release for absorption or in target sites (Katsuma, Watanabe, Kawai, Takemura, & Sako, 2006; Sawada, Kondo, Nakashima, Sako, & Hayashi, 2004), and prevention of drug degradation in the stomach or blood (Lai, Bae, Yoshida, Kataoka, & Kwon, 2010; Shimizu et al., 2003). Other formulations for high safety have also been researched to reduce blood concentration and systemic side effects (Maeda et al., 2011; Sako, Nakashima, Sawada, & Fukui, 1996; Yoshida, Nakanishi, et al., 2009), prevent gastric drug release and damage (Babish, Tripp, Howell, & Bland, 2010; Jain, Jindal, & Singh, 2005), and produce higher drug concentration in target sites than normal tissues (Adams, Lavasanifar, & Kwon, 2003; Maeda & Matsumura, 1989; Maeda, Wu, Sawa, Matsumura, & Hori, 2000). Oral disintegrating tablets (Mizumoto Tamura, Kawai, Kajiyama, & Itai, 2008a,b), and taste-masking formulations (Kurimoto et al., 2005; Tasaki, Yoshida, Maeda, Katsuma, & Sako, 2009; Yoshida et al., 2008a,b; Yoshida, Tasaki, et al., 2009), are patient-friendly and improve drug adherence.

Stimuli-responsive polymers are used in the systems and formulations for interaction with the surface of the body, organs, tissues, and cells to improve efficacy, safety, and adherence. Stimuli-responsive polymers are categorized into two groups: (1) polymers responsive to the body's internal stimuli and (2) polymers responsive to external stimuli. The first group contains pH-, temperature-, ion-, redox-, enzyme-, and biomolecule-sensitive polymers. Concentrations of various molecules (protons, ions, glutathione, enzymes, glucose, etc.) are different between tissues in the body (Table 10.1). Body temperature is higher than room temperature, except in the tropics. These changes of concentrations and temperature in the body are used for many drug delivery and diagnostic systems.

Polymers responsive to the body's external stimuli contain polymers sensitive to temperature, near-infrared light, ultrasound, magnetic fields, light, and electrics fields. These kinds of energy, which are supplied from outside of the body, are used for drug delivery systems.

Switchable and Responsive Surfaces and Materials for Biomedical Applications. http://dx.doi.org/10.1016/B978-0-85709-713-2.00010-9

Table 10.1 **PH and ionic concentration in each site in the body**

	Amount (L)	pH	Na^+ (mEq/L)	K^+ (mEq/L)	Cl^- (mEq/L)	HCO_3^- (mEq/L)
Saliva	1.5	6.0–7.0	30	20	30	15
Gastric fluid	2.5	1.0–3.5 (fasting)	50 (H^+, 90)	10	110	0
Bile	0.5	7.8	140	5	105	40
Pancreatic fluid	0.7	8.0–8.3	140	5	60	90
Small-intestinal fluid	1.5	7.5–8.0	120	5	110	35
Plasma	3	7.4	140	5	100	30
Interstitial fluid	10		150	5	110	30
Intracellular fluid	30		10	160	115	30

In this chapter, recent studies on drug delivery and diagnostic systems using stimuli-responsive polymers are summarized from the viewpoint of interactions between the systems and the surface of the body, organs, tissues, and cells. Progress and prospect of this field are discussed, and the author's brief comments on future trends are given.

10.2 Polymers sensitive to body's internal stimuli

10.2.1 pH-sensitive polymers

Many drugs interact with the surface of the tongue and the oral cavity, bind to receptors of taste buds, and induce unpleasant qualities such as bitterness, sourness, or saltiness, or cause oral numbness (Hashimoto et al., 2002; Keast & Breslin, 2005; Kumazawa, Nomura, & Kurihara, 1988; Nanda, Kandarapu, & Garg, 2002; Sugao, Yamazaki, Shiozawa, & Yano, 1998). Cationic polymers that dissolve pH-sensitively can prevent drugs from binding to the surface of the tongue and the oral cavity at pH 5.8–7.4 (Randale et al., 2010), and release drugs in the stomach at pH 1–3.5 (Table 10.1). Microspheres of aminoalkyl methacrylate copolymer (Eudragit E) containing sumatriptan succinate (Sheshala, Khan, & Darwis, 2011) and donepezil hydrochloride (Yan, Woo, Kang, Yong, & Choi, 2010) were prepared by spray-drying technique. The microspheres suppressed drug release in phosphate buffer pH 7.4 (Sheshala et al., 2011) or simulated salivary fluid (Yan et al., 2010) for 1 min, and suppressed drugs binding to the surface of the tongue and their bitter tastes. Both formulations released drugs rapidly in acidic buffers, and showed similar plasma concentration–time profiles to the marketed products (Sheshala et al., 2011, Yan et al., 2010). Pellets containing quinine sulfate (Kayumba et al., 2007) and granules containing promethazine (Haware, Chaudhari, Parakh, & Bauer-Brandl, 2008) were also prepared using

Eudragit E, and released drugs faster at low pH than at neutral pH (Kayumba et al., 2007, Haware et al., 2008). Coating of Eudragit E on particles containing atorvastatin for oral disintegrating tablets has been reported to suppress drugs binding to the surface of the tongue (Tasaki et al., 2011).

Microspheres of polyvinylacetal diethylaminoacetate (AEA) containing trimebutine maleate suppressed drug release at pH 6.8, and rapidly released drug at pH 1.2 (Hashimoto et al., 2002). A sensation test in healthy volunteers showed that the AEA microspheres suppressed drugs binding to the surface of the oral cavity and their bitter tastes (Hashimoto et al., 2002). In another work, AEA was coated on sidenafil-montmorillonite (Sdn-MMT) nanohybrid for taste-masking (Lee et al., 2012). The cationic Sdn molecules bound to the inner surface of the inorganic clay material, MMT. The interaction between Sdn and MMT was strong, and prevented drugs from being released at all pH values from the nanohybrid containing no AEA. In contrast, AEA-coated nanohybrid showed suppression of Sdn release in simulated salivary fluid, and rapid drug release at pH 1.2, because the cationic AEA at the low pH promoted dissociation of the drug from the MMT. Plasma concentration–time profiles of the system were similar to the marketed products in beagle dogs (Lee et al., 2012). These cationic polymers are widely used in many taste-masking products of bitter drugs, and very useful for high adherence of patients.

Nano-sized polymers and particles can be accumulated in the tumor tissues more efficiently than in normal tissues by enhanced permeability and retention (EPR) effect after their intravenous administration (Matsumura & Maeda, 1986). Leaky neovascular vessels in the tumors comparing to tight endothelial junctions in the normal blood vessels, and lack of lymphatic drainage in the tumors, enhance nanoparticular tumor accumulation. Various nanoparticular drug-delivery systems have been researched and validated in preclinical and clinical trials. Moreover, it has been reported that pH decreases to 6.5–7.0 in the solid tumors and inflammation sites compared to extracellular pH around 7.4 of blood and normal tissues. Polymeric mixed micelles (PHSM$^{\text{pop-up Tat}}$) were prepared with two block copolymers of poly(L-lactic acid)-b-poly(ethylene glycol)-b-poly(L-hystidine)-Tat peptide and poly(L-hystidine)-b-poly(ethylene glycol) to utilize the tumoral low pH (Lee et al., 2008, Figure 10.1(a)). The hydrophobic poly(L-lactic acid) (PLA) and the poly(L-hystidine) [p(His)] in the first and second polymers composed cores of the micelles, because p(His) had imidazole groups with pK_a around 7 and were hydrophobic at pH 7.4. The poly(ethylene glycol) (PEG) formed a shell layer surrounding the micelle for long blood circulation. The Tat peptide (GRKKRRQRRRPQ) is one of the cationic cell-penetrating peptides (CPPs) that interact with cellular membranes and facilitate cellular uptake of various cargo. To prevent the peptide from interacting with the surface of normal cells and blood proteins, the p(His) of the first polymer was hydrophobic and composed the core, and the Tat peptides were shielded by the PEG shell layers at blood pH. The p(His) changes to be cationic and hydrophilic at the tumoral acidic pH, the p(His) and the Tat peptide move onto the surface of the micelles, and the Tat peptide can interact with the surface of the tumor cells. The micelles were taken up by cultured cells much more efficiently at pH 6.8 than pH 7.4. The PHSM$^{\text{pop-up Tat}}$ loading doxorubicin showed pH-dependent drug release, about four times faster drug release at pH 6.0–6.4 than at pH 7.0–8.0. Higher cytotoxic effect of the micelles at pH 6.4–6.8 was also reported than at pH

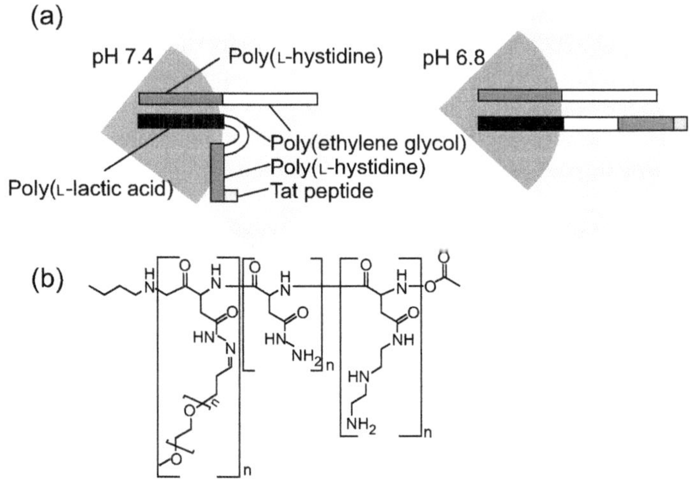

Figure 10.1 Structures of polymers and polymeric drug delivery systems sensitive to pH. (a) pH-sensitive PHSM$^{\text{pop-up Tat}}$ (Lee et al., 2008). (b) pH-sensitive ply[Asp(Hyd–PEG)]-block-p[Asp(diaminoethane)] (Lai et al., 2010).

7.2–7.4. The micellar doxorubicin showed four to nine times lower IC_{50} than free doxorubicin in drug-resistant cell lines at pH 6.8, which indicated the Tat peptide showed a pH-dependent cell-penetrating effect between the surfaces of the micelles and cells (Lee et al., 2008). The PHSM$^{\text{pop-up Tat}}$ loading doxorubicin also showed much higher efficacy against mice solid tumors than free doxorubicin (Lee et al., 2008).

Polyethyleneimines (PEIs) have many cationic charges, interact with pDNA, and adhere on the surface of the cells. However, the cationic PEIs interact with not only the target cells but also normal cells and blood proteins, and cannot show the EPR effect due to the interaction with blood vessels, the aggregation in blood, and short blood circulation. PEG modification with the PEIs is useful for long blood circulation and accumulation in the tumors by the EPR effect; however, reducing their association with the surface of tumor cells, cellular uptake, and endosomal escape, which is called the PEG dilemma. PEI–PEG conjugated via hydrazone bonds (PEI–Hyd–PEG) has been reported to solve the PEG dilemma by using tumoral acidic pH (Sawant et al., 2012). PEI–Hyd–PEG was added into polyplexes between PEI-dioleoylphosphatidylethanolamine conjugate and pDNA. Addition of PEI–Hyd–PEG lowered transfection efficiency at pH 7.4, probably caused by the PEG coating on the polyplexes. In contrast, transfection efficiency was much higher at pH 5 than at pH 7.4 (Sawant et al., 2012). This system has both advantages of PEGylation and PEIs on the surface of the particles and cells using the tumoral low pH.

Nano-sized polymers and particles adhere on the surface of tumor cells, and are taken up into tumor cells by endocytosis or macropinocytosis. In the endocytic pathway, pH gradually decreases from 7.4 (physiological pH) to 6 (endosomal pH) and to 5 (lysosomal pH). For intracellular release of pDNA, poly[Asp(Hyd–PEG)]-block-p[Asp(diaminoethane)] (Figure 10.1(b)) has been reported to show high transfection efficiency (Lai et al., 2010). The cationic diaminoethane side chains interacted with pDNA and formed nanoparticular polyplexes. The PEGs were conjugated with the polymer backbone via hydrazone bonds; therefore PEGs were more rapidly released at pH 5 than at pH 7.4. Surface of the polyplexes changed from neutral at pH 7.4 to cationic after PEG release at pH 5, and showed highly efficient transfection (Lai et al., 2010). These results were probably caused by the hydrolysis of hydrazone bonds and PEG release in the endocytic pathway enhancing interaction between the cationic polyplexes and endosomal membrane and promoting their endosomal escape. These pH-sensitive polymers are very useful for controlling the interaction between the drug delivery nanoparticles and biosurfaces, and are expected to show promising results in clinical trials in future.

10.2.2 Temperature-sensitive polymers

Body temperature 37 °C is higher than room temperature except in the tropics. Temperature-sensitive polymers have been used and reported for drug delivery responding to the temperature change between before and after administration into the body. Chitosan is the deacetylated derivative of chitin, a neutral component of shrimp and crab shells. Chitosan has low toxicity, biocompatibility, biodegradability, and adhesivity to tissues. Thermal gelation of an aqueous solution of chitosan and β-glycerophosphate at body temperature, and its medical application, have been reported (Hoemann et al., 2007; Kim et al., 2010; Ruel-Gariépy & Leroux, 2004; Shive et al., 2006). BST-CarGel® is a liquid solution of chitosan and β-glycerophosphate having physiological pH and osmolarity for cartilage repair (Ruel-Gariépy & Leroux, 2004; Shive et al., 2006). In a popular method to repair damaged cartilage, tiny fractures are created in the underlying bone by microfracture technique for the release of pluripotent mesenchymal stem cells from the bone marrow. It was recently proven that BST-CarGel showed higher efficacy than the microfracture technique in phase III clinical trials. Chitosan promoted formation of clot between BST-CarGel® and whole blood while simultaneously impeding clot retraction, probably caused by electrostatic interaction between cationic charges of chitosan and anionic phospholipids on the surface of the erythrocytes and platelets (Hoemann et al., 2007). Chitosan also increased the adhesivity and residency of the clot to the cartilage regions, because of interaction between their cationic charges and anionic glycosaminoglycans of the cartilage (Ruel-Gariépy & Leroux, 2004; Shive et al., 2006).

Block copolymers consisting of poly(lactic acid-co-glycolic acid) (PLGA) and PEG have temperature-sensitive gelation properties. Thermal gelation of low molecular weight PLGA–PEG–PLGA copolymers (ReGel® technology) is used for intratumoral injection against solid tumors (Fowers, Baudys, Rathi, & Shih, 2003; Ruel-Gariépy & Leroux, 2004; Vukelja et al., 2007; Zentner et al., 2001). The solution with low viscosity below room temperature can easily be injected via a 25-gauge needle (Fowers et al., 2003). The solution rapidly becomes a stiff insoluble gel depot with high viscosity at body temperature (Fowers et al., 2003). OncoGel™ (paclitaxel in ReGel®) was degraded over 4–6 weeks from the surface of the gel (Zentner et al., 2001). Paclitaxel was released throughout the 6-week period with a half-life of about 3 weeks. The first release mechanism was diffusion from the surface of the gel, followed by a combination of diffusion and erosion of the depot. Preclinical studies using mice models showed similar or superior efficacy with a 10-fold lower dose of OncoGel™ compared to the maximum tolerated systemic dose of free paclitaxel. In a phase I clinical study, none of the patients had any drug-related serious adverse events; OncoGel™-related adverse events were extremely minor and were limited to erythema and pain on injection in two patients each (Vukelja et al., 2007). Systemic levels of paclitaxel were detectable only in 3.3% of the samples analyzed (range: 0.53–0.71 ng/mL). OncoGel delivered intralesionally at doses up to 2.0 mg paclitaxel/cm^3 tumor volume was well tolerated, and paclitaxel remained localized at the injection site, confirming design principles to minimize systemic exposure (Vukelja et al., 2007). In a small phase IIa study against esophageal cancer, 70% of patients had a reduction in tumor volume when OncoGel was used in combination with radiotherapy. Biopsy samples did not contain tumor cells in almost 40% of patients. A phase IIb study in 124 patients with operable esophageal cancer is ongoing.

In situ–forming hydrogel has been useful to improve drug retention on the corneal surface and obtain high bioavailability and efficacy in ophthalmic drug delivery (Nanjawade, Manvi, & Manjappa, 2007). *In situ*–forming hydrogels have both advantages—accuracy and facility of administration of solutions and prolonged resistance time of gels. Pluronic F127 is one of the poloxamers, and consists of polyoxyethylene units (70%) and polyoxypropylene units (30%). At a concentration of 15% or higher in aqueous solution, Pluronic F127 is transformed from a low-viscosity solution to a semisolid gel upon heating to a temperature higher than 23 °C. Formulations containing Pluronic F127 and timolol maleate showed 2.5-fold higher bioavailability than drug solution after instilling into the conjunctival sac of albino rabbits (El-Kamel, 2002). The tear film is kept on the ocular surface by mucin interactions. The mucins concentration is the highest near the surface of the globe, and it gradually decreases as the tear/air interface is approached. Moreover, some different types of mucins (MUC1, MUC16, MUC5-AC, MUC4, and MUC7) occupy different positions and perform different functions on the ocular surface. It is proposed that poloxamers have mucomimetic property by their hydrophobic and hydrophilic sequences (Nanjawade et al., 2007). Poloxamers might interact with proteins on the surface of the globe, and efficiently remain on the corneal surface. These results indicated that temperature-sensitive polymers can interact with the surface of the tissue in the administered sites, and release drugs gradually to increase efficacy, reduce side effects, and keep administration easy.

10.2.3 Ion-sensitive polymers

Ion-exchange resins are insoluble polymers containing a backbone of cross-linked polystyrene and side chains of ion-active groups. The ion-exchange resins are widely used for prevention of drugs from binding to the surface of the tongue and the oral cavity and suppression of drugs' unpleasant tastes. The saliva volume in the oral cavity is a small amount, i.e., a few mL, and the resins maintain drugs bound and prevent drugs from binding to the surface of the tongue. In the gastrointestinal tracts, large amount of fluids (200–400 mL) promote drug release from the resin–drug complexes. An ion-exchange resin with carboxylic acid groups (Amberlite CG-50) is used for taste-masking of pseudoephedrine in chewable Rondec tablets (Borodkin & Sundberg, 1971). In the product, the resin–drug complexes were further coated with a layer containing ethylcellulose and hydroxypropylmethylcellulose for suppression of drug release in saliva. The particles prevented drugs from interacting with the surface of the tongue, suppressed unpleasant tastes, and released drug rapidly in simulated gastric juice for high bioavailability.

Several polymers having a lower critical solution temperature (LCST) can be used for ion-sensitive drug delivery systems. Highly water-soluble substances such as electrolytes and sugars remove water molecules from the polymers having the LCSTs and decrease their LCSTs, which is known as the salting-out effect. A "salting-out taste-masking system" (Figure 10.2(a)) prevented drugs from binding to the surface of the oral cavity by multi-coating layers on particles containing drugs (Tasaki et al., 2009; Yoshida et al., 2008a,b; Yoshida, Nakanishi, et al., 2009; Yoshida, Tasaki, et al., 2009). This system consisted of particle-containing drugs, a salting-out layer containing salts and water-soluble polymers, and a water-penetration-control layer containing water-insoluble materials. The water-penetration-control layer controls rate of saliva intake into the particles in the mouth. Salts dissolve in the saliva, and prevent the water-soluble polymers having LCST from dissolving via the salting-out effect. The insolubilized water-soluble polymers suppress drug release and drugs binding to the surface of the oral cavity. In the gastrointestinal tract, most of the salts

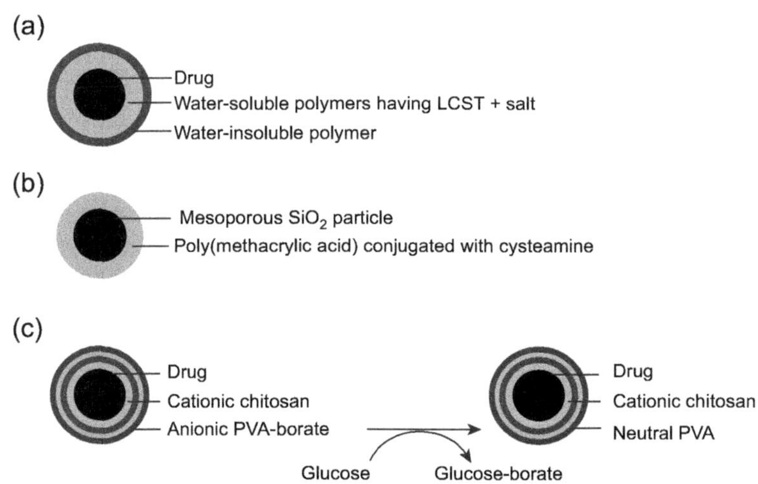

Figure 10.2 Structures of polymeric drug delivery systems sensitive to ions, redox, and biomolecules. (a) Ion-sensitive salting-out taste-masking system (Yoshida et al., 2008a,b; Yoshida, Nakanishi, et al., 2009; Yoshida, Tasaki, et al., 2009). (b) Redox-sensitive PMA-SH(disulfide) (Yan et al., 2011; Zelikin, 2006). (c) Biomolecule-sensitive microparticles having gel multilayers of chitosan and polyvinylalcohol (PVA)-borate (Manna & Patil, 2010).

dissolve out of the system and the water-soluble polymers dissolve; therefore drugs are rapidly released. This system generated long lag times, and subsequently released drug rapidly. The main mechanism behind the drug-release-control of the system was proved to be the salting-out effect (Tasaki et al., 2009; Yoshida, Nakanishi, et al., 2009; Yoshida, Tasaki, et al., 2009). This salting-out taste-masking system prevented solifenacin succinate from binding to the surface of the tongue and the oral cavity, suppressed its unpleasant tastes, and is used for an oral disintegrating tablet of Vesicare OD.

10.2.4 Redox-sensitive polymers

Some viruses, such as Sindbis virus and HIV, are thought to require reduction of critical disulfide bonds in viral envelope glycoproteins, which occurs at the surface of the target cell, to achieve membrane fusion and delivery of the viral RNA into cells (Torres & Gait, 2012). Similarly, hepatitis C virus envelope glycoproteins contain reduced cysteine residues essential for viral entry. Such findings indicate that exofacial thiols on the cellular surface may mediate a natural mechanism for cellular internalization of extracellular compounds. Protein disulfide isomerase (PDI) acts mainly as a reducing agent at the cell surface. Thiol-modified polymers have been reported to enter into cells efficiently mediated by the PDIs. On the other hand, glutathione concentrations are about 1000 times higher inside cells (0.5–10 mM) than outside cells (several μM) (Saito, Swanson, & Lee, 2003). Disulfide bonds in proteins are reduced by the PDIs or the intracellular glutathiones. Nanocarriers consisting of polymers with reducible groups such as disulfide bonds are expected to be stable outside cells, degraded by the PDIs on the surface of cells or the glutathione inside cells, and release drugs on the cellular surface or intracellular-specifically.

Thiolated gelatin was prepared by reaction of arginine groups on gelatin with 2-iminothiolanes (Kommareddy & Amiji, 2005). Aqueous solution of the thiolated gelatin was mixed with ethanol, and nanoparticles with a size of about 200 nm were achieved. The thiolated gelatin nanoparticles containing pDNA showed significantly more effective transfection than their equivalent nonthiolated gelatin nanoparticles. The thiolation might enhance adhesion of the nanoparticles on the surface of cells, and therefore efficient cellular uptake. Moreover the thiolated gelatin nanoparticles showed glutathione-dependent drug release, faster drug release at high glutathione concentration than in a buffer containing no glutathione.

Chitosan thiomer (chitosan-NAC) was synthesized by the conjugation of anime groups of chitosan and N-acetylcysteines (Loretz, Thaler, & Bernkop-Schnürch, 2007, Yin et al., 2009). Polyplexes between chitosan-NAC and pDNA, about 100 nm in size, showed 2.5 times more efficient transfection than their nonthiolated counterparts (Loretz et al., 2007). Polyplexes between chitosan-NAC and insulin showed 2.5 and 12 times higher permeability in excited rat ileum, comparing to polyplexes of nonmodified chitosan and free insulin (Yin et al., 2009). Uptake of the chitosan-NAC polyplexes into Peyer's patches in rat intestine, and hygroglycemic effects after their oral and ileal administration to rats, were also higher than nonmodified chitosan counterparts and free insulin (Yin et al., 2009). These results indicated the thiolation-enhanced mucoadhesion to the surface of cell and intestines, and efficient uptake of the nanoparticles.

Thiolated poly(methacrylic acid) polymer was synthesized, and microcapsules were prepared through LbL assembly on mesoporous SiO_2 particles (Yan, Wang, Heath, Nice, & Caruso, 2011; Zelikin, Quinn, & Caruso, 2006, Figure 10.2(b)). Thiolated poly(methacrylic acid) microcapsules stabilized by disulfide bonds (PMA-SH(disulfide)) or by thioether bonds (PMA-SH(thioether)) showed better cellular adhesion in HeLa cells than equivalent poly(methacrylic acid) microcapsules stabilized by amide bonds (PMA(amide)) (Yan et al., 2011). When microcapsules were incubated with cells (microcapsule-to-cell amount ratio of 100:1), 77%, 63%, and 27% of cells were associated with PMA-SH (disulfide), PMA-SH (thioether), and PMA (amide) microcapsules. A membrane-impermeable reagent, 2,2′-dinitro-5,5′-dithiobenzoic acid (DTNB), blocks exofacial thiols and inhibits disulfide-thiol exchange at the cell surface. Pretreatment of PMA-SH (disulfide) microcapsules with DTNB and their coincubation with cells decreased microcapsule–cell association. These results indicated that the thiolation of the polymers enhanced the interaction between thiols on the microparticles and cell-surface thiols.

The cell-surface thiols can be used as a biomarker to predict skin sensitization (Suzuki, Hirota, Hagino, Itagaki, & Aiba, 2009). THP-1 cells (human monocytic cells, activated macrophages) change intracellular signaling and cell-surface thiols in response to various chemicals. Thiolated polymers and their particles have the possibility of being used as skin-sensitization biomarkers.

10.2.5 Enzyme-triggered polymers

Extracellular matrix (ECM) provides structural support to cells in the body. In tumor tissues, overexpressed matrix metalloproteases (MMPs) degrade ECM around tumor cells, which can promote proliferation, invasion, and metastasis of the tumor cells. MMP14 (MT1-MMP) is anchored to cellular membranes, and activates MMP2 for degradation of ECM on cellular surfaces. The activated MMP2 recognizes and hydrolyzes specific peptide sequences of ECM components such as collagen. These MMP2 cleavable peptides have been incorporated into polymer chains, and used for tumor surface-specific drug delivery.

Methotrexate was conjugated to dextran via an MMP2 cleavable peptide (PVGLIG) (Chau, Dang, et al., 2006; Chau, Padera, et al., 2006). The conjugates showed much longer half-lives in blood, and about 100 times higher tumor concentration than free methotrexate after intravenous injection to mice (Chau, Dang, et al., 2006a). The conjugate significantly suppressed tumor progression, and free methotrexate showed no efficacy after intraperitoneal injection to subcutaneous tumor-bearing mice (Chau, Padera, et al., 2006b). In another work, a generation 5 PEGlylated polylysine dendrimer was conjugated to methotrexate via the MMP2 cleavable peptide (Kaminskas et al., 2011). The dendrimers showed long circulation in blood, and higher efficacy against subcutaneous tumor of mice comparing to free methotrexate. In these systems, the polymer–drug conjugates might release the drug on the surface of tumor cells by cleavage of the peptides.

In CPPs for enhancing cellular uptake of cargos, two popular cationic CPPs, Tat peptide and oligoarginine, have some cationic lysines or arginines in their sequences. The polycationic sequences have advantages of interacting with anionic surface of cells, and induce endocytosis. However, it was reported that nonarginine CPP bound to the local vasculature just after intravenous injection, due to the electrostatic interaction between polycationic peptides and anionic surface of the blood vessels (Aguilera, Olson, Timmers, Jiang, & Tsien, 2009). Moreover, aggregates between the polycationic CPP and anionic plasma proteins such as albumin are formed, and are trapped in the liver or embolize blood vessels in the lung, and induce respiratory collapse and acute toxicity (Aguilera et al., 2009). To reduce the nonspecific adhesion and acute toxicity of CPPs, polycationic nonarginines were conjugated with anionic poly(glutamic acid) via the MMP2 cleavable peptide (PLGLAG) (Aguilera et al., 2009). The conjugates showed much longer half-lives in blood and more uniform distribution to normal tissues, and reduced acute toxicity comparing to nonarginines after intravenous injection. The conjugates increased tumor accumulation, were cleaved by MMP2, and then increased adhesivity on the cellular surface. These results indicated that activation of the polymers using the MMP2 cleavable peptide was useful for higher efficacy and lower side effects. In various diseases, many kinds of enzymes are also overexpressed or activated on cellular surfaces. Polymers responding to the enzymes have probability of being used for future drug delivery and diagnostic systems specific to the disease.

10.2.6 Biomolecule-sensitive polymers

Integrin $\alpha_V\beta_3$ receptors are overexpressed on the surface of vascular endothelial, smooth muscle, and tumor cells in several tumors, such as gliomas and ovarian cancers, and are associated with angiogenesis and tumor metastasis. A tripeptide sequence, Arg–Gly–Asp (RGD), is recognized and bound by integrin $\alpha_V\beta_3$. Many polymers have been conjugated to the RGD peptide, and used for tumor-targeting drug delivery.

Conjugate of polyglutamic acid (PGA) with paclitaxels (PTX) and the RGD peptides showed efficient tumor accumulation and high efficacy against mice tumors (Eldar-Boock et al., 2011). PGA–RGD conjugates accumulated about 1.5 times more efficiently in tumors than PGA. PGA–PTX–RGD conjugates more strongly suppressed tumor progression than PGA–PTX conjugates and free PTX. From the view of body weight change, PGA–PTX–RGD conjugates were safer than PGA–PTX conjugates and free PTX. Opaxio, a PGA–PTX conjugate, is currently undergoing phase III clinical trials and shows promising results (Langer et al., 2008; O'Brien et al., 2008; Paz-Ares et al., 2008). Modification of Opaxio with the RGD peptide might help to obtain future products with better efficacy and lower side effects by the interaction with integrin $\alpha_V\beta_3$ on the cellular surface.

In another study, a hairpin peptide was conjugated with the RGD peptide and methotrexate, and binding of the RGD peptide to integrin $\alpha_V\beta_3$ was designed to induce conformation change of the hairpin and drug release (Kotamraj, Russu, Jasti, Wu, & Li, 2011). A 12 amino acid peptide, RWQYV^{D}PGKFTVQ (^{D}P was *d*-Pro), stably folded into an antiparallel β-hairpin conformation in an aqueous environment. When the RGD peptide bound to integrin $\alpha_V\beta_3$, the hairpin conformation changed to the straight chain probably by the higher binding constant of the straight chain to integrin $\alpha_V\beta_3$, or steric hindrance from integrin $\alpha_V\beta_3$. An endopeptidase cannot hydrolyze the hairpin conformation, but can do the straight chain; therefore drug can be released only after binding of the peptides to integrin $\alpha_V\beta_3$. Coincubation of the conjugates with cells expressing integrin $\alpha_V\beta_3$ showed about 1.6 times faster drug release compared to the conjugates without cells (Kotamraj et al., 2011). Biomolecule-sensitive conformation changes of proteins are widely used for their activation in the body, but artificial design and control of the conformation changes are very difficult in our technologies at present. More development of our science and technologies will enable these approaches to control interaction with target cells and deliver drugs site-specifically.

Glucose transporters are overexpressed on the surface of tumor cells. Glucose analogue [^{18}F] fluoro-2-deoxy-D-glucose (FDG) is clinically used with positron emission tomography for detecting tumors, because FDG is efficiently taken up into tumor cells, about seven times more efficiently in lung cancers and four times more efficiently in liver cancers than in normal tissues (Nolop et al., 1987; Feng, Wang, Gao, & Shen, 2007). The tumor-specific elevation of glucose metabolism can be used for tumor-specific drug delivery. Microparticles having gel multilayers of chitosan and polyvinylalcohol (PVA)-borate (Figure 10.2(c)) have been reported to release drug responding to glucose concentration (Manna & Patil, 2010). The PVAs form complexes with borax molecules, and hydrogels by the physical cross-linking through hydrogen bonds. In the presence of glucose, bolate ions prefer to form complexes with glucose molecules rather than PVA hydroxyl groups (Cui, Ward Muscatello, & Asher, 2009). In the multilayered system, glucose induced the change from anionic PVA–bolate complex layers to neutral PVA layers, electrostatic repulsion among the cationic chitosan layers, and rapid degradation of the multilayer. The microparticles showed about four times faster drug release at 25 mM glucose than at 5 mM glucose (Manna & Patil, 2010). These idea and results can be used for glucose-triggered polymeric tumor-targeting systems. Polymeric systems responding to biomolesules on the surface of the organs, tissues, and cells have the possibility of changing future medication.

10.3 Polymers sensitive to the body's external stimuli

10.3.1 Temperature-sensitive polymers

In hyperthermia therapy, tumors are exposed to slightly higher temperatures to damage and kill cancer cells, or to make cancer cells more sensitive to drugs. Polymers having LCSTs or glass transition temperatures around 38–40 °C are used for heat-sensitive drug release in the hyperthermic tumor tissues (Wu, Shen, Banerjee, & Zhou, 2010). Elastin-like polypeptides (ELPs) have LCST values below which they stay in solution, and above which they form aggregates due to collapse of the hydrophobic structure (Yamaoka et al., 2003). ELP1 was composed of 150 repetitive units of Val-Prol-Gly-Xaa-Gly, in which Xaa was a V:G:A ratio of 5:3:2 (59 kDa), and had LCST close to 40 °C. In one study, the ELP1 was conjugated with a CPP of SynB1 and doxorubicins via hydrazone bonds (Moktan, Perkins, Kratz, & Raucher, 2012). The SynB1 peptide enhances adhesion on the surface of cells and cellular uptake of its cargo (Rousselle et al., 2001), and induces cleavage of the acid-labile hydrazone linker and dozorubicin release in acidic endosomes/lysosomes. Immediately after intravenous injection of the conjugates to tumor-bearing mice, the tumors were heated with infrared light that raised the core tumor temperature to 42 °C. Injection of the conjugates and the hyperthermia showed about two times higher drug concentration in tumors, and more significant suppression of tumor progression, compared to only injection of the conjugates and injection of free drugs with or without hyperthermia (Moktan et al., 2012). These results indicated that temperature-sensitive polymers can enhance efficacy of hyperthermia on the surface of tumors.

Nanoparticles modified with thermoresponsive polymers and tumor-targeting ligands were reported to show temperature-sensitive binding on cells and cellular uptake (Mastrotto et al., 2011). A thermoresponsive polymer, poly(N-isopropylacrylamide-co-acrylamide) (pNIPAM-AM), with a cysteine terminus, has an LCST of 37 °C. Gold nanoparticles decorated with pNIPAM-AM and thiolated biotin, bound to avidin-coated plates about 15 times more efficiently at 40 °C than at 34 °C. The particles were expected to display chain-extended polymer brushes at their surfaces at 37 °C (T < LCST), thus hiding the targeting ligands in the particles. When the temperature was increased above 37 °C (T > LCST), the polymer collapsed into poorly solvated globules, exposed the ligands to the surface of particles, and interacted with the surface of cells. Gold nanoparticles decorated with pNIPAM-AM and thiolated folic acid internalized into KB cells 30 times more efficiently at 40 °C than at 34 °C (Mastrotto et al., 2011). Temperature-sensitive polymers were very useful for controlling adhesion of ligands and particles to the surface of cells responding to the high temperature of hyperthermia.

In another work, biodegradable block copolymers were synthesized using cyclic carbonates as functional monomers, and their thermosensitive polymeric micelles were prepared (Kim et al., 2011, Figure 10.3(a)). In the copolymer, the first hydrophobic block had ethylene groups as side chains, and formed the core of the micelle. The second hydrophilic-balanced block had side chains of PEGs and lauryl chains, and formed the corona of the micelle. The copolymer had 36 °C of LCST, was stable *in vitro*, and was degraded enzymatically *in vivo*. Size of the polymeric micelle changed from 41 nm below the LCST to 276 nm above the LCST. The micelles containing paclitaxel showed the faster drug release, and more efficient cellular uptake at 37 °C than at 32 °C (Kim et al., 2011). Poorly solvated globules of thermosensitive polymers above LCSTs might enhance adhesion to the surface of cells and cellular uptake.

These results indicated that temperature-sensitive polymers can control interaction between nanoparticular drug delivery systems and the cellular surface, as the pNIPAM are used in cell culture dishes for cell sheet engineering (Yang et al., 2006).

10.3.2 Near-infrared-light-sensitive polymers

Near-infrared (NIR) light between 750 and 1000 nm can penetrate several millimeters up to centimeter depths of tissue. Polymeric NIR-light sensitive drug-delivery systems have been reported by combination of polymers and gold nanoparticles that absorb NIR light and increase temperature. Biodegradable PLGA microspheres (1–15 μm) containing paclitaxel and hollow gold nanoparticles (about 35 nm) were prepared by

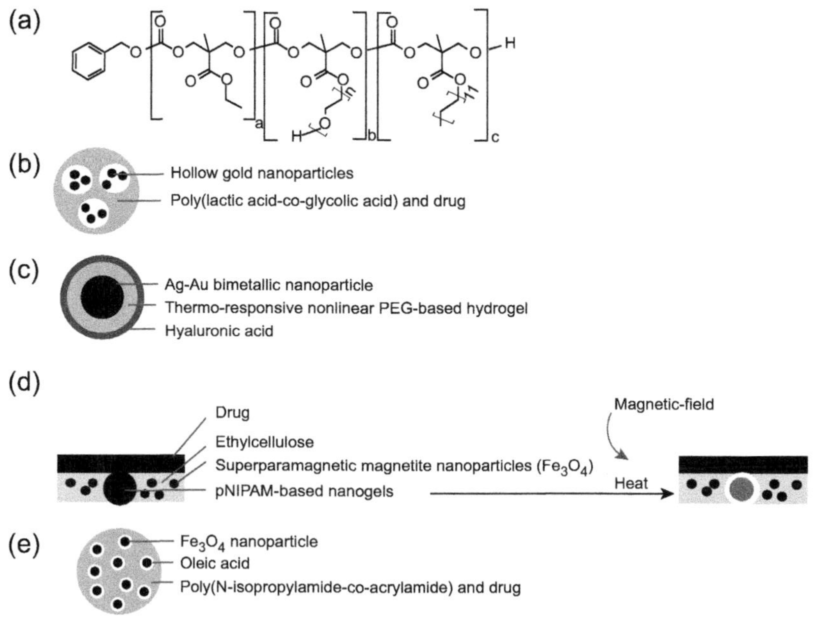

Figure 10.3 Structures of polymers and polymeric drug delivery systems sensitive to temperature, near-infrared light, and magnetic fields. (a) Temperature-sensitive biodegradable block copolymer (Kim et al., 2011). (b) NIR-sensitive PLGA microspheres containing hollow gold nanoparticles (You et al., 2010). (c) NIR-sensitive hybrid nanogels (Wu et al., 2010). (d) Magnetic-field-sensitive composite membrane (Schild, 1992). (e) Magnetic-field-sensitive magnetic composite particles (Yao et al., 2011).

the water-in-oil-in-water ($W_1/O/W_2$) double-emulsion solvent evaporation method (You, Shao, Wei, Gupta, & Li, 2010, Figure 10.3(b)). Lipophilic paclitaxel was dispersed in PLGA at the molecular level. The gold nanoparticles were probably dispersed in the W_1 water phase inside the microspheres. Faster drug release, higher cytotoxicity, and more efficient suppression of tumor progression after intratumoral injection of the microspheres to mice were obtained with NIR irradiation comparing to without the irradiation. The NIR irradiation induced rapid elevation of the temperature of the gold nanoparticles and the microspheres. By the temperature increase, the polymer chains of PLGA became more flexible, and the drug was more rapidly released from the PLGA matrix (You et al., 2010). Tumors on the surface of organs are expected to be regressed by these NIR-responsive polymeric systems.

In another system, hybrid nanogels consisted of Ag–Au bimetallic nanoparticle as core, thermoresponsive nonlinear PEG-based hydrogel as shell, and hyaluronic acid chains on the surface of the nanoparticles as targeting ligands (Wu et al., 2010, Figure 10.3(c)). The hybrid nanogels released drug thermosensitively. When temozolomide solubion was inculated with cells, NIR irradiation for 5 min had no effect on the cytotoxicity. The hybrid nanogels showed higher cytotoxicity with NIR irradiation than without the irradiation. Comparing the effects of photothermal, chemo, and combined chemo-photothermal treatments, the efficacy of the hybrid nanogels was proved to be chemo-photothermal synergistic (Wu et al., 2010). Combination of thermosensitive polymers and metals are useful for NIR-responsive drug treatment on the surface of the body, organs, and tissues.

10.3.3 Ultrasound-sensitive polymers

Ultrasound can be forced efficiently into deep tissue, and is widely used in clinical diagnostic imaging. Many polymeric systems have been reported for site-specific drug release using energy from ultrasound (Antipina & Sukhorukov, 2011; Du, Jin, Zhou, & Zhao, 2011; Nakatsuka et al., 2011; Park, 2011). Ultrasound-sensitive nanodroplets were prepared by filling with perfluoropentane into polymeric micelles of PEG–PLGA containing doxorubicin (Park, 2011). Since the boiling point of perfluoropentane was 29 °C, the nanodroplets were prepared at room temperature or lower, and would likely transform into nanobubbles at body temperature. The nanobubbles with ultrasound treatment showed faster drug release and higher efficacy against progression of mice tumor than without ultrasound. The PEG–PLGA adhered on the surface of the microbubbles, and enabled long circulation of the microbubbles in blood after intravenous injection. The ultrasound-sensitive particles should have a good balance of stability because the particles should be disrupted by ultrasound and be stable in the blood. Polymers are very important for these drug delivery systems to regulate interaction between the surface of the particles and proteins or cellular surfaces.

Microbubbles having a shell of poly(acrylic acid) and phospholipids have been reported as another ultrasound-sensitive system (Du et al., 2011). Poly(acrylic acid) had no interaction with liposomes at neutral pH, but bound to the surface of liposomes at acidic pH below its pK_a of 4.5. From these results, design of suitable polymers and control of interaction between the particles and cellular surface are expected to advance the ultrasound-sensitive drug delivery systems.

10.3.4 Magnetic-field-sensitive polymers

Magnetic resonance imaging (MRI) is a popular noninvasive diagnostic procedure. Exposure to alternating magnetic fields raises the temperature of magnetic colloids. Combinations of temperature-sensitive polymers and the magnetic colloids are used for magnetic-field-sensitive drug delivery (Hernández et al., 2010; Hoare et al., 2009; Yao, Chen, Ai, Wang, & Huang, 2011). A magnetic-field-sensitive membrane consisting of ethylcellulose (the membrane support), superparamagnetic magnetite nanoparticles (Fe_3O_4, the triggering entity), and thermosensitive pNIPAM-based nanogels (the switching entity) was reported (Schild, 1992, Figure 10.3(d)). Magnetic triggering increased the temperature of the device, thermosensitive pNIPAM contracted and formed pores in the membrane, and drugs were permeated through the membrane (Hoare et al., 2009). When the magnetic field was turned off, the device was cooled, the pNIPAM expanded and buried the membrane's pores, and drug permeation was stopped. This membrane showed low biocompatibility under the skin of rats. Microscopically, there was acute and early chronic inflammation around the implant. The sections showed a mature fibrous capsule with macrophages and occasional foreign body giant cells at the material–tissue interface. Surface modification of the membrane with biocompatible polymers, or the membrane using biocompatible polymers, might have possibility of magnetic-field-responsive drug release on the surface of organs and skin.

Another magnetic-field-sensitive composite particle was synthesized by emulsion-free polymerization of NIPAAm and acrylamide in the presence of oleic acid-modified Fe_3O_4 nanoparticles (Yao et al., 2011, Figure 10.3(e)). The copolymer had an LCST of 40 °C, and the size of the composite particles changed temperature-responsively. The composite particles released drug faster at a magnetic field than at no magnetic field. The MRI is a safe diagnostic procedure and many hospitals have MRI equipment; therefore the magnetic-field-sensitive drug delivery and diagnostic systems are promising for the future. Biocompatibility of the systems and magnetic-field-sensitive control of adhesion on the surface of the target organs, tissues, and cells probably becomes more important for these systems.

10.3.5 Light-sensitive polymers

Photodynamic therapy is used clinically to treat cancers, age-related macular degeneration, and skin disease. Polymeric systems with light-sensitive cleavable bonds have been reported for site-specific drug release (Agasti et al., 2009; Choi et al., 2010; Dvir, Banghart, Timko, Langer, & Kohane, 2010; Fischer et al., 2000; Maeda, Titani, & Sekiguchi, 1994; Shamay, Adar, Ashkenasy, & David, 2010). A photosensitive copolymeric system using photo-cleavable protecting groups on CPPs has been reported (Shamay et al., 2010). Photo-cleavable 6-nitroveratrylcarbonyl (Nvoc) groups protected three lysine residues on CPPs derived from penetratin (Ac-KRRMK^{Nvov}W$K^{Nvoc}$$K^{Nvoc}$) (Fischer et al., 2000). The photoactivatable CPPs and drugs were conjugated on N-(2-hydroxypropyl)methacrylamide copolymers. Normal cationic CPPs have no cell specificity and are taken up into not only target cells but also normal cells. However, this system can suppress uptake into normal cells, and deliver the conjugated drugs site-specifically by irradiation of UV light on the surface of targeting sites. In fact, these copolymers showed stronger binding affinity on the cellular surface after UV light irradiation, compared to the experiments in the dark (Shamay et al., 2010). Cellular uptake of the copolymers were 10 times more efficient with UV light than in the dark. This light-sensitive polymeric system is expected to have much smaller side effects in normal tissues, and be very useful for interacting with the surface of organs or skin for drug delivery.

Another photoactivatable CPP system has also been reported for light-responsive cellular uptake (Dvir et al., 2010). A peptide sequence of YIGSR is crucial for adhesion to integrin β1 on the cell membrane of stromal and endothelial cells. The biological activity of YIGSR peptide can be greatly attenuated by mutation or deletion of tyrosine (Maeda et al., 1994). Photo-cleavable 4,5-dimethoxy-2-nitrobenzyl (DMNB) group protected the tyrosine of the peptide, and the peptides were conjugated with carboxylated polystylene nanoparticles. Before light irradiation, the DMNB groups protected interaction between the nanoparticles and cell surface. Light irradiation induced detachment of DMNBs and two to five times more efficient adhesion of the nanoparticles on the cellular surface.

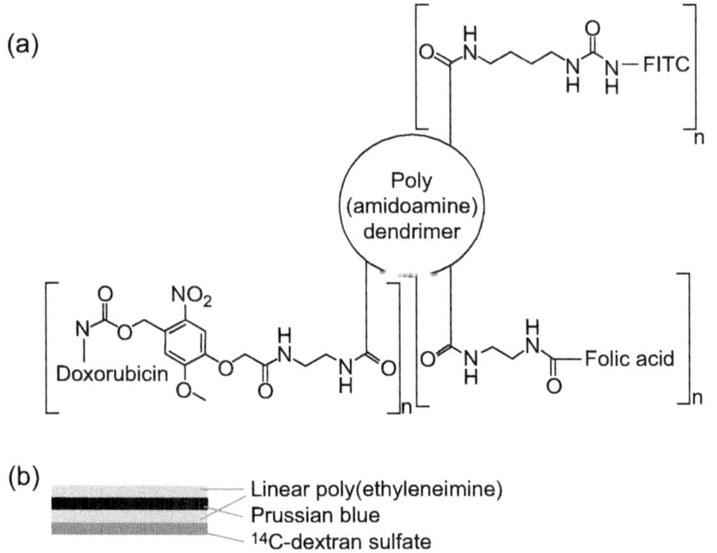

Figure 10.4 Structures of polymers and polymeric drug delivery systems sensitive to light and electric fields. (a) Light-sensitive polyamidoamine dendrimes conjugated with doxorubicin via photo-cleavable ortho-nitrobenxyl linkages (Choi et al., 2010). (b) Electric-field-sensitive (Wood et al., 2008).

Many light-sensitive polymers have been reported for photo-responsive drug release. Fifth-generation polyamidoamine dendrimers were conjugated with doxorubicin via photo-cleavable *ortho*-nitrobenxyl (*o*-NB) linkages, and with folic acids as targeting ligands (Choi et al., 2010, Figure 10.4(a)). The conjugates were inactive prior to UV irradiation, but inhibited cell growth after exposure to UV light as a function of irradiation time. In another system, 5-fluorouracil was conjugated to PEG chain via *o*-NB linkage and the conjugates were attached on gold nanoparticles (Agasti et al., 2009). The polymer-modified nanoparticles also showed UV light-specific cytotoxicity. Light-sensitive conformation change of polymers has also been researched (Asanuma, Liang, et al., 2001; Asanuma, Shirasuka, et al., 2001; Asanuma, Takarada, et al., 2001; Asanuma, Liang, Yoshida, & Komiyama, 2000; Yamazawa, Liang, Yoshida, Asanuma, & Komiyama, 2000; Asanuma, Ito, et al., 1999; Asanuma, Yoshida, et al., 1999). Photo-isomerization of azobenzenes and spiropyrans has been used for conformation change of nucleic acids and proteins. These light-sensitive polymers can be used for site-specific drug delivery and diagnostic systems on the surface of the body and organs, and high efficacy and low side effects of medication.

10.3.6 *Electric-field-sensitive polymers*

Neurons communicate using electrical signals and chemical messengers. Electric-field-sensitive polymers have been reported to control drug release responding to electrical signals like neurons (Leprince, Dogimont, Magnin, & Demoustier-Champagne, 2010; Luo & Cui, 2009). Nanoscale electroactive thin films with the thickness of 4.2 nm consisted of tetralayers of linear PEI/Prussian blue (PB)/linear PEI/^{14}C-dextran sulfate (^{14}C-DS) (Wood et al., 2008, Figure 10.4(b)). The PB was a nontoxic inorganic iron hexacyanoferrate compound having negative charge, and could be switched to oxidation state of Prussian brown having neutral charge by applying a potential of +1.25 V. The films were rapidly destabilized in the presence of the oxidizing potential, and released the model drug of ^{14}C-DS (Wood et al., 2008). When the potential was removed, the films were restabilized and suppressed drug release. This system can be used for drug delivery, tissue engineering, and medical diagnostics responding to electrical trigger on the surface of the body, organs, and tissues.

Conducting polymer of polypyrrole (PPy) has been used for electric-field-sensitive drug delivery systems (Leprince et al., 2010; Luo &Cui, 2009). In a study, drugs were loaded into pores of sponge-like nanoporous PPy film, and a thin layer of PPy was coated on the top of the film (Luo & Cui, 2009). Fluorescein and dexamethasone were used as model drugs, and their drug release from the films was controlled electrically by voltage application of −2.0 and −0.5 V. In addition to the significant change of drug release rates by the voltage application, these films could load large amount of drugs for future applications. In another work, PPy and dexamethasone were coated on a Pt nanopillar brushes on a gold surface (Leprince et al., 2010). The drug release from the composites was also controlled electrically. In the future, these electric-field-sensitive polymers might be used as artificial neurons and control biofunctions via interactions on cellular surfaces.

10.4 Future trends

Stimuli-responsive drug delivery is very important for providing higher efficacy, lower side effects, and higher adherence. Stimuli-sensitive polymers can control the interaction between drug delivery systems and the surface of the body, organs, tissues, and cells. Many formulations using the polymers responsive to the body's internal stimuli, such as pH, ions, and temperature, have been launched and developed in clinical trials. However, the body's internal stimuli of gastrointestinal pH, ionic concentrations, and endosomal pH have intra- and interindividual variability. For example, gastric pH in many patients is 1.0–3.5 (Table 10.1); however, patients with achlorhydria have gastric pH >4.0. Gastric pH is 1.0–3.5 in fasted condition (Table 10.1), and 4–5 in fed condition. In the oral cavity, pH is 6.0–7.0 in fasted condition (Table 10.1) and about 4 in fed condition. Therefore kinds of polymers, amounts of polymers, and dosing conditions should be carefully considered to develop and launch products for many patients. As another example, some kinds of cells might have more acidic endosomes/lysosomes than others (Hinton et al., 2009; Lafourcade, Sobo, Kieffer-Jaquinod, Garin, & van der Goot, 2008). It is very important to understand differences between each cell type in the body for targeting drug delivery systems, such as tumor, liver, and spleen cells. Amounts of enzymes and biomolecules in the body and target sites also probably have intra- and interindividual variability. The concentrations and expression of the biomolecules might correlate with symptoms and malignancy. As recent trends of drug–diagnostic co-development, stimuli-sensitive polymeric drug delivery systems might have to be used in clinical trials after diagnosis of the amount of biomolecules in each patient.

In contrast, the body's external stimuli have high certainty, because the stimuli are irradiated from outside of the body, and their irradiation sites, strength, and terms can be controlled freely. But the polymeric systems responding to the external stimuli have some hurdles. The first hurdle is the method for irradiation of the stimuli to the target sites. Target sites in the depth of the body cannot be irradiated with heat, NIR, and light from the surface of the body. Ultrasound, magnetic fields, and electric fields can penetrate into the depth of the body, but their precise focus on the small target sites without irradiation to surrounding normal sites is necessary and difficult. The second hurdle is the necessity of special equipment for the irradiation. Specification of the target sites, generation of the stimuli, and precise focus of the stimuli need highly functionalized equipment. Hospitals do not want to buy the expensive equipment, only after the medications were proved to be significantly more effective than the existing therapies. Proof of safety of the stimuli and polymers is also very important for clinical use. Therefore selection of target diseases, selection of the best stimuli for the purpose, and optimization of polymers and systems responding to the stimuli will become important for future applications.

One of the largest advantages of stimuli-sensitive polymers is that the polymers can be optimized by a combination of monomers, changing functional groups, and controlling copolymerization ratio and types (block, graft, or random) for responding to various stimuli. Academic researchers have much intelligence and challenging spirits about synthesis and evaluation of novel polymers based on new concepts and approaches. Moreover, progress of fundamental science is clarifying disease-specific molecules on the cellular surface and interactions on biosurfaces. However, clinically meaningful benefits for many patients are essential to develop and launch novel polymeric systems. In an example of clinical trials of anticancer drugs, novel systems must prolong survival time of patients meaningfully compared to the existing best medication, not statistically significant survival time, and not suppressing tumor progression. Novel systems also have high risk concerning the safety of the polymers such as cytotoxicity, antigenicity, hemolysis, genotoxicity, and inflammation of the polymers. For practical and realistic research and development of new systems, pharmaceutical companies have much experience on evaluation of the clinical efficacy, safety, selection of appropriate compounds, and indications. Collaboration of academic researchers and pharmaceutical companies will be essential to produce the best stimuli-sensitive polymers and the best drug delivery systems, and deliver the excellent products to many patients.

References

Adams, M. L., Lavasanifar, A., & Kwon, G. S. (2003). Amphiphilic block copolymers for drug delivery. *Journal of Pharmaceutical Sciences*, *92*(7), 1343–1355.

Agasti, S. S., Chompoosor, A., You, C. C., Ghosh, P., Kim, C. K., & Rotello, V. M. (2009). Photoregulated release of caged anticancer drugs from gold nanoparticles. *Journal of the American Chemical Society*, *131*(16), 5728–5729.

Aguilera, T. A., Olson, E. S., Timmers, M. M., Jiang, T., & Tsien, R. Y. (2009). Systemic in vivo distribution of activatable cell penetrating peptides is superior to that of cell penetrating peptides. *Integrative Biology (Cambridge)*, *1*(5–6), 371–381.

Alani, A. W., Rao, D. A., Seidel, R., Wang, J., Jiao, J., & Kwon, G. S. (2010). The effect of novel surfactants and Solutol HS 15 on paclitaxel aqueous solubility and permeability across a Caco-2 monolayer. *Journal of Pharmaceutical Sciences*, *99*(8), 3473–3485.

Antipina, M. N., & Sukhorukov, G. B. (2011). Remote control over guidance and release properties of composite polyelectrolyte based capsules. *Advanced Drug Delivery Reviews*, *63*(9), 716–729.

Asanuma, H., Ito, T., Yoshida, T., Liang, X., & Komiyama, M. (1999). Photo-regulation of the formation and dissociation of DNA-duplex by using the cis-trans isomerization of azobenzene. *Angewandte Chemie International Edition*, *38*, 2393–2395.

Asanuma, H., Liang, X., Yoshida, T., & Komiyama, M. (2000). Photo-control of Triple-Helix formation by using oligo(thymidine) bearing azobenzene. *Angewandte Chemie International Edition*, *39*, 1316–1318.

Asanuma, H., Liang, X., Yoshida, T., & Komiyama, M. (2001). Photo-control of DNA-duplex formation by use of oligonucleotide carrying an azobenzene. *ChemBioChem*, *2*, 39–44.

Asanuma, H., Shirasuka, K., Yoshida, T., Takarada, T., Liang, X., & Komiyama, M. (2001). Spiropyran as a regulator of DNA hybridization with reversed switching mode to that of azobenzene. *Chemistry Letters*, 108–109.

Asanuma, H., Takarada, T., Yoshida, T., Liang, X., & Komiyama, M. (2001). Enantioselective incorporation of azobenzenes into oligodeoxyribonucleotide for the effective photoregulation of duplex formation. *Angewandte Chemie International Edition*, *40*, 2671–2673.

Asanuma, H., Yoshida, T., Ito, T., & Komiyama, M. (1999). Photo-responsive oligonucleotides carrying azobenzene at the 2'-position of the ribonucleotide. *Tetrahedron Letters*, *40*, 7995–7998.

Babish, J., Tripp, M., Howell, T., & Bland, J. S., Metaproteomics, LLC. (2010). Anti-inflammatory pharmaceutical compositions for reducing inflammation and the treatment or prevention of gastric toxicity. US patent application US7811610. October 19, 2010.

Borodkin, S., & Sundberg, D. P., Abbott Lab. (1971). Coating resin-drugs for chewable tablets. US patent application US3594470. July 20, 1971.

Chau, Y., Dang, N. M., Tan, F. E., & Langer, R. (2006). Investigation of targeting mechanism of new dextran-peptide-methotrexate conjugates using biodistribution study in matrix-metalloproteinase-overexpressing tumor xenograft model. *Journal of Pharmaceutical Sciences*, *95*(3), 542–551.

Chau, Y., Padera, R. F., Dang, N. M., & Langer, R. (2006). Antitumor efficacy of a novel polymer-peptide-drug conjugate in human tumor xenograft models. *International Journal of Cancer*, *118*(6), 1519–1526.

Choi, S. K., Thomas, T., Li, M. H., Kotlyar, A., Desai, A., & Baker, J. R., Jr. (2010). Light-controlled release of caged doxorubicin from folate receptor-targeting PAMAM dendrimer nanoconjugate. *Chemical Communications (Cambridge)*, *46*(15), 2632–2634.

Cui, Q. W., Muscatello, M. M., & Asher, S. A. (2009). Photonic crystal borax competitive binding carbohydrate sensing motif. *Analyst*, *134*(5), 875–880.

Du, L., Jin, Y., Zhou, W., & Zhao, J. (2011). Ultrasound-triggered drug release and enhanced anticancer effect of doxorubicin-loaded poly(D,L-lactide-co-glycolide)-methoxy-poly (ethylene glycol) nanodroplets. *Ultrasound in Medicine & Biology*, *37*(8), 1252–1258.

Dvir, T., Banghart, M. R., Timko, B. P., Langer, R., & Kohane, D. S. (2010). Photo-targeted nanoparticles. *Nano Letters*, *10*(1), 250–254.

El-Kamel, A. H. (2002). In vitro and in vivo evaluation of Pluronic F127-based ocular delivery system for timolol maleate. *International Journal of Pharmaceutics*, *241*(1), 47–55.

Eldar-Boock, A., Miller, K., Sanchis, J., Lupu, R., Vicent, M. J., & Satchi-Fainaro, R. (2011). Integrin-assisted drug delivery of nano-scaled polymer therapeutics bearing paclitaxel. *Biomaterials*, *32*(15), 3862–3874.

Feng, Z., Wang, Z., Gao, C., & Shen, J. (2007). Template polymerization to fabricate hydrogen-bonded poly(acrylic acid)/poly(vinylpyrrolidone) hollow microcapsules with a pH-mediated swelling? Deswelling property. *Chemistry of Materials*, *19*(19), 4648–4657.

Fischer, P. M., Zhelev, N. Z., Wang, S., Melville, J. E., Fahraeus, R., & Lane, D. P. (2000). Structure-activity relationship of truncated and substituted analogues of the intracellular delivery vector Penetratin. *Journal of Peptide Research*, *55*, 163–172.

Fowers, K. D., Baudys, M., Rathi, R., & Shih, C. (2003). Thermally reversible Gelling materials for safe & Versatile depot delivery. *Drug Development Technology*, *3*, 58–64.

Hashimoto, Y., Tanaka, M., Kishimoto, H., Shiozawa, H., Hasegawa, K., Matsuyama, K., et al. (2002). Preparation, characterization and taste-masking properties of polyvinylacetal diethylaminoacetate microspheres containing trimebutine. *Journal of Pharmacy and Pharmacology*, *54*, 1323–1328.

Haware, R. V., Chaudhari, P. D., Parakh, S. R., & Bauer-Brandl, A. (2008). Development of a melting tablet containing promethazine HCl against motion sickness. *AAPS PharmSciTech*, *9*(3), 1006–1015.

Hernández, R., Sacristán, J., Asín, L., Torres, T. E., Ibarra, M. R., Goya, G. F., et al. (2010). Magnetic hydrogels derived from polysaccharides with improved specific power absorption: potential devices for remotely triggered drug delivery. *Journal of Physical Chemistry B*, *114*(37), 12002–12007.

Hinton, A., Sennoune, S. R., Bond, S., Fang, M., Reuveni, M., Sahagian, G. G., et al. (2009). Function of a subunit isoforms of the V-ATPase in pH homeostasis and in vitro invasion of MDA-MB231 human breast cancer cells. *Journal of Biological Chemistry*, *284*(24), 16400–16408.

Hoare, T., Santamaria, J., Goya, G. F., Irusta, S., Lin, D., Lau, S., et al. (2009). A magnetically triggered composite membrane for on-demand drug delivery. *Nano Letters*, *9*(10), 3651–3657.

Hoemann, C. D., Sun, J., McKee, M. D., Chevrier, A., Rossomacha, E., Rivard, G. E., et al. (2007). Chitosan-glycerol phosphate/blood implants elicit hyaline cartilage repair integrated with porous subchondral bone in microdrilled rabbit defects. *Osteoarthritis Cartilage*, *15*(1), 78–89.

Jain, R., Jindal, K.C., & Singh, S., Panacea Biotec ltd. (July 28, 2005). Compositions Comprising Glycosaminoglycan and Nonsteroidal Anti-Inflammatory Drug. International patent application WO2005067943.

Kaminskas, L. M., Kelly, B. D., McLeod, V. M., Sberna, G., Boyd, B. J., Owen, D. J., et al. (2011). Capping methotrexate α-carboxyl groups enhances systemic exposure and retains the cytotoxicity of drug conjugated PEGylated polylysine dendrimers. *Molecular Pharmaceutics*, *8*(2), 338–349.

Katsuma, M., Watanabe, S., Kawai, H., Takemura, S., & Sako, K. (2006). Effects of absorption promoters on insulin absorption through colon-targeted delivery. *International Journal of Pharmaceutics*, *307*(2), 156–162.

Kayumba, P. C., Huyghebaert, N., Cordella, C., Ntawukuliryayo, J. D., Vervaet, C., & Remon, J. P. (2007). Quinine sulphate pellets for flexible pediatric drug dosing: formulation development and evaluation of taste-masking efficiency using the electronic tongue. *European Journal of Pharmaceutics and Biopharmaceutics*, *66*(3), 460–465.

Keast, R. S. J., & Breslin, P. A. S. (2005). Bitterness suppression with zinc sulfate and Na-cyclamate: a model of combined peripheral and central neural approaches to flavor modification. *Pharmaceutical Research*, *22*, 1970–1977.

Kim, S., Nishimoto, S. K., Bumgardner, J. D., Haggard, W. O., Gaber, M. W., & Yang, Y. (2010). A chitosan/beta-glycerophosphate thermo-sensitive gel for the delivery of ellagic acid for the treatment of brain cancer. *Biomaterials*, *31*(14), 4157–4166.

Kim, S. H., Tan, J. P., Fukushima, K., Nederberg, F., Yang, Y. Y., Waymouth, R. M., et al. (2011). Thermoresponsive nanostructured polycarbonate block copolymers as biodegradable therapeutic delivery carriers. *Biomaterials, 32*(23), 5505–5514.

Kommareddy, S., & Amiji, M. (2005). Preparation and evaluation of thiol-modified gelatin nanoparticles for intracellular DNA delivery in response to glutathione. *Bioconjugate Chemistry, 16*(6), 1423–1432.

Kotamraj, P., Russu, W. A., Jasti, B., Wu, J., & Li, X. (2011). Novel integrin-targeted binding-triggered drug delivery system for methotrexate. *Pharmaceutical Research, 28*(12), 3208–3219.

Kumazawa, T., Nomura, T., & Kurihara, K. (1988). Liposomes as model for taste cells: receptor sites for bitter substances including N–C=S substances and mechanism of membrane potential changes. *Biochemistry, 27*, 1239–1244.

Kurimoto, I., Kasashima, Y., Kawai, H., Takaishi, Y., Katsuma, M., Ohi, H., et al. Yamanouchi pharmaceutical Co., Ltd. (May 6, 2005). Drug-containing coated microparticle for orally disintegrating tablet. International patent application WO2005039542.

Lafourcade, C., Sobo, K., Kieffer-Jaquinod, S., Garin, J., & van der Goot, F. G. (2008). Regulation of the V-ATPase along the endocytic pathway occurs through reversible subunit association and membrane localization. *PLoS One, 3*(7), e2758.

Lai, T. C., Bae, Y., Yoshida, T., Kataoka, K., & Kwon, G. S. (2010). pH-sensitive multi-PEGylated block copolymer as a bioresponsive pDNA delivery vector. *Pharmaceutical Research, 27*(11), 2260–2273.

Langer, C. J., O'Byrne, K. J., Socinski, M. A., Mikhailov, S. M., Leśniewski-Kmak, K., Smakal, M., et al. (2008). Phase III trial comparing paclitaxel poliglumex (CT-2103, PPX) in combination with carboplatin versus standard paclitaxel and carboplatin in the treatment of PS 2 patients with chemotherapy-naive advanced non-small cell lung cancer. *Journal of Thoracic Oncology, 3*(6), 623–630.

Lee, J. H., Choi, G., Oh, Y. J., Park, J. W., Choy, Y. B., Park, M. C., et al. (2012). A nanohybrid system for taste masking of sildenafil. *International Journal of Nanomedicine, 7*, 1635–1649.

Lee, E. S., Gao, Z., Kim, D., Park, K., Kwon, I. C., & Bae, Y. H. (2008). Super pH-sensitive multifunctional polymeric micelle for tumor pH(e) specific TAT exposure and multidrug resistance. *Journal of Controlled Release, 129*(3), 228–236.

Leprince, L., Dogimont, A., Magnin, D., & Demoustier-Champagne, S. (2010). Dexamethasone electrically controlled release from polypyrrole-coated nanostructured electrodes. *Journal of Materials Science Materials in Medicine, 21*(3), 925–930.

Loretz, B., Thaler, M., & Bernkop-Schnürch, A. (2007). Role of sulfhydryl groups in transfection? A case study with chitosan-NAC nanoparticles. *Bioconjugate Chemistry, 18*(4), 1028–1035.

Luo, X., & Cui, X. T. (2009). Sponge-like nanostructured conducting polymers for electrically controlled drug release. *Electrochemistry Communications, 11*(10), 1956.

Maeda, H., & Matsumura, Y. (1989). Tumoritropic and lymphotropic principles of macromolecular drugs. *Critical Reviews Therapeutic Drug Carrier System, 6*(3), 193–210.

Maeda, A., Shinoda, T., Ito, T., Baba, K., Oku, N., & Mizumoto, T. (2011). Evaluating tamsulosin hydrochloride-released microparticles prepared using single-step matrix coating. *International Journal of Pharmaceutics, 408*, 84–90.

Maeda, T., Titani, K., & Sekiguchi, K. (1994). Cell-adhesive activity and receptor-binding specificity of the laminin-derived YIGSR sequence grafted onto staphylococcal protein A. *Journal of Biochemistry, 115*(2), 182–189.

Maeda, H., Wu, J., Sawa, T., Matsumura, Y., & Hori, K. (2000). Tumor vascular permeability and the EPR effect in macromolecular therapeutics: a review. *Journal of Controlled Release, 65*(1–2), 271–284.

Manna, U., & Patil, S. (2010). Glucose-triggered drug delivery from borate mediated layer-by-layer self-assembly. *ACS Applied Materials and Interfaces*, *2*(5), 1521–1527.
Mastrotto, F., Caliceti, P., Amendola, V., Bersani, S., Magnusson, J. P., Meneghetti, M., et al. (2011). Polymer control of ligand display on gold nanoparticles for multimodal switchable cell targeting. *Chemistry Communications (Cambridge)*, *47*(35), 9846–9848.
Matsumura, Y., & Maeda, H. (1986). A new concept for macromolecular therapeutics in cancer chemotherapy: mechanism of tumoritropic accumulation of proteins and the antitumor agent smancs. *Cancer Research*, *42*(12 Pt1), 6387–6392.
Mizumoto, T., Tamura, T., Kawai, H., Kajiyama, A., & Itai, S. (2008a). Formulation design of an oral, fast-disintegrating dosage form containing taste-masked particles of famotidine. *Chemistry and Pharmaceutical Bulletin (Tokyo)*, *56*(7), 946–950.
Mizumoto, T., Tamura, T., Kawai, H., Kajiyama, A., & Itai, S. (2008b). Formulation design of taste-masked particles, including famotidine, for an oral fast-disintegrating dosage form. *Chemical and Pharmaceutical Bulletin (Tokyo)*, *56*(4), 530–535.
Moktan, S., Perkins, E., Kratz, F., & Raucher, D. (2012). Thermal targeting of an acid-sensitive doxorubicin conjugate of elastin-like polypeptide enhances the therapeutic efficacy compared with the Parent compound in vivo. *Molecular Cancer Therapy*, *11*(7), 1547–1556.
Nakatsuka, M. A., Lee, J. H., Nakayama, E., Hung, A. M., Hsu, M. J., Mattrey, R. F., et al. (2011). Facile one-pot synthesis of polymer-phospholipid composite microbubbles with enhanced drug loading capacity for ultrasound-triggered therapy. *Soft Matter*, *2011*(7), 1656–1659.
Nanda, A., Kandarapu, R., & Garg, S. (2002). An update on taste masking technologies for oral pharmaceuticals. *Indian Journal of Pharmaceutical Sciences*, *64*, 10–17.
Nanjawade, B. K., Manvi, F. V., & Manjappa, A. S. (2007). In situ-forming hydrogels for sustained ophthalmic drug delivery. *Journal of Controlled Release*, *122*(2), 119–134.
Nolop, K. B., Rhodes, C. G., Brudin, L. H., Beaney, R. P., Krausz, T., Jones, T., et al. (1987). Glucose utilization in vivo by human pulmonary neoplasms. *Cancer*, *60*(11), 2682–2689.
O'Brien, M. E., Socinski, M. A., Popovich, A. Y., Bondarenko, I. N., Tomova, A., Bilynsky, B. T., et al. (2008). Randomized phase III trial comparing single-agent paclitaxel Poliglumex (CT-2103, PPX) with single-agent gemcitabine or vinorelbine for the treatment of PS 2 patients with chemotherapy-na?ve advanced non-small cell lung cancer. *Journal of Thoracic Oncology*, *3*(7), 728–734.
Park, K. (2011). Shockwave-ruptured nanopayload carriers (SHERPAs) for ultrasound-triggered drug release. *Journal of Controlled Release*, *155*(3), 343.
Paz-Ares, L., Ross, H., O'Brien, M., Riviere, A., Gatzemeier, U., Von Pawel, J., et al. (2008). Phase III trial comparing paclitaxel poliglumex vs docetaxel in the second-line treatment of non-small-cell lung cancer. *British Journal of Cancer*, *98*(10), 1608–1613.
Randale, S. A., Dabhi, C. S., Tekade, A. R., Belgamwar, V. S., Gattani, S. G., & Surana, S. J. (2010). Rapidly disintegrating tablets containing taste masked metoclopramide hydrochloride prepared by extrusion-precipitation method. *Chemical and Pharmaceutical Bulletin (Tokyo)*, *58*(4), 443–448.
Rousselle, C., Smirnova, M., Clair, P., Lefauconnier, J. M., Chavanieu, A., Calas, B., et al. (2001). Enhanced delivery of doxorubicin into the brain via a peptide-vector-mediated strategy: saturation kinetics and specificity. *Journal of Pharmacology and Experimental Therapeutics*, *296*(1), 124–131.
Ruel-Gariépy, E., & Leroux, J. C. (2004). In situ-forming hydrogels–review of temperature-sensitive systems. *European Journal of Pharmaceutics and Biopharmaceutics*, *58*(2), 409–426.

Saito, G., Swanson, J. A., & Lee, K. D. (2003). Drug delivery strategy utilizing conjugation via reversible disulfide linkages: role and site of cellular reducing activities. *Advanced Drug Delivery Reviews*, *55*(2), 199–215.

Sako, K., Nakashima, H., Sawada, T., & Fukui, M. (1996). Relationship between gelation rate of controlled-release acetaminophen tablets containing polyethylene oxide and colonic drug release in dogs. *Pharmaceutical Research*, *13*, 594–598.

Sawada, T., Kondo, H., Nakashima, H., Sako, K., & Hayashi, M. (2004). Time-release compression-coated core tablet containing nifedipine for chronopharmacotherapy. *International Journal of Pharmaceutics*, *280*(1–2), 103–111.

Sawant, R. R., Sriraman, S. K., Navarro, G., Biswas, S., Dalvi, R. A., & Torchilin, V. P. (2012). Polyethyleneimine-lipid conjugate-based pH-sensitive micellar carrier for gene delivery. *Biomaterials*, *33*(15), 3942–3951.

Schild, H. G. (1992). Poly(*N*-isopropylacrylamide): experiment, theory and application. *Progress in Polymer Science*, *17*, 163–249.

Shamay, Y., Adar, L., Ashkenasy, G., & David, A. (2010). Light induced drug delivery into cancer cells. *Biomaterials*, *32*(5), 1377–1386.

Sheshala, R., Khan, N., & Darwis, Y. (2011). Formulation and optimization of orally disintegrating tablets of sumatriptan succinate. *Chemistry and Pharmaceutical Bulletin (Tokyo)*, *59*(8), 920–928.

Shimizu, T., Nakano, Y., Morimoto, S., Tabata, T., Hamaguchi, N., & Igari, Y. (2003). Formulation study for lansoprazole fast-disintegrating tablet. I. Effect of compression on dissolution behavior. *Chemistry and Pharmaceutical Bulletin (Tokyo)*, *51*(8), 942–947.

Shive, M. S., Hoemann, C. D., Restrepo, A., Hurtig, M. B., Duval, N., Ranger, P., et al. (2006). BST-CarGel: In situ chondroInduction for cartilage repair. *Operative Techniques in Orthopaedics*, *16*(4), 271–278.

Sugao, H., Yamazaki, S., Shiozawa, H., & Yano, K. (1998). Taste masking of bitter drug powder without loss of bioavailability by heat treatment of wax-coated microparticles. *Journal of Pharmaceutical Sciences*, *87*, 96–100.

Suzuki, M., Hirota, M., Hagino, S., Itagaki, H., & Aiba, S. (2009). Evaluation of changes of cell-surface thiols as a new biomarker for in vitro sensitization test. *Toxicology In Vitro*, *23*(4), 687–696.

Tasaki, H., Ishii, T., Kasashima, Y., Kojima, R., Umejima, H., Kondo, H., et al. (October 6, 2011). Astellas Pharma Inc. Orally disintegrating tablet. International patent application WO2011121824.

Tasaki, H., Yoshida, T., Maeda, A., Katsuma, M., & Sako, K. (2009). Effects of physicochemical properties of salting-out layer components on drug release. *International Journal of Pharmaceutics*, *376*(1–2), 13–21.

Torres, A. G., & Gait, M. J. (2012). Exploiting cell surface thiols to enhance cellular uptake. *Trends in Biotechnology*, *30*(4), 185–190.

Vukelja, S. J., Anthony, S. P., Arseneau, J. C., Berman, B. S., Cunningham, C. C., Nemunaitis, J. J., et al. (2007). Phase 1 study of escalating-dose OncoGel (ReGel/paclitaxel) depot injection, a controlled-release formulation of paclitaxel, for local management of superficial solid tumor lesions. *Anticancer Drugs*, *18*(3), 283–289.

Wood, K. C., Zacharia, N. S., Schmidt, D. J., Wrightman, S. N., Andaya, B. J., & Hammond, P. T. (2008). Electroactive controlled release thin films. *Proceedings of the National Academy of Sciences of the United States of America*, *105*(7), 2280–2285.

Wu, W., Shen, J., Banerjee, P., & Zhou, S. (2010). Core-shell hybrid nanogels for integration of optical temperature-sensing, targeted tumor cell imaging, and combined chemo-photothermal treatment. *Biomaterials*, *31*(29), 7555–7566.

Yamaguchi, H., & Tominaga, T. (March 15, 2007). Astellas Pharma Inc. Microparticle of hardly-soluble substance having enteric base material adsorbed on the surface of the substance. International patent application WO2007029660.

Yamaoka, T., Tamura, T., Seto, Y., Tada, T., Kunugi, S., & Tirrell, D. A. (2003). Mechanism for the phase transition of a genetically engineered elastin model peptide (VPGIG)40 in aqueous solution. *Biomacromolecules*, *4*, 1680–1685.

Yamashita, K., Nakate, T., Okimoto, K., Ohike, A., Tokunaga, Y., Ibuki, R., et al. (2003). Establishment of new preparation method for solid dispersion formulation of tacrolimus. *International Journal of Pharmaceutics*, *267*(1–2), 79–91.

Yamazawa, A., Liang, X., Yoshida, T., Asanuma, H., & Komiyama, M. (2000). Photo-Regulation of DNA polymerase reaction by oligonucleotides bearing azobenzene. *Angewandte Chemie International Edition*, *39*, 2356–2357.

Yang, J., Yamato, M., Nishida, K., Ohki, T., Kanzaki, M., Sekine, H., et al. (2006). Cell delivery in regenerative medicine: the cell sheet engineering approach. *Journal of Controlled Release*, *116*(2), 193–203.

Yan, Y., Wang, Y., Heath, J. K., Nice, E. C., & Caruso, F. (2011). Cellular association and cargo release of redox-responsive polymer capsules mediated by exofacial thiols. *Advanced Materials*, *23*(34), 3916–3921.

Yan, Y. D., Woo, J. S., Kang, J. H., Yong, C. S., & Choi, H. G. (2010). Preparation and evaluation of taste-masked donepezil hydrochloride orally disintegrating tablets. *Biological and Pharmaceutical Bulletin*, *33*(8), 1364–1370.

Yao, A., Chen, Q., Ai, F., Wang, D., & Huang, W. (2011). Preparation and characterization of temperature-responsive magnetic composite particles for multi-modal cancer therapy. *Journal of Materials Science Materials in Medicine*, *22*(10), 2239–2247.

Yin, L., Ding, J., He, C., Cui, L., Tang, C., & Yin, C. (2009). Drug permeability and mucoadhesion properties of thiolated trimethyl chitosan nanoparticles in oral insulin delivery. *Biomaterials*, *30*(29), 5691–5700.

Yoshida, T., Kurimoto, I., Yoshihara, K., Umejima, H., Ito, N., Watanabe, S., et al. (2012). Aminoalkyl methacrylate copolymers for improving the solubility of tacrolimus I: evaluation of solid dispersion formulations. *International Journal of Pharmaceutics*, *428*(1–2), 18–24.

Yoshida, T., Nakanishi, K., Maeda, A., Sako, K., Kasashima, Y., Kondo, H., et al. (2009a). Astellas Pharma Inc. Pharmaceutical composition containing lipophilic IL-2 production inhibitor. International patent application WO/2009/054463. July 14, 2010.

Yoshida, T., Tasaki, H., Maeda, A., Katsuma, M., Sako, K., & Uchida, T. (2008a). Mechanism of controlled drug release from a salting-out taste-masking system. *Journal of Controlled Release*, *131*(1), 47–53.

Yoshida, T., Tasaki, H., Maeda, A., Katsuma, M., Sako, K., & Uchida, T. (2008b). Optimization of salting-out taste-masking system for micro-beads containing drugs with high solubility. *Chemistry and Pharmaceutical Bulletin (Tokyo)*, *56*(11), 1579–1584.

Yoshida, T., Tasaki, H., Maeda, A., Katsuma, M., Sako, K., & Uchida, T. (2009b). Salting-out taste-masking system generates lag time with subsequent immediate release. *International Journal of Pharmaceutics*, *365*(1–2), 81–88.

You, J., Shao, R., Wei, X., Gupta, S., & Li, C. (2010). Near-infrared light triggers release of Paclitaxel from biodegradable microspheres: photothermal effect and enhanced antitumor activity. *Small*, *6*(9), 1022–1031.

Zelikin, A. N., Quinn, J. F., & Caruso, F. (2006). Disulfide cross-linked polymer capsules: en route to biodeconstructible systems. *Biomacromolecules*, *7*(1), 27–30.

Zentner, G. M., Rathi, R., Shih, C., McRea, J. C., Seo, M. H., Oh, H., et al. (2001). Biodegradable block copolymers for delivery of proteins and water-insoluble drugs. *Journal of Controlled Release*, *72*, 203–215.

Application of responsive polymers in implantable medical devices and biosensors

11

J. Li, M. Stachowski, Z. Zhang
Teleflex Medical, Cambridge, MA, USA

11.1 Introduction

Implantable devices are man-made implants partly or totally introduced into the human body and intended to remain after the procedure. These devices reside within the human body either temporarily or permanently, for diagnostic, monitoring, or therapeutic purposes. Examples include catheters, stents, drug delivery systems, pacemakers, implantable cardiac defibrillators, and neurostimulators. Currently more than 25 million US citizens rely on them for life-critical functions (Daniel, 2008). The market was valued at \$43.1 billion in 2011 and is expected to increase 8% annually over 7 years to be worth \$73.9 billion by 2018 (Implantable Medical, 2013).

Considering that biological structures, such as skin, muscles, and blood vessels, are responsive to both external and internal stimuli, for decades people have attempted to design synthetic polymers that could match the "smartness" of natural materials (Hoffman, 2013). Stimuli-responsive polymers are polymers that respond with noticeable physical or chemical changes in their behaviors with a small change of environmental stimuli variations. They are also called "environmentally sensitive," "smart," or "intelligent" polymers. For example, some thermo-responsive polymers (also known as temperature-sensitive) such as poly(*N*-isopropyl acrylamide) (PNIPAAm), undergo a sharp coil to globule transition at the critical temperature, below which it changes from a hydrophilic to a hydrophobic state. Correspondingly, the shape, the mechanical properties, and/or the affinity to cells and biomolecules can also change under the transition temperature. Besides temperature, other stimuli that could trigger changes in polymer properties include but are not limited to pH, light, ionic strength, electric field, magnetic field, solvents, and the presence of chemicals and enzymes. Due to their unique properties, the stimuli-responsive polymers have shown many potential applications as implantable medical devices including: (1) facilitating the deployment or removal of the devices with minimum damage to the host tissue such as application on catheters, stents, sutures, and wound dressings; (2) supporting the function of current devices or treating ailments, such as responsive contact lenses or aneurysm removal devices; (3) delivering drugs or controlling infections corresponding to environmental changes or molecular stimuli, such as implantable insulin injectors; or (4) monitoring physiological factors or biomolecules, such as implantable glucose monitors.

Switchable and Responsive Surfaces and Materials for Biomedical Applications. http://dx.doi.org/10.1016/B978-0-85709-713-2.00011-0

Compared with other *in vitro* biomedical applications, there are significant technical and regulatory hurdles in applying responsive polymers as implants. Biocompatibility, a measurement of how compatible a device is with a biological system, is probably the most complex issue of introducing a new polymer to an implant. The biocompatibility of medical devices and responsive polymers will be first addressed in the following section. In this chapter, responsive polymers are reviewed as various implants, divided by application areas which include cardiovascular devices, ophthalmic devices, surgical devices, dental and orthopedic devices, respiratory devices, gastrointestinal devices, urogenital devices, drug delivery devices, and implantable biosensors.

11.2 Biocompatibility of medical devices and responsive polymers

Biocompatibility of a medical device is, by definition, a measurement of how compatible a device is with a biological system. In terms of implantable materials, biocompatibility has been defined as "the ability of a material to perform with an appropriate host response in a specific application" (Williams, 1987). Various foreign body responses occur as a material is implanted and the *in vivo* environment poses a significant challenge. For a typical subcutaneous implant, the foreign body response is schemed in Figure 11.1(a) (Anderson, 2001; Bridges & Garcia, 2008). The surface of the device is first covered by proteins and biomolecules as a response to the implanted foreign material. The implantation stimulates neutrophils in response to inflammation signals. The sequentially differentiated macrophage cells are activated to secrete inflammatory mediators and fuse to form multinucleated foreign body giant cells (FBS). Finally, fibroblast cells access the site and develop a collagenous fibrous capsule around the implant to isolate the foreign body from the natural tissue (Anderson, Rodriguez, & Chang, 2008). For an intravascular implant, such as an intravenous catheter or a stent, the foreign body reaction can be different (Figure 11.1(b)) (Gorbet & Sefton, 2004; Ratner, 2007). Blood protein adsorption is the first event involved in blood–materials interactions, activating the clotting factors and triggering a cascade of blood coagulation, an intrinsic pathway. Injury also activates an extrinsic pathway through tissue factor expression by cells. The whole cascade generates an enzyme called thrombin, which converts the soluble protein fibrinogen to an insoluble fibrin. The fibrin acts as the foundation of the blood clot by providing a scaffold for the entrapment of platelets, blood cells, and other plasma proteins. Once a fibrin clot is formed, the clot itself can promote more clotting. While very simplified, the above mechanisms indicate that foreign body responses are a progressive process and vary significantly with materials, time, and implant sites. It has always been a significant challenge for biomedical scientists and engineers to find a material that could be biocompatible in most *in vivo* situations, especially as a long-term implant.

As a real medical device is implanted, the injury during the implantation and involvement of microbial burden makes the material–body interactions more complex.

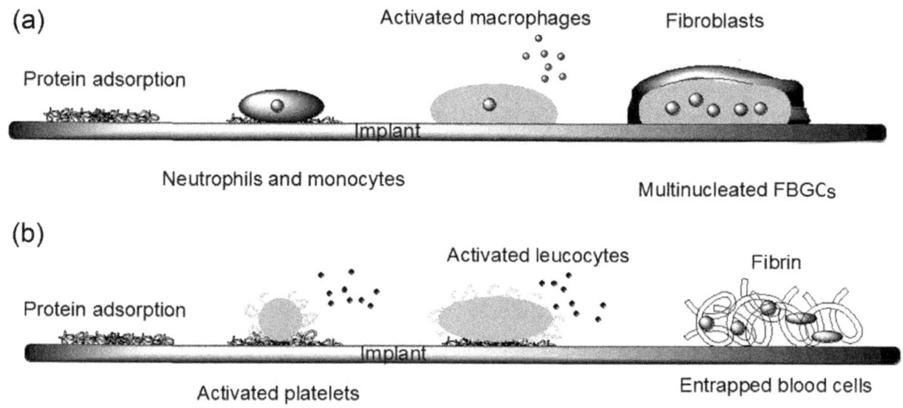

Figure 11.1 (a) Sequential events of foreign body responses of a subcutaneous implant (Anderson, 2001; Bridges & Garcia, 2008): (1) protein adsorption takes place on implant surfaces; (2) neutrophils and monocytes emigrate from the vasculature and adhere to the layer of adsorbed proteins; (3) differentiated macrophages become activated, secreting a variety of inflammatory mediators, and often fuse into multinucleated foreign body giant cells (FBGCs); and (4) fibroblasts infiltrate the site and generate a collagenous fibrous capsule around the implant. (b) Sequential events of foreign body responses of an intravascular implant (Gorbet & Sefton, 2004; Ratner, 2007): (1) protein adsorption takes place on implant surfaces; (2) platelets adhere and became activated; (3) leukocytes adhere and become activated, both generating clotting factors such as thrombin; and (4) thrombin converts soluble protein fibrinogen to an insoluble fibrin, providing a scaffold for the entrapment of platelets, blood cells, and other plasma proteins.

Insertion of a cardiovascular catheter, as an example, involves a set of interacted foreign body reactions associated with infection risk (Jacobs, 2003). Within minutes of the insertion of a catheter into a vein, the patient's body begins to react to the presence of a foreign body and to the damage to the endothelial lining of the vein. The insertion site and vessel wall may become inflamed, and a fibrin layer begins to form on the catheter surface. Within 24 h, a layer of platelets and white blood cells is trapped within the fibrin layer on which a thrombus can form. Moreover, a catheter that has just entered the bloodstream may carry some level of microorganisms as introduced through the skin. At the time of insertion, bacteria attachment on catheter surfaces is initiated simultaneously. Once the blood and catheter make contact, blood components facilitate the adhesion and colonization of microorganisms. Extracellular polysaccharide produced by some bacteria, together with the adherent cells known as biofilm, acts as a barrier to host immune responses and protection from antimicrobial therapies (Jacobs, 2003). For a cardiovascular catheter, both device-induced thrombi and infection can pose a significant threat to the health and safety of patients.

Responsive polymers have unique biocompatibility issues when applied as implants since the foreign body interactions may impact both their responsiveness and stability. Protein and cell fouling are especially crucial for implantable diagnostic devices that monitor-responsive signals continuously *in vivo* (Gifford et al., 2006). Capsulation and thrombus formation around the devices can result in decreased function and ultimately device failure. Implantation also provides an unfriendly environment that impacts stability of responsive surfaces and polymers. Macrophages generate free radicals that can degrade polymers oxidatively. For example, polyethers such as polyethylene glycol (PEG) are susceptible to free radicals and oxidative degradation *in vivo*. Polyesters can be hydrolyzed enzymatically or be self-catalyzed to degrade. In addition to the *in vivo* environment, the process of sterilization, storage, and implantation that are required for medical devices provide further challenges to responsive polymers. Generally, a few paths can be taken to improve the biocompatibility of responsive polymers as medical devices: (1) formulate and design-responsive polymers based on known implantable materials, such as polyurethanes, silicones, and polyesters; (2) introduce moieties with anti-fouling properties, such as hydrogels, PEG, and betaine polymers; (3) make polymers degradable *in vivo*; and/or (4) incorporate eluting agents to reduce the inflammatory response or thrombus formation. The following sections will provide examples of design-responsive polymers as implantable devices in different areas.

To mitigate the risk of applying a device to the human body, a series of tests such as those guided by the International Organization of Standards (ISO), are required to be performed on the device prior to human use. Biocompatibility evaluation of medical devices, based on the ISO 10993, is performed to determine the potential harmful effects from contact of the component materials with the body. Biocompatibility tests also evaluate the fitness and function of the devices under an implant environment. It should be noted that requirements of biocompatibility vary considerably based on the device function and design, so most regulatory evaluations are device-orientated rather than material-orientated. Usually, materials that have been proven to be "biocompatible" in one device have to be reevaluated for a different application.

11.3 Cardiovascular devices

Stimuli-responsive polymers have many applications in cardiovascular devices. Among them, a special type of polymer called shape memory polymers (SMPs) are widely investigated and used. SMPs are stimuli-responsive polymers that modify their shape according to changes in external stimuli, mostly temperature. SMPs are able to be deformed to a new shape and retain this temporary shape, only to revert to their original (permanent) shape when heated to a transition temperature, which may be a glass transition temperature (Tg) or a melting temperature. The shape memory effect results from the structure and morphology of the polymer in combination with the processing and programming technology. One example of SMPs was demonstrated by Osaka, et al. in 1995 (Osada & Matsuda, 1995), using the hydrogels prepared from acrylic acid and *N*-stearyl acrylate with *N*,*N'*-methylenebisacrylamide as a cross-linker to form a straight bar. After swelling in water, heated to 50 °C, and coiled and cooled to room temperature, the gel was rigid and retained its coiled structure due to the existence of the crystalline aggregate structure from the long alkyl chains. Upon heating again to 50 °C, the packing of side chains having turned amorphous, the gel became soft and recovered its original straight shape in 15 s.

Toward biomedical applications, a variety of biodegradable SMPs have been synthesized, including poly(ethylene terephthalate) (PET) and PEG copolymers (Park et al., 2004), oligo(ε-caprolactones) (Lendlein, Schmidt, Schroeter, & Langer, 2005), ethylene-vinyl acetate copolymers (Li, Zhu, Zhang, Zhao, & Xu, 1999), copolymer of oligo(ε-caprolactone) dimethacrylate and *N*-butyl acrylate (Lendlein, Schmidt, & Langer, 2001), *t*-butyl acrylate and di(ethylene glycol) dimethacrylate (Yakacki et al., 2007), composite of poly(D,L-lactide) with hydroxyapatite (Zheng, Zhou, Li, & Weng, 2006), and widely used polyurethane-based polymers (Alteheld, Feng, Kelch, & Lendlein, 2005; Ping, Wang, Chen, & Jing, 2005). The biocompatibility of polyurethanes has been well demonstrated for use as catheters (Gutowska et al., 1995) and heart valves (Butterfield, Wheatley, Williams, & Fisher, 2001; Dabagh, Abdekhodaie, & Khorasani, 2005). Metcalfe et al. (2003) performed *in vitro* and *in vitro* studies using an SMP polyurethane for endovascular interventions, and demonstrated that it was less thrombogenic and noncytotoxic. In another study, a series of SMP polyurethanes from Mitsubishi Heavy Industries, including MM5510, MM4520, MM5520, MM6520, and MM7520, have been investigated for their thermomechanical properties for biomedical applications (Baer, Wilson, Matthews, & Maitland, 2007). Some of these were popularly used for the fabrication of medical devices, such as stents and clot extractors (Baer et al., 2009; Baer et al., 2007).

There are two different approaches to use SMPs in medical devices; the generally preferred approach is a design decision that depends on the targeted body system (Gall et al., 2005). In the first approach, the transition temperature of SMPs is tailored to a low temperature up to about 37 °C. Therefore, the shape recovery is able to be activated by the body's thermal energy to simply allow the fast transition at body temperature. However, these kinds of SMPs have to be held in cold storage, typically below room temperature, to avoid unwanted deployments before use. Another disadvantage is that the mechanical properties are highly dependent on the Tg and cannot be altered in the device design. Therefore, it would be impossible to design an extremely stiff device when Tg is close to body temperature, which limits them in terms of applications. In the second approach, the transition temperature is higher than 37 °C. It has the advantage of room temperature storage, but the main disadvantage of this approach is the need to heat the polymer to induce the shape recovery, which may locally damage cells at temperatures of about 5 °C above body temperature. Therefore, some local heating bursts may be used to minimize exposure to elevated temperature and circumvent cell damage.

The research on SMPs in the last two decades has enabled better control of a polymer's transition temperature, moduli, and recovery ratios, etc. Serrano and Ameer (2012) reviewed recent advances of SMPs, covering introduction of SMPs, design of SMPs for biomedical uses, and biomedical applications of SMPs.

Minimally invasive surgery could reduce surgical intervention–involved complications and postsurgery costs. The shape memory properties of SMPs have made them great candidates for making devices for minimally invasive surgery. A medical device made of SMPs in an initially compressed and temporary shape could be inserted into the body through a small incision to expand on demand into its functional, permanent shape (for example, sheet, stent, and so on) at body temperature. For cardiovascular applications, SMPs are mainly used as clot extraction devices (Hartman et al., 2007; Maitland, Metzger, Schumann, Lee, & Wilson, 2002; Metzger, Wilson, Schumann, Matthews, & Maitland, 2002; Small et al., 2007a; Small, Wilson, Benett, Loge, & Maitland, 2005; Wilson, Small Iv, Benett, Bearinger, & Maitland, 2005), in aneurysm embolization (Hampikian, Heaton, Tong, Zhang, & Wong, 2006; Maitland et al., 2007; Metcalfe et al., 2003; Rodriguez et al., 2012; Small et al., 2007b) and as vascular stents (Chen et al., 2007; Gall et al., 2005; Wache, Tartakowska, Hentrich, & Wagner, 2003; Xue, Dai, & Li, 2010; Yakacki et al., 2007).

11.3.1 Clot extraction devices

Thromboembolic vascular occlusion (formation or lodging of a blood clot) is the primary cause for stroke, which is the third leading cause of death and disability in the United States. Researchers at Lawrence Livermore National Laboratory have developed novel microactuator devices based on SMPs to mechanically remove blood clots. In 2002 (Metzger et al., 2002), they performed thermomechanical studies on two polyurethane SMPs, MP4510 and MP 5510 (Mitsubishi Heavy Industries Ltd.), to provide information on conditions for device activation and estimate the recovery forces that the device is able to deliver during the transition from its straight (temporary) to coiled (permanent) shape. It was suggested that the device could successfully grab and extract a clot in a physiological system. They then designed laser-activated SMP devices mounted on a diffusing fiber to be delivered in the straight form through a catheter (Maitland et al., 2002). Three devices were designed including two conventional ischemic stroke devices (coil and umbrella) and one device for releasing embolic coils (microgripper). The optical fiber coupled to the SMP (MP5510, transition temperature at 55 °C) provided heating in the range of 65–85 °C to activate the transition of the SMP. Later, they performed thermal imaging and computer simulation to characterize the optical and photothermal behavior of the laser-activated intravascular thrombectomy device, built from their SMP and an infrared diode laser coupled together by optical fiber. The deployment of the device *in vitro* successfully captured an artificial blood clot within a water-filled bifurcated vessel model under flow (Small et al., 2005).

Recently, a new type of polyurethane-based SMPs were designed and synthesized from hexamethylene diisocynate (HDI), *N,N,N'N'*-tetrakis(hydroxypropyl)ethylenediamine (HPED), and triethanolamine (TEA), showing moderately high recovery forces, excellent optical properties and sharper glass transitions, which translate to fast and well-controlled actuation (Wilson et al., 2005). With the new synthesized SMPs, they further developed an electromechanical microactuator made of SMP and shape memory nickel–titanium alloy (nitinol) for an endovascular device (Small et al., 2007a). After the current-induced electro-resistive heating, the device changes from a straight rod shape into a corkscrew shape (Figure 11.2) for clot removal. They successfully demonstrated its applicability in a water-filled silicone neurovascular model and found the potential thermal damage only localized in the artery wall area adjacent to the device. Furthermore, an *in vivo* test was performed with this extraction device for removal of induced vascular occlusions in the common carotid artery in tested rabbits. By angiography among the five cases, only one failed to restore the blood flow (Hartman et al., 2007).

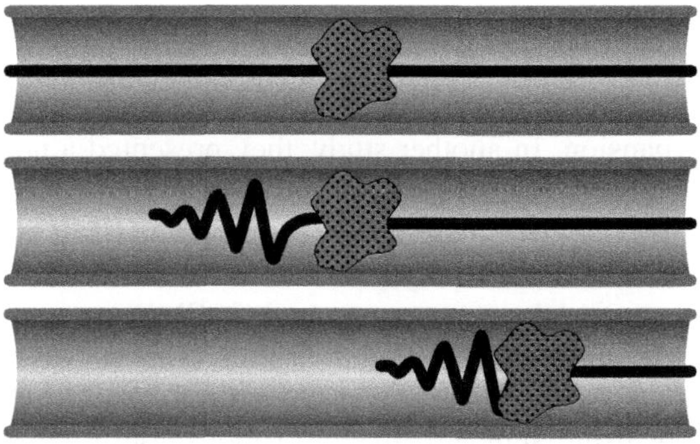

Figure 11.2 Schematic representation of the SMP microactuator for endovascular thrombectomy. (a) The microactuator is delivered distally through the blood clot in its secondary straight rod form; (b) then the microactuator is heated to transform into its primary corkscrew form; and (c) the deployed microactuator is finally retracted to capture the clot. *Source*: Reprinted with permission Small et al. (2007a).

11.3.2 Aneurysm occlusion devices

Endovascular procedures have gained popularity in the treatment of arterial aneurysm, but the drawbacks, including significant incidence of residual lesions, deficient healing at the neck, or recurrences (Raymond & Roy, 1997; Raymond, Desfaits, & Roy, 1999), make the search for new and more effective occlusion materials more and more important. Shape memory polyurethanes have been studied as candidates due to their excellent biocompatibility yet poor thrombogenicity.

Metcalfe et al. (2003) tested cold hibernated elastic memory (CHEM) polyurethane for the treatment of lateral wall aneurysms on the carotid arteries of dogs. The SMP polyurethane was made in open cellular (foamy) structures with the glass transition temperature of 60 °C. After 3 weeks, *in vivo* experiments showed that the CHEM foam led to improved aneurysm embolization, and a thick neointima formation by favoring the ingrowth of cells. However, residual necks and recurrences were observed in some cases due to the incomplete filling of the vascular lumen.

Endovascular coiling is preferred to surgical clipping for the treatment of intracranial aneurysms, which affect between 2% and 6% of the world population (Ahuja et al., 1993; Horowitz, Samson, & Purdy, 1997). Hampikian et al. (2006) evaluated the feasibility of using a novel block copolymer polyurethane, Calomer™, to replace platinum in aneurysm coils. The thermomechanical and radiographic properties were tested for both SMP polyurethane and a composite made of SMP with 3% metal filler. The intracranial aneurysm helical coils made from the SMP composite were successfully deployed inside an aneurysm model with simulated fluid forces. The scientists at Lawrence Livermore National Laboratory also developed a laser-activated SMP foam device for aneurysm occlusion (Maitland et al., 2007). Their SMP was synthesized from HDI, HPED, and TEA, similar to that used in the clot removal device (Wilson et al., 2005). In their preliminary *in vitro* basilar aneurysm model study, the

SMP foams could be laser deployed with the influence of flow rate. At no flow, there was fast and full expansion with overheating at the aneurysm wall; low flow rate led to slow, full expansion with minimal temperature increase; high flow rate resulted in incomplete expansion. In another study, they presented a novel prototype device, based on SMPs, for endovascular embolization of fusiform aneurysms (Small et al, 2007a,b). It consisted of two main components: (1) an SMP stent made of stainless steel dip-coated with MM-5520 polyurethane (Mitsubishi Heavy Industries) for maintaining an open lumen in the parent artery and (2) SMP foam attached to the stent for embolization of the aneurysm. They demonstrated this device as a potential tool for treating non-necked fusiform aneurysm. In order to allow for visualization during vascular treatment using standard imaging (fluoroscopy), opacification of an SMP device for intracranial aneurysm was developed recently by the same research group (Rodriguez et al., 2012). Opacified SMP foam was made by the incorporation of tungsten during the synthesis of SMP polyurethane from HDI, HPED, and TEA. By adding 4% of tungsten particles, the doped SMP foam exhibited increased mechanical properties without significantly changing transition temperature and inflammatory response *in vivo*, compared to the original SMP foam. This suggested more practical devices for intracranial aneurysm.

11.3.3 Stents

Cardiovascular stents are synthetic material scaffolds used to expand and/or support blood-carrying vessels. The first clinical application of a metallic stent was performed in 1986 by Sigwart, Puel, Mirkovitch, Joffre, & Kappenberger (1987). Since then, stents have been used in a majority of the percutaneous coronary interventions, performed because they can be delivered via minimally invasive surgery resulting in rapid recovery time and with less surgical risk. So far, stents have been made of metals such as stainless steel, tantalum, nitinol, cobalt alloy, and platinum iridium; however, they suffer from limited flexibility, stiffness mismatch, compliance mismatch, thrombogenicity, intimae proliferation, etc. (Serrano & Ameer, 2012; Wache et al., 2003; Yakacki et al., 2007). Therefore, a polymer coating on a metal stent is usually used to improve its biocompatibility, and/or load drugs to improve restenosis complications. With their shape memory characteristic, SMPs appear to be important candidates for a new generation of vascular stents with the possibility to overcome the above-mentioned drawbacks from metallic stents, as well as lower total cost. Furthermore, an SMP stent can be packed into various extreme shapes without permanent deformation after deployment. An SMP stent deployed from two different stored shapes has been demonstrated (Gall et al., 2005).

For stent uses, a polyurethane SMP was molded into a primary shape and the size of the target artery, then crimped into a small secondary shape and mounted on a delivery system, inserted into an artery, and navigated to the lesion site, where it could finally be thermally actuated to recover its primary shape (Baer et al., 2007). With a polyurethane SMP, Wache et al. (2003) not only performed elongation, fixation, and recoil tests to demonstrate its feasibility as an alternative to the conventional stent but also conducted its drug loading and release study for drug-eluting capability. From cross-linked poly(t-butyl acrylate)-poly(ethylene glycol) dimethacrylate, Yakacki et al. (2007) designed several SMP stents including solid and 50% perforated stents targeted to be recovered at body temperature. The recovery and mechanical properties of the SMP were tailored by its glass transition and rubbery modulus independently. They have found that the full shape recovery time was highly dependent on the glass transition temperature, degree of cross-linking, and deformation (storage) temperature.

Development of rapid self-expandable stents has been conducted (Chen et al., 2007; Lauto et al., 2001; Meng et al., 2006; Tamai et al., 2000; Venkatraman, Tan, Joso, Boey, & Wang, 2006; Xue et al., 2010). In 2000, Tamai et al. fabricated the self-expandable coronary stent from biodegradable poly-L-lactic acid (PLLA) (Tamai et al., 2000). Its self-expansion was able to be achieved in 0.2 s at 70 °C, 13 s at 50 °C, or 20 min at 37 °C. With a total of 25 stents implanted, neither thrombosis nor major cardiac event was observed within 30 days. Later, Venkatraman et al. (2006) developed a biodegradable stent from PLLA and poly-D-L-lactide-glycolide (PLGA). This temperature-responsive stent was able to self-expand at 37 °C; however, its complete expansion could not be shorter than 8 min. Fabricated from cross-linked chitosan blended with glycerol and poly(ethylene oxide), (PEO) SMP, the biodegradable stent was able to be rapidly recovered from crimp to full expansion (permanent) status at 37 °C by hydration after 150 s by Chen et al. (2007). Their preliminary *in vivo* tests in a rabbit artery (abdominal aorta) showed 100% patency. After 24 h, neither migration of the implanted stent nor thrombus formation occurred. However, the ideal self-expansion time is less than 1 min to benefit the *in vivo* deployment and prevent migration after insertion (Venkatraman et al., 2006). This was achieved by another degradable, fast self-expandable stent by Xue et al. (2010). In their work, the SMP of block copolymers containing hyperbranched three-arm poly(ε-caprolactone) (PCL) as a switching segment and poly[(R)3-hydrobutyrate-co-(R)-3-hydroxyvalerate] (PHBV) as the hard segment, was designed, synthesized, and characterized. The obtained SMP showed a transition temperature of 39–40 °C, suitable for thermal switching at body temperature. Their stent from the copolymer with 25% of PHBV exhibited an excellent shape memory property with complete self-expansion at 37 °C for only 25 s.

The SMP (chitosan blending with glycerol and PEO) used for self-expandable stent (Chen et al., 2007) and was also used to make a sirolimus-eluting stent by Chen et al. (2009) for the treatment of restenosis. In their report, a hydrophobic heparin coating was made on the stent to prevent thrombosis, as well as to offer a diffusion barrier for the drug. The obtained stent kept its shape memory effect with controlled release of sirolimus in a nearly linear way. The successful sirolimus-eluting stent was proven when implanted into rabbit infrarenal abdominal aortas by significantly reducing neointimal formation.

Other than direct heating, other stimuli, such as light (Maitland et al., 2002), electric current (Koerner, Price, Pearce, Alexander, & Vaia, 2004), or magnetic field (Buckley et al., 2006), have been used as indirect heating sources to deploy an SMP. Lendlein et al. first revealed that ultraviolet (UV) light could be used to deploy recovery of shape memory polymers at ambient temperature (Lendlein, Jiang, Junger, & Langer, 2005). On the other hand, Huang (Huang, 2010; Huang, Yang, Liu, & Phee, 2007; Yang, Huang, Li, Lee, & Li, 2004) reported a thermo-moisture-responsive

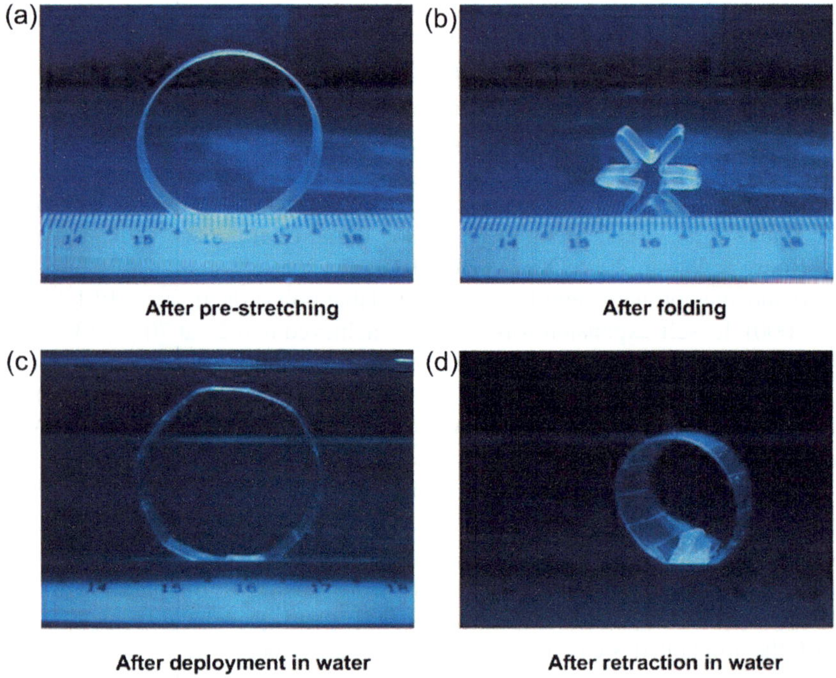

Figure 11.3 Retraction of a pre-deformed polyurethane SMP stent in water. (a) Round shape after pre-stretching; (b) star shape after folding; (c) expanded after deployment in water; and (d) smaller size after retraction in water.

Source: Reprinted with permission from Huang et al. (2007).

polyurethane SMP. When immersed in water, the SMP can absorb moisture to significantly reduce its Tg up to 15 °C. Therefore, the shape recovery can be triggered by immersing the SMP in ambient water to reach its Tg instead of heating the material. Therefore, this moisture-driven shape recovery material opens the door for new potential applications. For example, an expandable but retractable stent was demonstrated with MM3520 polyurethane (Mitsubishi Heavy Industries) in Figure 11.3 (Huang et al., 2007). Thin film polyurethane SMP was created and wrapped into a round shape (a), then mechanically folded into a star shape (b) with a much smaller diameter for delivery to the location by a catheter; after deployment in water, it expanded into another shape (c) in a standard mechanical way; finally after a period, it retracted to a significantly smaller size (d), due to shape recovery by water absorption. Therefore, it is easy to remove the stent.

11.3.4 Temporary endovascular occlusion devices

Another example of stimuli-responsive polymers in a vascular application is a commercialized thermosensitive polymer, LeGoo, used for temporary endovascular occlusion. To temporarily occlude the vascular branches for hemostasis in vascular surgery, free dissection and clamping of the branches or occluding balloons introduced into branch ostia can be accomplished. However, they suffer from some drawbacks, especially leading to extra operative trauma and time. LeGoo potentially offers a simple, rapid, yet reliable solution to temporary arterial occlusion with less blood loss.

LeGoo is a purified thermal sensitive copolymer (Poloxamer 407) in saline solution by Pluromed (Woburn, MA, acquired by Sanofi in May 2012). Poloxamer 407 is a triblock copolymer of poly(ethylene oxide)–poly(propylene oxide)–poly(ethylene oxide) (PEO–PPO–PEO) with PPO molecular mass of 4000 g/mol and a 70% PEO content. It is a nontoxic, nonabsorbed, nonmetabolized polymer, and does not lead to an embolism. At a concentration over 12% in water, Poloxamer 407 solution undergoes a rapid transition from liquid at room temperature to a solid gel after being exposed to physiological temperature, and redissolves upon cooling, exhibiting good thermo-sensitivity. Therefore, intravascular injection of LeGoo results in a vascular plug suitable for use against temporary vascular occlusion. The gel plug erodes in blood as the polymer is soluble. The dissolution time of the gel depends on the amount of the injected polymer, the concentration of polymer, and the local temperature. LeGoo has been injected during coronary artery bypass surgery by cannula introduced via an arteriotomy, called the "cannulation method" (Bouchot, Berger, Berne, Brunotte, & Brenot, 2010; Rastan et al., 2010; Wimmer-Greinecker et al., 2011). Also in aortic surgery to temporarily stop bleeding from the arterial ostia, the method of injecting LeGoo by a device that arrests blood flow, for example, a balloon, called "a retrograde method," was also studied (Dregelid, 2012). In this way, the plug could form without injecting LeGoo as far into the vessel as required by the cannulation method, reducing the risk of side-branch embolization. Dregelid's preliminary *in vitro* study suggested that the injection of this thermosensitive polymer into the arterial ostia by retrograde method is feasible.

One concern in coronary occlusion during off-pump coronary artery bypass (OPCAB) is endothelial damage to the target vessel. Gucu et al. (2013) performed an *in vivo* study on rats by occlusion of the aorta with LeGoo, with a microvascular clamp, and with no occlusion as a control. Among the three groups, it was found that Poloxamer 407 gel as vascular occlusion led to no or minimal damage on the endothelium. In a different study, Bouchot, Aubin, Carrier, Cohn, and Perrault (2006) reported that neither functional nor structural damage of the endothelium of the coronary arteries, or ischemic injury to myocardium, was found when Poloxamer 407 was utilized for temporary occlusion. The results have shown that LeGoo is a quick, safe, and effective temporary occluder in OPACB surgery.

11.4 Drug delivery/infection control coatings and devices

Catheters are medical devices that can be inserted in the body to treat diseases or perform a surgical procedure. They can be inserted into a body cavity, duct, or vessel to function as drainage, administration of fluids or gases, access by surgical instruments, and also to perform a wide variety of other tasks depending on the type of catheter.

Implantable devices, including catheters, are associated with a definitive risk of bacterial and fungal infections. For example, catheter-associated urinary tract infection (CAUTI) is a particularly intransigent problem and is the most common source of hospital-acquired (nosocomial) infections, with roughly 450,000 cases occurring per year in the United States alone (Scott and Roberts, 2007). Biofilms play a pivotal role in health care–associated infections, especially those related to the implantation of medical devices such as various catheters and orthopedic implants. To meet the challenge, some new techniques have surfaced. For example, a nonleaching sulfobetaine surface was fabricated on a vascular catheter by Semprus Biosciences (Cambridge, MA, now part of Teleflex), to reduce bacteria attachment as well as thrombus formation due to the antifouling properties of the zwitterionic polymers (Smith et al., 2012).

Currently, the incorporation of antimicrobial agents in devices is still widely used to prevent the development of biofilm-related infections. Among them, the development of responsive polymers for drug delivery is ever more promising as it offers unprecedented control over drug dosing and can provide the ability to give sustained or pulsatile release mode of action.

11.4.1 Catheter-based drug release coatings

While they can be used as catheter materials, stimuli-responsive polymers are preferred as catheter coatings for drug release. In 1995, Gutowska et al. demonstrated the possibility of PNIPAAm coating on polyurethane intravenous catheters (Vialon®) for heparin release triggered to improve antithrombogenic properties (Gutowska et al., 1995). To improve the coating formed on a catheter, Biomer®, a hydrophobic multiblock copolyether–urethane–urea, was used with better binding strength as well as enhanced mechanical properties. The coating was prepared by dip-coating in *N,N'*di-methylacetamide (DMAC) followed by coprecipitation in hot water. The loading of heparin was performed at lower temperature at which PNIPAAm has a high swelling ratio to enable more drug loading. At body temperature above the LCST of PNIPAAm, *in vitro* release of heparin was rapid in the initial several minutes (squeezed out by the collapsed polymer network) followed by slow release controlled by diffusion through the collapsed matrix at 1 $\mu g/cm^2$ per hour up to 6h. Compared with unmodified catheters, *in vivo* experiments in dogs for 1.5h showed the heparin release from the Biomer®/PNIPAAm-coated catheters exhibited reduced thrombus formation.

Stimuli-responsive polymers used as coatings for controlled release of antibiotics are reported (Contreras-García, Bucio, Concheiro, & Alvarez-Lorenzo, 2010, Contreras-García, Alvarez-Lorenzo, Taboada, Concheiro, & Bucio 2011; Ruiz et al., 2008). The groups in Mexico and Spain have demonstrated the possibility of functionalization of a polypropylene (PP) surface with PNIPAAm-based smart polymers to control the release of different drugs for reducing the risk of biofilm formation. PP was chosen as it is widely used as a material for a variety of medical devices including catheters, hernia meshes, and sutures; however, it has also been shown to be easily colonized by bacteria (Anjum, Gulrez, Singh, & Gupta, 2006; Cosson et al., 2003; Freytag, Thies, König, & Welte, 2003). Using a radiation technique, various functional polymers were grafted onto PP, either by polymer brushes, polymer networks, or interpenetrating networks (Muñoz-Muñoz, Ruiz, Alvarez-Lorenzo, Concheiro, & Bucio, 2009; Ruiz et al., 2008). To reduce biofilm formation from gram negative methicillin-resistant *Staphylococcus aureus* (MRSA), vancomycin was released from PP grafted with interpenetrating networks of PNIPAAm and PAAc. AAc was chosen to control the load of drug due to the electric interaction with the weakly basic vancomycin. The existence of PAAc enabled the modified materials to possess dual-responsive properties. The obtained materials could sustain the release of vancomycin for over 8 h at a release rate adequate for preventing bacterial growth. In other publications, control release of negatively charged drugs, such as nalidixic acid to inhibit *Escherichia coli* or nonsteroidal anti-inflammatory drugs of diclofenac and ibuprofen, *N*-(3-aminopropyl) methacrylamide hydrochloride (APMA) was used along with NIPAAm grafted on PP by a preradiation method (Contreras-García et al., 2010, 2011). The *in vitro* tests showed that the controlled release of these drugs was successful.

11.4.2 Drug delivery devices

SMPs were also studied for controlled drug delivery. For example, Wache et al. presented a vascular endoprosthesis from an SMP polyurethane (Wache et al., 2003). This drug-loaded self-expandable stent could also work as a drug delivery system at the same time. Sustained release of drugs from biodegradable SMPs has been investigated (Nagahama, Ueda, Ouchi, & Ohya, 2009; Wischke, Neffe, Steuer, & Lendlein, 2009). Nagahama et al. (2009) reported that SMPs based on the star-shaped branched CL cross-linked with HDI exhibited strongly temperature-sensitive shape recovery within the range of 37–39 °C. With 10 wt% theophylline (drug) loading, the resultant system successfully released the drug over a month in a phosphate buffer solution at 37 °C, without an initial burst release. A complete shape memory copolyester urethane was synthesized from a star-shaped tetrahydroxy telechelics of oligo[(*rac*-lactide)-co-glycolide)] cross-linked with trimethyl-1,6-hexamethylene diisocyanate (TMDI) (Wischke et al., 2009). Drugs with different hydrophobicities, such as enoxacin, nitrofurantoin, and ethacridine lactate, were able to be loaded into the polymer network without changing shape memory functionality. A predictable release of about 90% of the loaded drug in 80 days was observed. The same research group also designed and developed biodegradable SMP for controlled release of incorporated hydrophilic and hydrophobic drugs (Neffe, Hanh, Steuer, & Lendlein, 2009). Their SMP networks were prepared by UV-curing [oligo(ε-caprolactone)-co-glycolide] dimethacrylate precursors with drugs loaded by swelling and *in situ* incorporation. This drug-release system was expected to be used as a ureteral stent with antibacterial properties.

11.5 Surgical devices

11.5.1 Sutures

An SMP is also a good candidate for smart sutures. A responsive suture was made of linear, phase separated multiblock copolymers from oligo(ε-caprolactone)diol and oligo(*p*-dioxanone)diol by Lendlein and Langer (2002). The existence of the

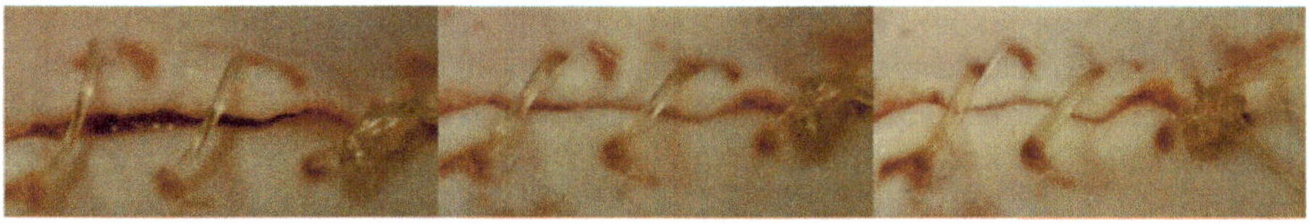

Figure 11.4 Degradable shape-memory suture for wound closure. The photos from the animal experiment show the shrinkage of the fiber with the temperature increasing from 20 °C (left) to 37 °C (middle) and to 41 °C (right).

Source: Reprinted with permission from Lendlein & Langer, (2002).

hydrolyzable ester in the SMP's structure allows biodegradation under physiological conditions. The suture made of a thermoplastic shape memory polymer has been programmed to allow stretching up to about 200% to form a loose knot in a temporary shape. When heated above its transition temp to 40 °C, the fiber shrank and tightened the knot within 20 s. This self-tightening of knots was to minimize the invasiveness of the surgery. This smart suture was further tested in a rat animal model by using it to close an abdominal wound with a standard surgical needle (Figure 11.4). Loosely sutured on the wound, the shape memory materials were stimulated when the temperature was increased to 41 °C. In the test, the fiber was elongated by 200% during programming and was able to generate a force of 1.6 N upon actuating the shape memory effect *in vitro*. The force exerted on the fiber to close the wound can be tailored by programming parameters such as elongation ratios or molecular parameters of the polymer such as the hard segment content.

In addition to copolymers, polymer blends have also been investigated to make SMPs. Zhang et al. reported a novel type of shape memory polymer blend from triblock copolymer of poly(styrene-butadiene-styrene) and poly(ε-caprolactone) (PCL) (Zhang, Wang, Zhong, & Du, 2009). One example of the blend polymer was able to automatically knot in 10 s at 70 °C in water. Their simple fabrication and fast recovery process implied the application as thermo-responsive sutures. Huang et al. also demonstrated the self-tightening of a wire from a thermo-moisture-responsive shape memory polyurethane in water at room temperature (Huang, 2010; Huang et al., 2007). They demonstrated that the tightening of a knot was slowly completed in about 1 h. Furthermore, the removal of a knot was also proven with the same SMP in water. This provided an alternative solution of using water to drive recovery of a suture when heating is not practical.

11.5.2 Wound dressings

Wound care is one of the most lucrative and rapidly expanding medical device market segments for both manufactures and providers. A dressing is applied to a wound in order to promote healing, protection, and/or prevent further harm. It was estimated that demand for wound dressings is expected to increase 6.6% annually to $5.7 billion in 2016. Wound dressing can be divided into two categories: dry and wet dressings. Since it was found that healing is faster in a moist environment than in a dry environment, the wet dressing has gained more attention and become more popular. Numerous wound dressing materials based on calcium alginates, hydrocolloids, and hydrogels along with supporting textile materials have been developed and are on the market.

Earlier in 1970, thermosensitive Pluronic F-127 (triblock copolymer of PEO–PPO–PEO) was used in the treatment of burns, as it forms a gel at body temperature and is easily removed by bathing when it is no longer needed (Schmolka, 1972). In 1989, Henry and Schmolka published a paper that reviewed the use of Poloxamer as burn wound dressings (Henry & Schmolka, 1989). In their summary, thermosensitive Poloxamer 407-based materials can be applied in liquid form to seek all of the depths and contours of the wound where they will then rapidly gel in place at body temperature. As a wound dressing, they have many advantages including low cost, good control in water loss, heat loss and electrolyte loss, good storage stability, as well as easy delivery and removal. Poloxamer 407 gels when combined with humectants for hydration qualities and with antibiotics to prevent infection have been considered to meet all of the requirements of an ideal temporary skin substitute. When occluded with a fluid layer in contact with the tissue surface, wounds reepithelialize more rapidly with less pain and inflammation (Caldwell, Bowser, & Crabtree, 1981).

In the past two decades, more studies have been conducted by incorporating stimuli-responsive polymers for a smart wound dressing. Besides the properties of good flexibility, gas permeability, durability, and ability to control water loss, a wound dressing is expected to be easy and comfortable to remove or change (Gupta, Agarwal, & Alam, 2010). Here, thermosensitive PNIPAAm is regarded as a good candidate for the purpose of being easily stripped off. During its use at physiological temperature, PNIPAAm in a dressing is dehydrated to form a shrinkage structure. To remove, the temperature is reduced below the LCST to make PNIPAAm in a swelling state to expand its chains; then it can be easily stripped off from the wound surface without damaging the new regenerated tissues (Chen et al., 2005; Reddy, Kano, Maruyama, Hadano, & Takahara, 2008; Wang, Chow, Tan, Zhang, & Yang, 2004; Wang, Chow, Phan, Lim, & Yang, 2006; Yang & Lin, 2004). Wound dressings based on PNIPAAm grafted on nonwoven fabric (such as PP) (Chen et al., 2005; Yang & Lin, 2004) or a semi-interpenetrating network (semi-IPN) of PNIPAAm with segmented polyurethane urea (Reddy et al., 2008) have also been prepared for this purpose.

Lin, Chen, Chen, Wu, and Chen (2000) performed an animal study on New Zealand rabbits for treatment of extensive skin injury by comparing several wound dressings: (1) control with dressing materials, (2) nonwoven fabric of polyethylene (PE), (3) nonwoven fabric grafted with PNIPAAm, and (4) a trilayer wound dressing from PNIPAAm-grafted PE, followed by UV grafting of bovine gelatin with glycosaminoglycans to serve as a matrix for infiltration of fibroblasts, macrophages, lymphocytes, and capillaries derived from the wound bed. After 6 weeks of operation, the (1) and (2) groups stayed in the proliferative phase, with no epidermis or dermis structure being traced in the section; however, group (3) healed completely in the maturation phase. The wound site in group (4) totally recovered with normal dermis and epidermis structure around it. The dressing materials of this group fell off automatically from the wound site without any damage to the skin. Thus, a PNIPAAm connection layer could adhere well with the tissue when the wound is in the moist condition and easily separate from the tissue once the wound is closed and in a dry condition.

Hydrogels as a wound dressing could provide the moist environment needed to accelerate wound healing. However, the presence of moisture may also accelerate the proliferation of microorganisms present in the wound site and increase the possibility of infection. Therefore, drugs have been loaded into the wound dressing with PNIPAAm to control release (Gupta, Mishra, & Saxena, 2008; Lin, Chen, & Run-Chu, 2001; Reddy et al., 2008; Wang et al., 2004, 2006). For an ideal wound dressing, one of its critical properties is the ability to promote skin regeneration. PNIPAAm, however, exhibited limited cell interaction (Wang, Chen, Zhao, Guo, & Zhang, 2009). Although PNIPAAm has been used in cell sheet engineering, the direct cell attachment to its surface may be insufficient for applications where tissue regeneration is expected. Therefore, there are studies to make composite dressings with other materials such as chitosan (Chen et al., 2005; Wang et al., 2009; Yang & Lin, 2004) and gelatin (Liu & Ito, 2002). Chitosan is regarded as a biocompatible compound that demonstrates anti-infection and acceleration of wound-healing capabilities. Research groups from Taiwan developed the incorporated chitosan onto nonwoven PP fabric grafted with acrylic acid (AAc) and NIPAAm by UV graft polymerization (Yang & Lin, 2004) or PNIPAAm-grafted PP via immobilization (Chen et al., 2005). Interestingly, these wound dressings also showed antimicrobial activity against such organisms as *Pseudomonas aeruginosa*, *S. aureus*, and *E. coli*. Jiang et al. (2012) chose acryloyl lysine to be introduced into the wound dressing as poly(L-lysine), to improve cell adhesion and proliferation in tissue engineering. In their wound dressing, acryloyl lysine was incorporated into PEG cross-linked PNIPAAm hydrogel. As expected, with the increase of lysine content from 0% to 10% of NIPAAm, the volume phase transition temperature could be tailored in the range of 32–40 °C. An antimicrobial drug, polyhexamethylene biguanide (PHMB), was loaded into the hydrogel at room temperature. The advantage of incorporating lysine into the wound dressing was proven by increased cell interaction *in vitro* and improved wound healing *in vivo* on Sprague Dawley rats. The drug-loaded wound dressing maintained its activity *in vitro* and reduced infection *in vivo*. Therefore, they provided a promising candidate for a wound dressing with the ability of promoting healing while delivering antimicrobial drugs to prevent infection.

11.6 Dental and orthopedic devices

The first SMP, polynorbornene, was developed by CdF Chime company, France, in 1984 and commercialized by Nippon Zeon Company of Japan under the trade name of Norsorex. In 1991, polynorbornene was reported by Nakasima et al. in orthodontic applications (Nakasima, Hu, Ichinose, & Shimada, 1991). The conventional elastic materials used in orthodontic treatment failed to provide a long-lasting, accurate force. With its transition temperature at 35 °C, close to body temperature, polynorbornene exhibited much less deformation at 37 °C over time compared with other orthodontic plastic modules. When applied intraorally, the SMP produced the force gradually and maintained it without any significant loss for 3–4 weeks, making it feasible for aligning human teeth in orthodontics.

One patent was found to use SMPs for fabricating a new type of fixed and removed orthodontic apparatus (Mather, Liu, & Burstone, 2006). The proposed orthodontic devices fabricated from SMPs include ligatures, self-ligating brackets, force modules, torque modules, removable aligner appliances, arch wires, etc. Compared to the traditional materials, the SMP orthodontic devices allowed for easier and more comfortable orthodontic operation for the orthodontist to insert into the patient's mouth, caused less pain to the patient, provided a more aesthetically appealing appliance during the treatment, as well as the ability to be produced at low cost.

Jung and Cho used SMP polyurethane as wire for correction of misaligned teeth (Jung & Cho, 2010). Their polyurethane block copolymer was synthesized from 4,4′-methylene bis(phenylisocyanate) (MDI), poly(ε-caprolactone)diol, and 1,4-butanediol. The obtained SMP wires exhibited transition temperatures at 36–42 °C with the hard segment content at 30–50%. The wires showed good shape recovery force for over 3 months. In their dental model, the polyurethane wire was fabricated by melt-spinning and used as an orthodontic appliance. When the appliance was heated above the transition temperature, the misaligned teeth were moved to the desired position by the shape recovery force. As the SMP wire is transparent to some degree, it provides aesthetic appearance over metals and shape memory alloys.

Jiang et al. reported a polypeptide coating on medical devices (dental and orthopedic) for pH-induced antibiotic capture and release to prevent implant infection (Jiang & Li, 2009). The pH-responsive drug delivery was based on nanocoatings from two biocompatible and weak polyelectrolytes of poly(L-lysine) (PLL) and poly(L-glutamic acid) (PLGA), which could be positively and negatively charged at low pH and high pH, respectively. The substrates can be glass, metal, ceramic, etc., which are commonly used as implants. The formation of this multilayer nanocoating was prepared by electrostatic self-assembly (also known as layer-by-layer assembly), with the advantage regardless of the surface area and shape of the substrates. The thickness of the nanocoatings was controlled by the number of the bilayers and the pH of the coatings when prepared. PLL is fully charged at pH below 8.0, but its net charge decreases with the increase of the pH above 8.0. On the other hand, PLGA is fully charged when the pH is above 6.0. Therefore, more PLL molecules are needed at a pH of 10 than that of 7 to match the surface of its neighboring PLGA layer, and the charge repulsion among PLGA is smaller at a pH of 10. This leads to the thicker bilayer being formed at a pH of 10 than a pH of 7. The uptake and release of drug from PLL/PLGA coatings were pH responsive. The capture of an antibiotic, for example, the negatively charged cefazolin, into the polypeptide nanocoatings was pH induced. Using the PLL/PLGA formed at high pH (Jacobs, 2003), more drug molecules of cefazolin could be incorporated into the nanocoatings at a lower pH (Anderson et al., 2008) than a higher pH (Gorbet & Sefton, 2004) as more PLL are available for binding as pH decreases. Similarly, if the positively charged drug is to be captured, PLL/PLGA coating should be formed at a lower pH (for example, 4) and captured at a higher pH. The release of cefazolin from the multilayer PLL/PLGA was also pH responsive: from pH 7 to 10, they all showed burst release and the release rate increased with pH, and then showed a sustained release could obtained up to 2 weeks. This PLL/PLGA nanocoating on stainless steel with cefazolin exhibited antibacterial activity against colonization of *S. aureus*, which has potential for reducing implant-associated infection.

Ge et al. reported a thermally responsive drug delivery device using a conducting polymer, polypyrrole (PPy), for an orthopedic implant (Ge et al., 2010). The simple system was fabricated by electric deposition of a drug-carried PPy film on the one side of a titanium foil. Titanium was chosen as it is a popular implantable material in orthopedic applications. The drug-doped PPy worked as a cathode with titanium as an anode, leading to a fast release of drug (over 60% released after 12 h) at room temperature without an external power supply. To have the drug released "on demand" in a body environment, a thin eicosane and poly(L-lactide) (PLLA)-blended film was spun coated on the opposite side to cover the naked Ti. Here biodegradable and biocompatible PLLA was used to help eicosane get a better blend film on Ti. At lower temperature, the eicosane-PLLA blend served as an insulator to inhibit the release of drug from PPy. On the other hand, at body temperature, eicosane starts to melt (with its melting point at 36.8 °C), whereas the PLLA exists as a solid to create a porous structure. Then the ions in solution could penetrate the film through the pores to reach Ti to work as a galvanic cell and trigger the massive drug release. With their thermo-responsive system, the model drug (adenosine triphosphate) was released about 82% after 12 h at 37 °C, compared to that of only 14% being released after 12 h at room temperature, suggesting an application of titanium-related orthopedic implants in the wound-healing process.

11.7 Ophthalmic devices

In ocular applications, stimuli-responsive polymers have been studied for controlled release of drugs for ophthalmic diseases. Conventional eye drops suffer from bioavailability due to drainage losses and are therefore not a sufficient method. As early as 1975, polyvinyl alcohol has been studied as a vehicle for ophthalmic drug delivery (Maichuk, 1975; Patton & Robinson, 1975). Since then, more polymers, including collagen (Bloomfield et al., 1977), gelatin (Nadkarni & Yalkowsky, 1993), pluronic F127 (Desai & Blanchard, 1998), PEG-based copolymer (Anumolu, Singh, Gao, Stein, & Sinko, 2009), polyHEMA (Hsiue, Guu, & Cheng, 2001), etc., have been studied for controlled delivery of gentamicin for infected corneas or pilocarpine for glaucoma. Later, thermosensitive PNIPAAm-based polymer was developed by Hsiue, Hsu, Yang, Lee, and Yang (2002) for controlled-release ophthalmic drops for glaucoma therapy. Epinephrine, as a model drug, was loaded into either linear PNIPAAm or a mixture of linear PNIPAAm with cross-linked PNIPAAm nanoparticles at room temperature for control release. It showed that neither of the two systems had a cytotoxic effect. Animal tests on male New Zealand white rabbits were evaluated. The intraocular pressure–lowering effect was found to last for 24 h with linear PNIPAAm and even 32 h for a mixture of linear PNIPAAm and nanoparticles of cross-linked PNIPAAm, compared to only 6 h for traditional ophthalmic drops.

Dual-stimuli-responsive polymers have been prepared (Verestiuc, Ivanov, Barbu, & Tsibouklis, 2004), and controlled-release drugs for ocular therapy from multistimuli-responsive hydrogels have been studied. Verestiuc et al. (2004) synthesized the dual-responsive hydrogels by combining the temperature-sensitive PNIPAAm with biodegradable pH-sensitive cationic polymer of chitosan. They demonstrated that the dual-responsive hydrogels could act as vehicles for controlled release of pilocarpine, which has been widely used in the treatment of chronic open-angle glaucoma and acute angle-closure glaucoma for over a hundred years. While widely used topically in the eye, pilocarpine suffers from low ocular bioavailability and short residence time in the eye due to its good water solubility. Therefore, the stimuli-triggered release as well as sustained release of pilocarpine would be helpful for glaucoma therapy. However, in their preliminary study the release of pilocarpine was fairly fast and most of the drug was released in the first 40 m. To meet this challenge, Casolaro, Casolaro, and Lamponi (2012) developed the dual-stimuli-responsive hydrogels for controlled pilocarpine ocular delivery. The hydrogel synthesized from *N*-acryloyl-L-valine exhibited pH sensitivity (from carboxyl group) and thermosensitivity (the existence of isopropyl and amido functionalities in structure close to PNIPAAm). For example, it deswelled when pH was below about 4 at 25 °C or above 35 °C at pH 4.37. The transition pH and temperature were also influenced by ion strength. After pilocarpine hydrochloride was loaded into the swelling hydrogel through electrostatic interaction, the release of pilocarpine was studied at pH 7.4 at two different temperatures of 25 °C and 36 °C. At both temperatures, the release profile showed burst release for the first several hours followed by sustained release lasting for over a week. At a higher temperature, pilocarpine release increased and was faster due to the shrinking of the gel at this temperature squeezing the drug from the network, in line with its temperature responsiveness. Additionally, the authors also showed the dual-responsive hydrogel had no cytotoxicity against the mouse fibroblast NIH3T3 cells, and the presence of pilocarpine strongly increased cell proliferation even after 2 days.

The use of stimuli-responsive polymer gels for multifocal contact lenses was patented by Guillon and Maissa (2012). Multifocal lenses are used to correct presbyopia and other disorders of accommodation by a gradient of increasing lens power, added to the wearer's correction for the other refractive errors. Currently, multifocal contact lenses (including contact lenses, intraocular lenses, implant lenses, inlay lenses, onlay lenses, or any other ophthalmic refractive devices) lack comfort, discontinuity on the surface of the lens, have issues focusing a distance and near image on the retina at the same time, or yield a poorer quality of retinal image, to name a few. To overcome these drawbacks, the authors provided the idea of contact lenses made at least partially from a responsive polymer gel with the quality of vision achieved with conventional single-vision contact lenses for both distance and near vision. Response to stimuli, such as temperature, pH, light, ion strength, shear force, electric field, magnetic field, or chemical trigger, could be provided by the ocular environment around the lens or by any artificial means within the contact lens, its shape, or refractive index changes. In one arrangement, a micro- or nano-chip is embedded in the contact lens, without being visible to the user, for monitoring changes, such as eye movement, interpupillary distance, or distance between the left and the right contact lens to emit a stimulus such as electric field or magnetic field.

11.8 Respiratory devices

Thermo-responsive polymers have also been used in endotracheal tubes (ETT). A commercialized example is the RÜSCH endotracheal tube from Teleflex, which allows the tube to soften at body temperature for greater patient comfort, while the smooth tube tip is molded and gently cupped to minimize trauma during intubation. It is made from a thermosensitive PVC via formulation. Since the thermosensitive PVC is expected to readily conform to a patient's anatomy for improved performance and patient comfort, the formulated thermosensitive PVC could also be used in an intermittent catheter, thoracic drainage catheters, etc.

Some responsive polymers, especially hydrogels, have poor mechanical properties such as low tensile strength and low Young's modulus that limit their practical applications, sometimes even as medical device coatings. Increasing the cross-link density, copolymerizing with other monomers, and incorporating with other composites are several methods that improve the polymer mechanical strength even though they might adversely lower the stimuli-response properties. Jones et al. (2012) synthesized semi-interpenetrating hydrogels of PNIPAAm or poly(NIPAAm-co-HEMA) containing PCL. The combination of the two polymers exhibited a thermosensitive response with a phase transition temperature at 32 °C and 34 °C, respectively, close to that of the PNIPAAm homopolymer. The existence of PCL in the network substantially improves the tensile strength and Young's modulus by up to 100 times, sufficient enough to be used as ETT materials. At the same time, metronidazole was used as a model drug in their study for release. Metronidazole is loaded during the polymer preparation, and its release from semi-IPN of PNIPAAm containing PCL at 37 °C above its transition temperature showed pulsed release with entire drug release within 1 h. Ventilator-associated pneumonia (VAP) is frequently associated with the use of ETT and has been shown to be due to bacteria adherence to the luminal surface of ETT (Adair et al., 1999).

VAP is common in civilian hospital intensive care units (ICUs), and an even more serious problem in combat military ICUs. One solution is to have sufficiently high concentrations of antimicrobial agents at the luminal surface of ETT devices. Typically, there is a limited volume of fluid at the luminal surface, presenting a thermodynamic resistance to drug diffusion from the bioactive devices. The prepared thermosensitive semi-IPN of PNIPAAm containing PCL may resolve this problem. With temperature of the inspired air exceeding the critical phase transition temperature of the polymer, the rapid release of a drug is activated by the huge reduction of the polymer network and is therefore not diffusion controlled release. This pulsed drug release modulated by temperature from this temperature-sensitive material provides a novel strategy for the prevention/treatment of VAP (Kenawy, Worley, & Broughton, 2007).

11.9 Gastrointestinal devices

A device using stimuli-responsive polymers for curbing appetite or treating diseases or disorders was patented by Lendlein and Langer (Lendlein & Langer, 2004). The small device is compressed or stretched in a temporary form and is able to be administered orally or rectally. With a stimulus such as temperature or pH, the device changes its shape to the permanent form to mechanically fix in the stomach, esophagus, or intestine. The device is made from biocompatible, stimuli-responsive polymers such as SMPs. This device can function as an implant for curbing appetite by reducing the volume of the stomach, esophagus, or intestine without interfering with the food flow through the gastrointestinal tract. Thus, it is used to help patients lose weight. In a different application, this device could deliver drugs to treat diseases and disorders in the stomach or intestine due to the change in pH or temperature. SMP materials used for the device can be biodegradable or nondegradable depending on the application. In another invention, Marco claimed a biodegradable self-inflating intragastric implant for curbing appetite (Marco, 2006). SMP was reported to be used in at least a portion of the expandable part in the device. In a deflated state before being administrated, it self-expands when in contact with gastric juices and/or is heated to about 37 °C.

For targeted delivery in the gastrointestinal tract, which has an acidic pH of 4 in the stomach and basic pH of 8 in the lower intestine, Jahren et al. designed pH-responsive chitosan hydrogels (Jahren, Butler, Adams, & Cameron, 2010). The chitosan hydrogels, with either a conventional cross-linker of glutaraldehyde or a naturally occurring and biocompatible cross-linker of genipin, reached a maximum swelling at pH 4 and declined in swelling with both reduced and elevated pH. The release profile of paracetamol from the hydrogels showed an above 50% release in 2 h.

11.10 Urogenital devices/urinary devices

Stimuli-responsive polymers are also used to make devices in the urogenital area. As applied in ETT tubes, formulated thermosensitive PVC can be used in intermittent catheters that soften at body temperature to minimize trauma during insertion. Lendlein and Langer (Lendlein & Langer, 2004) claimed a self-expanding device from SMPs suitable for insertion into the vagina or cervix. After deployment, it is used for contraception or drug delivery in the uterus or in urogenital applications. The biodegradable SMP network for controlled release incorporated hydrophilic and hydrophobic drugs (Neffe et al., 2009) and is expected to be used as a ureteral stent with antibacterial properties.

A device with shape memory technology and osmotic delivery has been developed to treat patients who suffer from a painful bladder condition known as interstitial cystitis/bladder pain syndrome (TARIS Biomedical Inc., MA). The device was made of a dual-lumen silicone catheter with one lumen loaded with lidocaine, which continuously leached out as contacted with urine. The second lumen of the device contains superelastic nitinol wire in a predefined form, which provides the bladder-retentive property of the device. The device has been implanted in patients for weeks and showed continuous retention and drug release in bladder (Nickel et al., 2012).

11.11 Insulin delivery devices for diabetes management

Diabetes has become a major cause of death in many countries. There is interest in the development of an insulin delivery system by glucose-responsive polymers. Starting from the mid-1980s, glucose-responsive insulin delivery systems based on pH-responsive polymers with entrapped or immobilized glucose oxidase (GOX) were developed by several groups (Albin, Horbett, & Ratner, 1985; Albin, Horbett, Miller, & Ricker, 1987; Ishihara & Matsui, 1986; Klumb & Horbett, 1992; Traitel, Cohen, & Kost, 2000). When the external glucose concentration increases to cause glucose to diffuse into the polymer membrane, GOX could catalyze the conversion of glucose into gluconic acid to lower the environmental pH, leading to the swelling of the polymers to increase the release of insulin. Therefore, these polybase pH-responsive polymers were copolymers with tertiary amines, for example, copolymer of *N*,*N*-dimethyl aminoethyl dimethacrylate (DEAEMA)-2-hydroethyl acrylate (HEA)-4-trimethylsilylstyrene (TMS) (Ishihara & Matsui, 1986) or DEAEMA-HEMA-tetraethylene glycol dimethacrylate (TEGDMA) (Albin et al., 1985, 1987; Klumb & Horbett, 1992; Traitel et al., 2000). *In vivo* experiments in rats demonstrated at least some of the entrapped insulin had still retained the activity to reduce blood glucose levels (Traitel et al., 2000). On the other hand, Kang and Bae (2003) reported a sulfonamide-based polyacid hydrogel that could show a delicate pH sensitivity under a physiological condition for insulin delivery. This novel hydrogel was prepared by polymerization of sulfadimethoxine monomer, *N*,*N*-dimethylacrylamide (DMAAm), methacryloyl GOX, and catalase with sucrose particles as porogen. Here catalase was used to remove the hydrogen peroxide produced by the glucose-involved reaction. In spite of being covalently attached to the hydrogel, the enzymes did not show any significantly reduced bioactivity. Instead, response to the glucose concentration changes between 0 and 300 mg/dl in PBS solution at pH 7.4 at 37 °C, the hydrogel system exhibited reversible swelling ratio changes without any noticeable hysteresis. It was suggested that this could be utilized as molecular valves in glucose-sensitive membranes for insulin delivery.

Based on the ability of phenylboronic acid (PBA) and its derivatives to form a stable complex with polyols, including glucose in aqueous media, the research group in Japan developed synthetic polymers containing PBA instead of biologically active GOD to reach the on–off release of insulin (Kataoka, Miyazaki, Bunya, Okano, & Sakurai, 1998; Shiino et al., 1995). With the addition of glucose at physiological pH, a greater release of insulin from amine-containing PBA gel was found than was in the PBA gel (Shiino et al., 1995). They further developed the glucose-responsive gel based on PNIPAAm derivatized with a definite fraction of PBA groups (Kataoka et al., 1998). It exhibited repeated on–off regulation of insulin release due to the change of external glucose in solution, and can be utilized in insulin delivery devices.

11.12 Biosensors

Implantable diagnostic devices can detect and monitor diseases, conditions, or infections. Such devices may be required to function reliably for hours, days, or even months in human bodies without regeneration or addition of reagents. Biosensors are diagnostic devices typically used for detection of biological analyses with a physicochemical detector (Wilson & Gifford, 2005). Responsive polymers have been demonstrated in an increasing number of diagnostic and biosensor applications (Adhikari & Majumdar, 2004; Deligkaris, Tadele, Olthuis, & van den Berg, 2010). However, virtually every implanted material initiates protein adsorption and triggers a foreign body reaction that finally walls off and isolates the implant, which has been described in previous sections. Even so, many researchers are optimistic that they can avoid foreign responses or manipulate them to their advantage (Service RF, 2002). Currently, only a limited number of diagnostic devices have been reported for long-term implantation. Compared to proteins and cells, small molecules, such as glucose, lactate, and cholesterol, can gain access to the detectors with less impact by fouling and encapsulation. With the widespread disease of diabetes requiring continuous monitoring of blood sugar levels, there is a particular interest in implantable biosensors as glucose sensors, which account for approximately 85% of the biosensor industry (Wang, 2007). Various materials have been designed to adapt to such glucose sensors application (Bratlie, York, Invernale, Langer, & Anderson, 2012; Wu, Wang, Yu, Wang, & Chen, 2011). This section summarizes the application of responsive polymers as implantable sensors with a particular focus on implantable glucose sensors.

11.12.1 Enzyme electrode-based biosensors

Most commercialized implantable glucose sensors are based on a glucose–enzyme reaction with their products detected by an amperometric sensor. The enzyme, usually a GOX, is immobilized on an electrode to provide a redox reaction with glucose and generate a current at the electrodes. With the immobilization of the GOX, the redox polymer layer can be viewed as a glucose-responsive system (Heo & Takeuchi, 2013; Wang, 2007). Commercialized implanted glucose sensors are usually implanted subcutaneously with a sensor life span of less than 7 days. In addition to subcutaneous implantation, glucose sensors integrated with intravascular catheters (Armour, Lucisano, McKean, & Gough, 1990) and contact lenses (Yao, Shum, Cowan, Lähdesmäki, & Parviz, 2011) have also been reported. Due to the discomfort, short life span, and the requirement for daily calibration of current implantable glucose sensors, there is still a tremendous need for improving blood glucose monitoring (Gerritsen, Jansen, & Lutterman, 1999).

The polymers that immobilize enzymes on the electrodes respond to analytes through an electron-transfer process. Different materials and their combinations have been applied on the electrodes to improve the sensitivity, biocompatibility, and lifetime of the biosensors. The redox enzymes can be either chemically bound within the cross-linked polymer network or physically embedded within the materials. Figure 11.5 shows a typical implantable glucose electrode with a sandwich structure designed

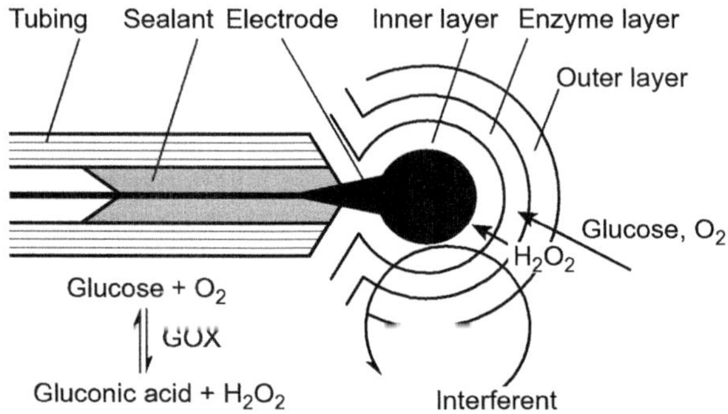

Figure 11.5 A typical implantable glucose electrode that is responsive to glucose concentration.

for immobilizing the enzyme (Clark & Duggan, 1982). Glucose diffuses through an outer layer to reach the immobilized enzyme, GOX, which is placed very close to the surface of a platinum electrode. The outer layer should have a function of allowing maximum passage of oxygen and retarding the passage of glucose to control glucose diffusion. This membrane must also be biocompatible and stable *in vivo*. The inner layer serves as the support for enzyme immobilization and also as the selective membrane for H_2O_2. The enzyme layer is where glucose is converted to the electoactive species H_2O_2.

Dialysis membranes from cellulose acetate (CA) are among the earliest substrates that have been used by Clark and Lyons (Clark & Lyons, 1962). To exclude interfering anions, a negatively charged perfluorinated ionomer Nafion™ membrane was alternatively deposited with CA on the electrode (Zhang et al., 1994). As an outer layer, Nafion™ also provides protection and improves biocompatibility. Several needle-type glucose sensors with Nafion™ layer remain functional for at least 10 days after subcutaneous implantation in dogs, without degradation of their sensitivity (Moussy, Harrison, O'Brien, & Rajotte, 1993; Moussy, Jakeway, Harrison, & Rajotte, 1994). Other examples of interference-excluding membranes include polydimethylsiloxane (PDMS) (Ertefai & Gough, 1989; Yang, Atanasov, & Wilkins, 1995), polyurethane (Moatti-Sirat et al., 1992; Pickup, Claremont, & Shaw, 1993; Yu, Long, Moussy, & Moussy, 2006), and modified CA (Qiu & Hu, 2013). These polymers allow for permeation of molecules having a similar molecular weight to the analyte; interferences of sizes larger than the analyte are excluded. Durability of the biosensors can be improved using these protective membranes. For example, an epoxy-enhanced polyurethane membrane was used as the outer protective membrane of the sensor and kept functioning in rats for 10–56 days (Yu et al., 2006). A glucose sensor, covered by a thin electrolyte layer, a protective layer of medical-grade PDMS, and a membrane of PDMS with wells for the immobilized enzymes located over certain electrodes was capable of performing long-term monitoring of tissue glucose concentrations by wireless telemetry (Gough, Kumosa, Routh, Lin, & Lucisano, 2010). The implanted sensor was functional in a pig model for more than 1 year, indicating significant progress in extending the life span of implantable glucose sensors (Gough et al., 2010).

An earlier study used CA as an inner layer to exclude molecules such as ascorbate, urate, and bilirubin (Clark & Duggan, 1982). The interference by small, electroactive compounds could be further reduced by incorporating conductive polymers. Conductive polymers have been coated on the implantable glucose sensor electrodes to provide efficient transfer of electric charge produced by the biochemical reaction to an electronic circuit. The conductive polymers, usually intrinsically conducting polymers with conjugated backbones, provide high electron affinity and are highly susceptible to chemical or electrochemical oxidation or reduction (Singh, 2012). Using enzymes during electrochemical polymerization, enzymes can be immobilized on the electrodes with conductive polymers. For example, enzymes entrapped within films such as polypyrrole (PPy), polyaniline, or polythiophene, prepared by electropolymerization from aqueous solutions, have been commonly used to prepare glucose electrodes. Like Nafion™, some conductive polymers can exclude interference of molecules with sizes larger than the analyte, such as overoxidized PPy (Rizzi, Centonze, & Zambonin, 2000), poly(*o*-phenylenediamine) (Malitesta, Palmisano, Torsi, & Zambonin, 1990; Sasso, Pierce, Walla, & Yacynych, 1990), and poly(quinone) (Arai, Shoji, & Yasumori, 2006; Kaku, Okamoto, Charles, Holness, & Karan, 1995). With some *in vivo* results for more than 10 days (Moussy et al., 1993, 1994), conductive polymers have shown activity, sensitivity, and selectivity, and possessed good durability on glucose electrodes.

Within the enzyme layer, GOX has been immobilized through cross-linked albumin, synthetic hydrogels (Guiseppi-Elie, 2010), and conductive polymers (Singh, 2012). Hydrogels are commonly used materials applied to the electrodes to immobilize enzymes and/or provide biocompatibility, permeability, and fouling resistance. PolyHEMA and its copolymers are among the earliest hydrogels to be applied on electrodes (Shaw, Claremont, & Pickup, 1991). To improve porosity, hydrophilicity, or biocompatibility, 3-dihydroxypropyl methacrylate (DHPMA) (Wang et al., 2008; Yu et al., 2008), *N*-vinyl pyrrolidone (Heineman, 1993), vinyl alcohol (Vaddiraju, Singh, Burgess, Jain, & Papadimitrakopoulos, 2009), ethylene glycol (Quinn, Pathak, Heller, & Hubbell, 1995), 2-methacryloyloxyethyl phosphorylcholine (MPC) (Chen et al., 1992), and carboxybetaine (Yang, Xue, Carr, Wang, & Jiang, 2011; Zhang et al., 2009) have all been applied. These hydrogels are ionically but not electronically conductive and usually demonstrate high interfacial impedances. Conductive polymers are incorporated into the hydrogel network to improve the stimuli responsiveness and reduce interfacial electrical impedances (Heller, 2006) (Guiseppi-Elie, 2010). Incorporating an osmium complex within the hydrogel network could also improve the redox efficacy (Kenausis, Taylor, Katakis, & Heller, 1996; Mano, Mao, & Heller, 2005).

11.12.2 pH-responsive polymer-based biosensors

Glucose reacts with GOX forming gluconic acid (GlucA), and thus decreases the pH of the environment. With the change in pH, the gel swells or collapses depending on the characteristics of the particular polymer of the system. As a biosensor, the hydrogel should respond fast to expand and contract abruptly with glucose concentration (Peppas and Bures, 2006). In the last few years, extensive work has been carried out with poly(methacrylic acid) (PMAA), PEG, poly(diethylamino ethyl methacrylate) (PDEAEMA), as well as copolymer hydrogels such as poly(methacrylic acid-g-ethylene glycol) (PMAA-g-EG) (Bell & Peppas, 1996). Interpolymer complexes are formed in these hydrogels because of hydrogen bonding between the hydrogens of the carboxylic groups of the PMAA and the oxygens on the ether groups of the PEG chains. At low pH values, there is sufficient protonation of the carboxylic acid groups causing complexes to form, which results in a collapse of the gel due to the increased hydrophobicity. Hydrogels with amine moieties also respond to the pH change. The decrease in pH value induces the ionization of the amine moiety, resulting in the swelling of the hydrogel. A typical glucose sensor with pH-sensitive hydrogel incorporates a fluorophore, an amine functional group, and glucose oxidase into a cross-linked network (McCurley, 1994). The gel can be localized on the end of an optic fiber or monitored using a transdermal light source. Due to the volume change, the fluorescence signal is monitored as a function of glucose concentrations. While many publications have investigated the possibility of using a pH-sensitive hydrogel for an implantable glucose sensor, few *in vivo* results have been reported. The hydrogel-based biomaterials still need further improvement in the signal linearity and response time. The complex *in vivo* condition, such as pH and temperature change due to inflammation after implantation, could also affect the stability.

11.12.3 Affinity-based biosensors

Affinity binding sensors monitor specific binding of analytes to the receptors. With absence of enzymatic, chemical, or electrode reactions, these sensors are believed to be promising for implantable monitoring since they are equilibrium driven rather than reaction driven and are thus expected to have higher signal stability and less interference. The development of affinity sensors is currently focused on enzyme/peptide-based, synthetic polymer-based, or surface-based substrates. The response of analytes has been monitored through various technologies, such as fluorescence signals (Mansouri & Schultz, 1984), viscosity changes (Huang et al., 2014), or surface-based signals such as Raman scattering (Stuart et al., 2006). For glucose sensors, glucose-specific protein concanavalin A (ConA) (Schultz, Mansouri, & Goldstein, 1982), boronic acid-based artificial glucose-specific polymers (Striegler, 2003), glucose-binding peptides (Siegrist et al., 2010), or glucose-sensitive surfaces (Stuart et al., 2006) have been investigated for *in vivo* monitoring.

For a ConA ratiometric fluorescence resonance energy transfer (FRET) sensor, the sensing element contains a ConA immobilized within a hydrogel, and a fluorescein labeled dextran is used as a competing ligand. Glucose in the external medium diffuses through the hydrogel into the sensing element and competes with dextran for binding to ConA. At equilibrium, the level of free fluorescein is measured via the optical fiber and is correlated to the concentration of glucose (Mansouri & Schultz, 1984). The glucose-dependent fluorescence change is based on FRET between near-infrared-compatible donor and quencher dyes that are chemically linked to dextran and ConA, respectively. An acute evaluation of transdermal sensors with an optical fiber–coupled setup has been conducted over 4 h, and a chronic *in vivo* evaluation of fully implanted sensors has been shown for up to 16 days (Ballerstadt, Polak, Beuhler, & Frye, 2004). An implantable fiber-coupled fluorescence affinity sensor (FAS) has been evaluated in hairless rats and pigs (BioTex Inc., Houston, TX). The chronic *in vivo* performance of the FAS in two different animal models over 3 days was clinically acceptable and comparable to other continuous glucose monitoring platforms (Ballerstadt, Evans, Gowda, & McNichols, 2007). A second-generation FAS was subcutaneously implanted in the abdomens of 12 people with diabetes, and its acute performance to excursions in blood glucose was monitored over 4 h (Dutt-Ballerstadt et al., 2012). One of the major benefits of the FAS is the absence of "autodestructive" side products and any device-related warm-up time after sensor reconnection (Ballerstadt et al., 2007).

Viscometric-affinity ConA sensors have also been developed for glucose monitoring in the subcutaneous tissue. Usually, viscosity variation of a sensitive fluid of a high-molecular-weight dextran is the core of the sensor. At low glucose concentration, dextran molecules are cross-linked by binding to ConA, forming a viscous solution. When the glucose concentration increases, dextran molecules are partially replaced by glucose molecules at the binding sites of ConA. As a result, the network ConA-dextran is weakened and the viscosity of the sensing solution decreases (Ballerstädt & Ehwald, 1994; Boss, Meurville, Sallese, & Ryser, 2009). A continuously working affinity sensor based on the glucose-dependent viscosity of a sensitive liquid containing dextran and ConA has been designed by arranging a microdialysis probe, two flow-resisting capillaries, and two pressure transducers in a linear flow system (Disetronic Medical Systems, now part of Roche Diabetes Care) (Beyer et al., 2001). In a clinical study, the sensor was implanted into subcutaneous tissue of the forearm and abdomen. Glucose concentration was examined for 8 h showing clinically acceptable results and signal stability (Diem et al., 2004).

Synthetic responsive polymers have been developed for implantable glucose sensors to avoid toxic and inflammatory effects *in vivo* from enzymes. Incorporated into linear copolymers (Huang, Li, Schultz, Wang, & Lin, 2009) or hydrogels (Alexeev, Das, Finegold, & Asher, 2004), phenylboronic acid (PBA) can bind diols through reversible boronate ester formation. Like ConA sensors, fluorescence-based PBA glucose sensors have been demonstrated *in vivo* (James, Sandanayake, Iguchi, & Shinkai, 1995). Other signals, such as optical changes due to swelling (Roy & Sumerlin, 2012), viscosity variation (Huang et al., 2009), and colorimetric changes (Alexeev et al., 2004) have also been monitored as a function of glucose change. Long-term *in vivo* tests or clinical tests have been reported through intravascular implantation (Gamsey, Suri, Wessling, & Singaram, 2006), subcutaneous implantation (Heo, Shibata, Okitsu, Kawanishi, & Takeuchi, 2011), and ocular implantation (Alexeev et al., 2004). For example, by binding boronic acid with a fluorophore, fluorescent signals can be modulated as a function of boronic acid–saccharide interaction through photoinduced electron transfer (PET) (James et al., 1995). The PET moieties were bound to hydrogels, and an optical fluorescence sensor was developed for continuous intravascular glucose sensor. A clinical trial showed that the optical fluorescence sensor can successfully measure blood glucose levels with a high degree of accuracy across 18 intensive care patients up to 100 h of continuous monitoring (Glysure, UK) (Mulavisala, Gopal, Crane, Mackenzie, & Barwell, 2012). In another development, a polyamide

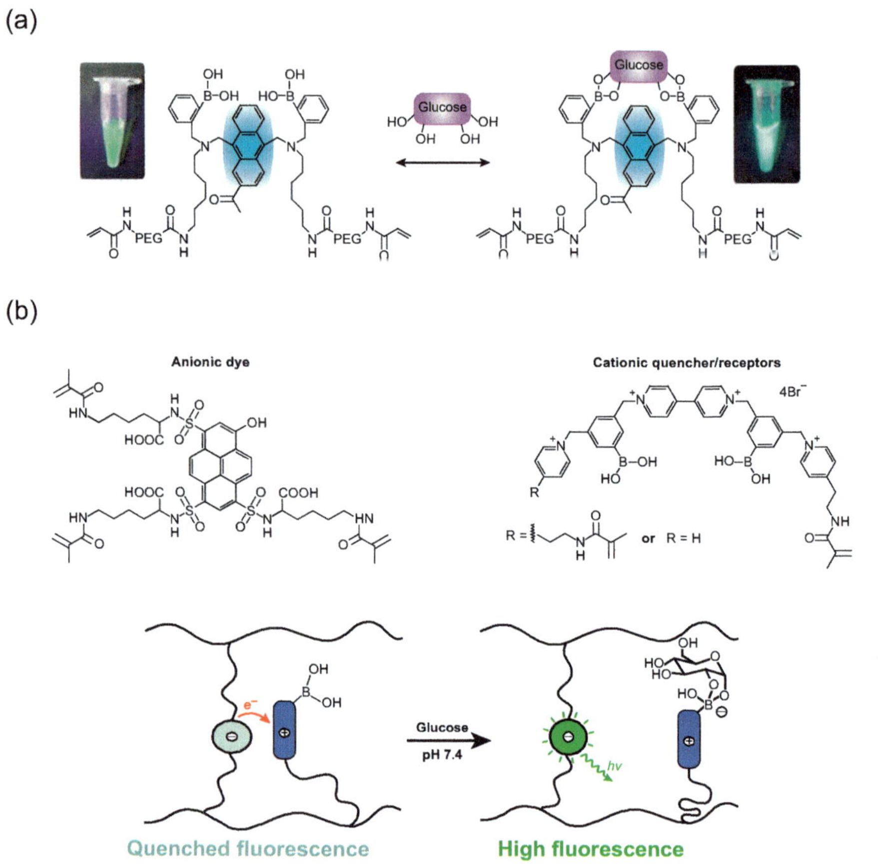

Figure 11.6 Two glucose-responsive polymers with phenylboronic acid (PBA) moieties and fluorophores. (a) In the absence of glucose molecules, the fluorescence of the anthracene is quenched by a photoinduced electron transfer (PET) that occurs from the unshared electron pair of the nitrogen atom to the anthracene. When glucose molecules bind to diboronic acid, a strong reaction between the nitrogen atom and a boron atom inhibits PET. As a result, the fluorescence of anthracene is higher than under glucose-free conditions (Heo et al., 2011). Reprinted with permission. (b) In the absence of glucose, a receptor quencher binds to a fluorophore and reduces the fluorescence emission. In the presence of glucose, the interaction between receptor quencher and fluorophore weakens, resulting in an increase in fluorescence intensity. The intensity of the fluorescence therefore depends on the concentration of glucose. *Sources*: Reprinted with permission from Gamsey et al., 2006; Romey et al., 2012.

hydrogel or a PEG-bonded polyacrylamide (PAM) was copolymerized with diboronic acid-based PET moieties (Figure 11.6(a)). The glucose-responsive hydrogel was implanted and the fluorescence signal was monitored transdermally for up to 140 days, showing the potential application for long-term continuous glucose monitoring (Heo et al., 2011). Another design includes immobilizing a fluorescent anionic dye and a viologen-based receptor appended with boronic acids into a polyHEMA hydrogel.

The fluorescence of the dye is modulated by the quenching efficiency of the receptor, which in turn is dependent on the glucose concentration. The hydrogels were tested for their ability to continuously and reversibly detect glucose over the course of several hours (Figure 11.6(b)) (Gamsey et al., 2006). The use of quenched fluorescence for glucose sensing has been reported in a pilot study in humans for 8 h. The sensors were intravascularly implanted and exhibited a continuous and direct measurement of plasma glucose in flowing blood (GluMetrics, CA) (Romey et al., 2012).

Label-free surface affinity-based biosensors, such as surface plasmon resonance (SPR) and quartz crystal microbalance (QCM), have shown high sensitivity *in vitro* in both small molecule and large molecule detection. However, their application as an implantable sensor is limited due to the complex *in vivo* environment. A surface affinity-based biosensor utilizing SERS as a detecting signal was developed to perform the first transcutaneous glucose sensing (Stuart et al., 2006). To achieve the *in vivo* stability and sensitivity, the SERS active surface, which has a silver film nanosphere (AgFON) substrate, was modified with nonfouling and interferant-exclusive self-assembled monolayers (SAMs). The glucose-responsive SAM is an alkane-thiolate tri(ethylene glycol) monolayer that acts as a partition layer, preconcentrating glucose near the SERS-active surface (Yonzon, Haynes, Zhang, Walsh, & Van Duyne, 2003). This SERS-active substrate has been implanted subcutaneously, and the glucose concentration was monitored continuously for more than 17 days with excellent accuracy and consistency (Ma et al., 2011).

11.13 Summary

Great progress has been made in the past several decades in developing implantable devices in different medical areas using responsive polymers and surfaces. Applications in cardiovascular, ophthalmic, surgical , dental, orthopedic, respiratory, gastrointestinal, urogenital, implantable, and drug delivery devices, and biosensors, have been reviewed in this chapter. Thanks to the unique responsiveness and rational design in terms of clinical needs, these polymers have been used to facilitate the deployment or removal of the devices with minimum damage to the host tissue, support the functions of current devices to treat ailments, deliver drugs, control infection, or monitor physiological factors or biomolecules.

Currently, most clinical success of responsive polymers still rests in short-term implantation and intervention. There is a tremendous need for developing long-term implantable devices that match the smartness and durability of human structures. Challenges are derived from every aspect of foreign body response and wound healing, such as protein adsorption, encapsulation, thrombus formation, and degradation. Modifying surfaces with inert and biocompatible polymers, formulating polymers with segments of both antifouling and responsive moieties, designing degradable polymers, or adding eluting agents that control inflammation and foreign body responses have been used to improve the biocompatibility of these devices. It is expected that more polymers, surfaces, and composites with accurate and programmable responsiveness, stability, and biocompatibility will be developed for implantable devices. To achieve this, many fundamental mechanisms of material–body responses are needed to be explored and understood. Moreover, the development of animal models with clinical relevance is also an indispensable link for device development.

References

Adair, C. G., et al. (1999). Implications of endotracheal tube biofilm for ventilator-associated pneumonia (Translated from English). *Intensive Care Medicine*, *25*(10), 1072–1076 (in English).

Adhikari, B., & Majumdar, S. (2004). Polymers in sensor applications. *Progress in Polymer Science*, *29*(7), 699–766.

Ahuja, A. A., et al. (1993). Platinum coil coatings to increase thrombogenicity: a preliminary study in rabbits. *American Journal of Neuroradiology*, *14*(4), 794–798.

Albin, G. W., Horbett, T. A., Miller, S. R., & Ricker, N. L. (1987). Theoretical and experimental studies of glucose sensitive membranes. *Journal of Controlled Release*, *6*(1), 267–291.

Albin, G., Horbett, T. A., & Ratner, B. D. (1985). Glucose sensitive membranes for controlled delivery of insulin: insulin transport studies. *Journal of Controlled Release*, *2*(0), 153–164.

Alexeev, V. L., Das, S., Finegold, D. N., & Asher, S. A. (2004). Photonic crystal glucose-sensing material for noninvasive monitoring of glucose in tear fluid. *Clinical Chemistry*, *50*(12), 2353–2360.

Alteheld, A., Feng, Y., Kelch, S., & Lendlein, A. (2005). Biodegradable, amorphous copolyester-urethane networks having shape-memory properties. *Angewandte Chemie International Edition*, *44*(8), 1188–1192.

Anderson, J. M. (2001). Biological responses to materials. *Annual Review of Materials Research*, *31*, 81–110.

Anderson, J. M., Rodriguez, A., & Chang, D. T. (2008). Foreign body reaction to biomaterials. *Seminars in Immunology*, *20*(2), 86–100.

Anjum, N., Gulrez, S. K. H., Singh, H., & Gupta, B. (2006). Development of antimicrobial polypropylene sutures by graft polymerization. I. Influence of grafting conditions and characterization. *Journal of Applied Polymer Science*, *101*(6), 3895–3901.

Anumolu, S. S., Singh, Y., Gao, D., Stein, S., & Sinko, P. J. (2009). Design and evaluation of novel fast forming pilocarpine-loaded ocular hydrogels for sustained pharmacological response. *Journal of Controlled Release*, *137*(2), 152–159.

Arai, G., Shoji, K., & Yasumori, I. (2006). Electrochemical characteristics of glucose oxidase immobilized in poly(quinone) redox polymers. *Journal of Electroanalytical Chemistry*, *591*(1), 1–6.

Armour, J. C., Lucisano, J. Y., McKean, B. D., & Gough, D. A. (1990). Application of chronic intravascular blood glucose sensor in dogs (Translated from English). *Diabetes*, *39*(12), 1519–1526 (in English).

Baer, G. M., et al. (2009). Thermomechanical properties, collapse pressure, and expansion of shape memory polymer neurovascular stent prototypes. *Journal of Biomedical Materials Research Part B: Applied Biomaterials*, *90B*(1), 421–429.

Baer, G., Wilson, T. S., Matthews, D. L., & Maitland, D. J. (2007). Shape-memory behavior of thermally stimulated polyurethane for medical applications. *Journal of Applied Polymer Science*, *103*(6), 3882–3892.

Ballerstädt, R., & Ehwald, R. (1994). Suitability of aqueous dispersions of dextran and concanavalin A for glucose sensing in different variants of the affinity sensor. *Biosensors and Bioelectronics*, *9*(8), 557–567.

Ballerstadt, R., Evans, C., Gowda, A., & McNichols, R. (2007). Fiber-coupled fluorescence affinity sensor for 3-day in vivo glucose sensing (Translated from English). *Journal of Diabetes Science and Technology*, *1*(3), 384–393 (in English).

Ballerstadt, R., Polak, A., Beuhler, A., & Frye, J. (2004). In vitro long-term performance study of a near-infrared fluorescence affinity sensor for glucose monitoring. *Biosensors and Bioelectronics*, *19*(8), 905–914.

Bell, C. L., & Peppas, N. A. (1996). Water, solute and protein diffusion in physiologically responsive hydrogels of poly (methacrylic acid-g-ethylene glycol) (Translated from English). *Biomaterials*, *17*(12), 1203–1218 (in English).

Beyer, U., et al. (2001). Recording of subcutaneous glucose dynamics by a viscometric affinity sensor (Translated from English). *Diabetologia*, *44*(4), 416–423 (in English).

Bloomfield, S. E., et al. (1977). Soluble artificial tear inserts. *Archives of Ophthalmology*, *95*(2), 247–250.

Boss, C., Meurville, E., Sallese, J.-M., & Ryser, P. (2009). Novel chemico-mechanical approach towards long-term implantable glucose sensing. *Procedia Chemistry*, *1*(1), 313–316.

Bouchot, O., Aubin, M., Carrier, M., Cohn, W., & Perrault, L. (2006). Temporary coronary artery occlusion during off-pump coronary artery bypass grafting with the new poloxamer P407 does not cause endothelial dysfunction in epicardial coronary arteries. *Journal of Thoracic Cardiovascular Surgery*, *132*, 1144–1149.

Bouchot, O., Berger, R. L., Berne, J. P., Brunotte, F., & Brenot, R. (2010). Clinical experience with a novel thermosensitive temporary coronary artery occluder (LeGoo). *Annals of Thoracic Surgery*, *89*(6), 1912–1917.

Bratlie, K. M., York, R. L., Invernale, M. A., Langer, R., & Anderson, D. G. (2012). Materials for diabetes therapeutics. *Advanced Healthcare Materials*, *1*(3), 267–284.

Bridges, A. W., & Garcia, A. J. (2008). Anti-inflammatory polymeric coatings for implantable biomaterials and devices (Translated from English). *Journal of Diabetes Science and Technology*, *2*(6), 984–994 (in English).

Buckley, P. R., et al. (2006). Inductively heated shape memory polymer for the magnetic actuation of medical devices. *IEEE Transactions on Biomedical Engineering*, *53*(10), 2075–2083.

Butterfield, M., Wheatley, D. J., Williams, D. F., & Fisher, J. (2001). A new design for polyurethane heart valves. *Journal of Heart Valve Disease*, *10*(1), 105–110.

Caldwell, F. T. J., Bowser, B. H., & Crabtree, J. H. (1981). The effect of occlusive dressings on the energy metabolism of severely burned children. *Annals of Surgery*, *193*(5), 579–591.

Casolaro, M., Casolaro, I., & Lamponi, S. (2012). Stimuli-responsive hydrogels for controlled pilocarpine ocular delivery. *European Journal of Pharmaceutics and Biopharmaceutics*, *80*(3), 553–561.

Chen, C.-Y., et al. (1992). A biocompatible needle-type glucose sensor based on platinum-electroplated carbon electrode (Translated from English). *Applied Biochemistry and Biotechnology*, *36*(3), 211–226 (in English).

Chen, K.-S., et al. (2005). Immobilization of chitosan gel with cross-linking reagent on PNIPAAm gel/PP nonwoven composites surface. *Materials Science and Engineering: C*, *25*(4), 472–478.

Chen, M.-C., et al. (2007). Rapidly self-expandable polymeric stents with a shape-memory property. *Biomacromolecules*, *8*(9), 2774–2780.

Chen, M.-C., et al. (2009). The characteristics and in vivo suppression of neointimal formation with sirolimus-eluting polymeric stents. *Biomaterials*, *30*(1), 79–88.

Clark, L. C., Jr., & Duggan, C. A. (1982). Implanted electroenzymatic glucose sensors (Translated from English). *Diabetes Care*, *5*(3), 174–180 (in English).

Clark, L. C., & Lyons, C. (1962). Electrode systems for continuous monitoring in cardiovascular surgery. *Annals of the New York Academy of Sciences*, *102*(1), 29–45.

Contreras-García, A., Alvarez-Lorenzo, C., Taboada, C., Concheiro, A., & Bucio, E. (2011). Stimuli–responsive networks grafted onto polypropylene for the sustained delivery of NSAIDs. *Acta Biomaterialia*, *7*(3), 996–1008

Contreras-García, A., Bucio, E., Concheiro, A., & Alvarez-Lorenzo, C. (2010). Polypropylene grafted with NIPAAm and APMA for creating hemocompatible surfaces that load/elute nalidixic acid. *Reactive and Functional Polymers*, *70*(10), 836–842.

Cosson, M., et al. (2003). Mechanical properties of synthetic implants used in the repair of prolapse and urinary incontinence in women: which is the ideal material? (Translated from English). *International Urogynecology Journal*, *14*(3), 169–178 (in English).

Dabagh, M., Abdekhodaie, M. J., & Khorasani, M. T. (2005). Effects of polydimethylsiloxane grafting on the calcification, physical properties, and biocompatibility of polyurethane in a heart valve. *Journal of Applied Polymer Science*, *98*(2), 758–766.

Daniel, H. (2008). In S. H. -B. Thomas, F. Kevin, K. Tadayoshi, & H. M. William (Eds.), *Security and privacy for implantable medical devices* (pp. 30–39).

Deligkaris, K., Tadele, T. S., Olthuis, W., & van den Berg, A. (2010). Hydrogel-based devices for biomedical applications. *Sensors and Actuators B: Chemical*, *147*(2), 765–774.

Desai, S. D., & Blanchard, J. (1998). In vitro evaluation of pluronic F127-based controlled-release ocular delivery systems for pilocarpine. *Journal of Pharmaceutical Sciences*, *87*(2), 226–230.

Diem, P., et al. (2004). Clinical performance of a continuous viscometric affinity sensor for glucose (Translated from English). *Diabetes Technology and Therapeutics*, *6*(6), 790–799 (in English).

Dregelid, E. (2012). Temporary vascular occlusion by rapid reverse phase polymer: a preliminary in vitro study of retrograde injection. *International Journal of Biomaterials*, *2012*, 1–7.

Dutt-Ballerstadt, R., et al. (2012). A human pilot study of the fluorescence affinity sensor for continuous glucose monitoring in diabetes (Translated from English). *Journal of Diabetes Science and Technology*, *6*(2), 362–370 (in English).

Ertefai, S., & Gough, D. A. (1989). Physiological preparation for studying the response of subcutaneously implanted glucose and oxygen sensors (Translated from English). *Journal of Biomedical Engineering*, *11*(5), 362–368 (in eng).

Freytag, C. C., Thies, F. L., König, W., & Welte, T. (2003). Prolonged application of closed in-line Suction catheters increases microbial colonization of the lower respiratory tract and bacterial growth on catheter surface (Translated from English). *Infection*, *31*(1), 31–37 (in English).

Gall, K., et al. (2005). Thermomechanics of the shape memory effect in polymers for biomedical applications. *Journal of Biomedical Materials Research Part A*, *73A*(3), 339–348.

Gamsey, S., Suri, J. T., Wessling, R. A., & Singaram, B. (2006). Continuous glucose detection using boronic acid-substituted viologens in fluorescent hydrogels: linker effects and extension to fiber optics. *Langmuir*, *22*(21), 9067–9074.

Ge, D., et al. (2010). A self-powered and thermally-responsive drug delivery system based on conducting polymers. *Electrochemistry Communications*, *12*(8), 1087–1090.

Gerritsen, M., Jansen, J. A., & Lutterman, J. A. (1999). Performance of subcutaneously implanted glucose sensors for continuous monitoring. *The Netherlands Journal of Medicine*, *54*(4), 167–179.

Gifford, R., et al. (2006). Protein interactions with subcutaneously implanted biosensors. *Biomaterials*, *27*(12), 2587–2598.

Gorbet, M. B., & Sefton, M. V. (2004). Biomaterial-associated thrombosis: roles of coagulation factors, complement, platelets and leukocytes. *Biomaterials*, *25*(26), 5681–5703.

Gough, D. A., Kumosa, L. S., Routh, T. L., Lin, J. T., & Lucisano, J. Y. (2010). Function of an implanted tissue glucose sensor for more than 1 year in animals. *Science Translational Medicine*, *2*(42), 42ra53.

Gucu, A., et al. (2013). Effects of temporary vascular occluder poloxamer 407 gel on the endothelium. *Journal of Cardiothoracic Surgery*, *8*(1), 16.

Guillon, M., & Maissa, C. A. (2012). *Multifocal contact lenses manufactured from a responsive polymer gel.* US 8,167,427 B2.

Guiseppi-Elie, A. (2010). Electroconductive hydrogels: synthesis, characterization and biomedical applications. *Biomaterials*, *31*(10), 2701–2716.

Gupta, B., Agarwal, R., & Alam, M. S. (2010). Textile-based smart wound dressings. *Indian Journal of Fibers and Textile Research*, *35*(2), 174–187.

Gupta, B., Mishra, S., & Saxena, S. (2008). Preparation of thermosensitive membranes by radiation grafting of acrylic acid/N-isopropyl acrylamide binary mixture on PET fabric. *Radiation Physics and Chemistry*, *77*(5), 553–560.

Gutowska, A., et al. (1995). Heparin release from thermosensitive polymer coatings: in vivo studies. *Journal of Biomedical Materials Research*, *29*(7), 811–821.

Hampikian, J. M., Heaton, B. C., Tong, F. C., Zhang, Z., & Wong, C. P. (2006). Mechanical and radiographic properties of a shape memory polymer composite for intracranial aneurysm coils. *Materials Science and Engineering: C*, *26*(8), 1373–1379.

Hartman, J., et al. (2007). Embolectomy in a rabbit acute arterial occlusion model using a novel electromechanical extraction device. *American Journal of Neuroradiology*, *28*(5), 872–874.

Heineman, W. (1993). Biosensors based on polymer networks formed by gamma irradiation crosslinking (Translated from English). *Applied Biochemistry and Biotechnology*, *41*(1–2), 87–97 (in English).

Heller, A. (2006). Electron-conducting redox hydrogels: design, characteristics and synthesis. *Current Opinion in Chemical Biology*, *10*(6), 664–672.

Henry, R. L., & Schmolka, I. R. (1989). Burn wound coverings and the use of poloxamer preparations. *Critical Reviews in Biocompatibility*, *5*(3), 207–220.

Heo, Y. J., Shibata, H., Okitsu, T., Kawanishi, T., & Takeuchi, S. (2011). Long-term in vivo glucose monitoring using fluorescent hydrogel fibers. *Proceedings of the National Academy of Sciences*, *108*(33), 13399–13403.

Heo, Y. J., & Takeuchi, S. (2013). Implantable biosensors: towards smart Tattoos: Implantable biosensors for continuous glucose monitoring. *Advanced Healthcare Materials*, *2*(1), 2.

Hoffman, A. S. (2013). Stimuli-responsive polymers: biomedical applications and challenges for clinical translation. *Advanced Drug Delivery Reviews*, *65*(1), 10–16.

Horowitz, M., Samson, D., & Purdy, P. (1997). Does electrothrombosis occur immediately after embolization of an aneurysm with Guglielmi detachable coils? *American Journal of Neuroradiology*, *18*(3), 510–513.

Hsiue, G.-H., Guu, J.-A., & Cheng, C.-C. (2001). Poly(2-hydroxyethyl methacrylate) film as a drug delivery system for pilocarpine. *Biomaterials*, *22*(13), 1763–1769.

Hsiue, G.-H., Hsu, S.-H, Yang, C.-C., Lee, S.-H., & Yang, I. K. (2002). Preparation of controlled release ophthalmic drops, for glaucoma therapy using thermosensitive poly-N-isopropylacrylamide. *Biomaterials*, *23*(2), 457–462.

Huang, W. (2010). Thermo-moisture responsive polyurethane shape memory polymer for biomedical devices. *The Open Medical Devices Journal*, *2*, 2–11.

Huang, X., et al. (2014). A differential dielectric affinity glucose sensor. *Lab on a Chip*, *14*, 294–301.

Huang, X., Li, S., Schultz, J. S., Wang, Q., & Lin, Q. (2009). A MEMS affinity glucose sensor using a biocompatible glucose-responsive polymer. *Sensors and Actuators B. Chemical*, *140*(2), 603–609.

Huang, W. M., Yang, B., Liu, N., & Phee, S. J. (2007). *Water-responsive programmable shape memory polymer devices*, pp. 64231S-64231S-64237.

Implantable medical devices market (reconstructive joint replacement, spinal implants, cardiovascular implants, dental implants, intraocular lens and breast implants)—U.S. Industry analysis, size, share, trends, growth and forecast 2012–2018. *Transparency market research*. (2013).

Ishihara, K., & Matsui, K. (1986). Glucose-responsive insulin release from polymer capsule. *Journal of Polymer Science Part C: Polymer Letters*, *24*(8), 413–417.

Jacobs, B. R. (2003). Central venous catheter occlusion and thrombosis. *Critical Care Clinics*, *19*(3), 489–514.

Jahren, S. L., Butler, M. F., Adams, S., & Cameron, R. E. (2010). Swelling and viscoelastic characterisation of pH-responsive chitosan hydrogels for targeted drug delivery. *Macromolecular Chemistry and Physics*, *211*(6), 644–650.

James, T. D., Sandanayake, K. R. A. S., Iguchi, R., & Shinkai, S. (1995). Novel saccharide-photoinduced electron transfer sensors based on the interaction of boronic acid and amine. *Journal of the American Chemical Society*, *117*(35), 8982–8987.

Jiang, B., et al. (2012). Investigation of lysine acrylate containing poly(N-isopropylacrylamide) hydrogels as wound dressings in normal and infected wounds. *Journal of Biomedical Materials Research Part B: Applied Biomaterials*, *100B*(3), 668–676.

Jiang, B., & Li, B. (2009). Polypeptide nanocoatings for preventing dental and orthopaedic device-associated infection: pH-induced antibiotic capture, release, and antibiotic efficacy. *Journal of Biomedical Materials Research Part B: Applied Biomaterials*, *88B*(2), 332–338.

Jones, D. S., et al. (2012). Novel semi-interpenetrating hydrogel networks with enhanced mechanical properties and thermoresponsive engineered drug delivery, designed as bioactive endotracheal tube biomaterials. *European Journal of Pharmaceutics and Biopharmaceutics*, *82*(3), 563–571.

Jung, Y., & Cho, J. (2010). Application of shape memory polyurethane in orthodontic (Translated from English). *Journal of Materials Science: Materials in Medicine*, *21*(10), 2881–2886 (in English).

Kaku, T., Okamoto, Y., Charles, L., Holness, W., & Karan, H. I. (1995). The effect of structure on poly(quinone) systems for amperometric glucose sensors. *Polymer*, *36*(14), 2813–2818.

Kang, S. I., & Bae, Y. H. (2003). A sulfonamide based glucose-responsive hydrogel with covalently immobilized glucose oxidase and catalase. *Journal of Controlled Release*, *86*(1), 115–121.

Kataoka, K., Miyazaki, H., Bunya, M., Okano, T., & Sakurai, Y. (1998). Totally synthetic polymer gels responding to external glucose Concentration: their preparation and application to on–off regulation of insulin release. *Journal of the American Chemical Society*, *120*(48), 12694–12695.

Kenausis, G., Taylor, C., Katakis, I., & Heller, A. (1996). 'Wiring' of glucose oxidase and lactate oxidase within a hydrogel made with poly(vinyl pyridine) complexed with [Os(4,4′-dimethoxy-2,2′-bipyridine)2Cl]+/2+. *Journal of the Chemical Society, Faraday Transactions*, *92*(20), 4131–4136.

Kenawy, E.-R., Worley, S. D., & Broughton, R. (2007). The chemistry and applications of antimicrobial polymers: a state-of-the-art review. *Biomacromolecules*, *8*(5), 1359–1384.

Klumb, L. A., & Horbett, T. A. (1992). Design of insulin delivery devices based on glucose sensitive membranes. *Journal of Controlled Release*, *18*(1), 59–80.

Koerner, H., Price, G., Pearce, N. A., Alexander, M., & Vaia, R. A. (2004). Remotely actuated polymer nanocomposites—stress-recovery of carbon-nanotube-filled thermoplastic elastomers. *Nature Materials*, *3*(2), 115–120.

Lauto, A., et al. (2001). Self-expandable chitosan stent: design and preparation. *Biomaterials*, *22*(13), 1869–1874.

Lendlein, A., Jiang, H., Junger, O., & Langer, R. (2005). Light-induced shape-memory polymers. *Nature*, *434*(7035), 879–882.

Lendlein, A., & Langer, R. (2002). Biodegradable, elastic shape-memory polymers for potential biomedical applications. *Science*, *296*(5573), 1673–1676.

Lendlein, A., & Langer, R. (2004). *Self-expanding device for the gastrointestinal or urogenital area*. WO 2004/073690 A1.

Lendlein, A., Schmidt, A. M., & Langer, R. (2001). AB-polymer networks based on oligo(ε-caprolactone) segments showing shape-memory properties. *Proceedings of the National Academy of Sciences*, *98*(3), 842–847.

Lendlein, A., Schmidt, A. M., Schroeter, M., & Langer, R. (2005). Shape-memory polymer networks from oligo(ε-caprolactone)dimethacrylates. *Journal of Polymer Science Part A: Polymer Chemistry*, *43*(7), 1369–1381.

Lin, F.-H., Chen, T.-M., Chen, K.-S., Wu, T.-H., & Chen, C.-C. (2000). An animal study of a novel tri-layer wound dressing material—non-woven fabric grafted with N-isopropyl acrylamide and gelatin. *Materials Chemistry and Physics*, *64*(3), 189–195.

Lin, S.-Y., Chen, K.-S., & Run-Chu, L. (2001). Design and evaluation of drug-loaded wound dressing having thermoresponsive, adhesive, absorptive and easy peeling properties. *Biomaterials*, *22*(22), 2999–3004.

Liu, H., & Ito, Y. (2002). Cell attachment and detachment on micropattern-immobilized poly(N-isopropylacrylamide) with gelatin. *Lab on a Chip*, *2*(3), 175–178.

Li, F., Zhu, W., Zhang, X., Zhao, C., & Xu, M. (1999). Shape memory effect of ethylene–vinyl acetate copolymers. *Journal of Applied Polymer Science*, *71*(7), 1063–1070.

Ma, K., et al. (2011). In vivo, transcutaneous glucose sensing using surface-enhanced spatially offset Raman spectroscopy: multiple rats, improved hypoglycemic accuracy, low incident power, and continuous monitoring for greater than 17 days. *Analytical Chemistry*, *83*(23), 9146–9152.

Maichuk, Y. F. (1975). Editorial: ophthalmic drug inserts. *Investigative Ophthalmology & Visual Science*, *14*(2), 87–90.

Maitland, D. J., et al. (2007). Prototype laser-activated shape memory polymer foam device for embolic treatment of aneurysms. *BIOMEDO*, *12*(3), 030504-030501-030504-030503.

Maitland, D. J., Metzger, M. F., Schumann, D., Lee, A., & Wilson, T. S. (2002). Photothermal properties of shape memory polymer micro-actuators for treating stroke. *Lasers in Surgery and Medicine*, *30*(1), 1–11.

Malitesta, C., Palmisano, F., Torsi, L., & Zambonin, P. G. (1990). Glucose fast-response amperometric sensor based on glucose oxidase immobilized in an electropolymerized poly(o-phenylenediamine) film. *Analytical Chemistry*, *62*(24), 2735–2740.

Mano, N., Mao, F., & Heller, A. (2005). On the parameters affecting the characteristics of the "wired" glucose oxidase anode. *Journal of Electroanalytical Chemistry, 574*(2), 347–357.

Mansouri, S., & Schultz, J. S. (1984). A miniature optical glucose sensor based on affinity binding. *Nature Biotechnology, 2*(10), 885–890.

Marco, D. (2006). *Biodegradable self-inflating intragastric implants and method of curbing appetite by the same*. WO 2006/082789 A2.

Mather, P. T., Liu, C., & Burstone, C. J. (2006). *Shape memory polymer orthodontic appliances, and methods of making and using the same*. WO 2006/071520 A2.

McCurley, M. F. (1994). An optical biosensor using a fluorescent, swelling sensing element. *Biosensors and Bioelectronics, 9*(7), 527–533.

Meng, B., et al. (2006). Study of biodegradable and self-expandable PLLA helical biliary stent in vivo and in vitro (Translated from English). *Journal of Materials Science: Materials in Medicine, 17*(7), 611–617 (in English).

Metcalfe, A., et al. (2003). Cold hibernated elastic memory foams for endovascular interventions. *Biomaterials, 24*(3), 491–497.

Metzger, M., Wilson, T., Schumann, D., Matthews, D., & Maitland, D. (2002). Mechanical properties of mechanical actuator for treating ischemic stroke (Translated from English). *Biomedical Microdevices, 4*(2), 89–96 (in English).

Moatti-Sirat, D., et al. (1992). Towards continuous glucose monitoring: in vivo evaluation of a miniaturized glucose sensor implanted for several days in rat subcutaneous tissue (Translated from English) *Diabetologia, 35*(3), 224–230 (in English).

Moussy, F., Harrison, D. J., O'Brien, D. W., & Rajotte, R. V. (1993). Performance of subcutaneously implanted needle-type glucose sensors employing a novel trilayer coating. *Analytical Chemistry, 65*(15), 2072–2077.

Moussy, F., Jakeway, S., Harrison, D. J., & Rajotte, R. V. (1994). In vitro and in vivo performance and lifetime of perfluorinated ionomer-coated glucose sensors after high-temperature curing. *Analytical Chemistry, 66*(22), 3882–3888.

Mulavisala, K. P., Gopal, P., Crane, B., Mackenzie, A., & Barwell, N. (2012). Accuracy of a continuous intravascular glucose monitor in ICU patients. In *32nd international symposium on intensive care and emergency medicine (ISICEM), Brussels, Belgium, March 20th–23rd 2012*.

Muñoz-Muñoz, F., Ruiz, J.-C., Alvarez-Lorenzo, C., Concheiro, A., & Bucio, E. (2009). Novel interpenetrating smart polymer networks grafted onto polypropylene by gamma radiation for loading and delivery of vancomycin. *European Polymer Journal, 45*(7), 1859–1867.

Nadkarni, S., & Yalkowsky, S. (1993). Controlled delivery of pilocarpine. 1. In vitro characterization of gelfoam® matrices (Translated from English). *Pharmaceutical Research, 10*(1), 109–112 (in English).

Nagahama, K., Ueda, Y., Ouchi, T., & Ohya, Y. (2009). Biodegradable shape-memory polymers exhibiting sharp thermal transitions and controlled drug release. *Biomacromolecules, 10*(7), 1789–1794.

Nakasima, A., Hu, J. R., Ichinose, M., & Shimada, H. (1991). Potential application of shape memory plastic as elastic material in clinical orthodontics. *European Journal of Orthodontics, 13*(3), 179–186.

Neffe, A. T., Hanh, B. D., Steuer, S., & Lendlein, A. (2009). Polymer networks combining controlled drug release, biodegradation, and shape memory capability. *Advanced Materials, 21*(32–33), 3394–3398.

Nickel, J. C., et al. (2012). Continuous intravesical lidocaine treatment for interstitial cystitis/bladder pain syndrome: safety and efficacy of a new drug delivery device. *Science Translational Medicine, 4*(143), 143ra100.

Osada, Y., & Matsuda, A. (1995). Shape memory in hydrogels. *Nature*, *376*(6537), 219.

Park, C., et al. (2004). Shape memory effect of poly(ethylene terephthalate) and poly(ethylene glycol) copolymer cross-linked with glycerol and sulfoisophthalate group and its application to impact-absorbing composite material. *Journal of Applied Polymer Science*, *94*(1), 308–316.

Patton, T. F., & Robinson, J. R. (1975). Ocular evaluation of polyvinyl alcohol vehicle in rabbits. *Journal of Pharmaceutical Sciences*, *64*(8), 1312–1316.

Peppas, N. A., & Bures, C. D. (2006). Glucose-responsive hydrogels. In G. E. Wnek & G. L. Bowlin (Eds.), *Encyclopedia of biomaterials and biomedical engineering*. New York, NY: Dekker. http://dx.doi.org/10.1081/E-EBBE-120042786.

Pickup, J. C., Claremont, D. J., & Shaw, G. W. (1993). Responses and calibration of amperometric glucose sensors implanted in the subcutaneous tissue of man (Translated from English). *Acta Diabetologica*, *30*(3), 143–148 (in English).

Ping, P., Wang, W., Chen, X., & Jing, X. (2005). Poly(ε-caprolactone) polyurethane and its shape-memory property. *Biomacromolecules*, *6*(2), 587–592.

Qiu, X., & Hu, S. (2013). "Smart" materials based on cellulose: a review of the preparations, properties, and applications. *Materials*, *6*(3), 738–781.

Quinn, C. P., Pathak, C. P., Heller, A., & Hubbell, J. A. (1995). Photo-crosslinked copolymers of 2-hydroxyethyl methacrylate, poly(ethylene glycol) tetra-acrylate and ethylene dimethacrylate for improving biocompatibility of biosensors. *Biomaterials*, *16*(5), 389–396.

Rastan, A. J., et al. (2010). Facilitated anastomosis using a reverse thermo-sensitive polymer for temporary coronary occlusion in off-pump minimally invasive direct coronary artery bypass surgery. *Interactive Cardiovascular and Thoracic Surgery*, *11*(5), 532–536.

Ratner, B. D. (2007). The catastrophe revisited: blood compatibility in the 21st Century. *Biomaterials*, *28*(34), 5144–5147.

Raymond, J., Desfaits, A. C., & Roy, D. (1999). Fibrinogen and vascular smooth muscle cell grafts promote healing of experimental aneurysms treated by embolization. *Stroke*, *30*(8), 1657–1664.

Raymond, J., & Roy, D. (1997). Safety and efficacy of endovascular treatment of acutely ruptured aneurysms. *Neurosurgery*, *41*(6), 1235–1246.

Reddy, T. T., Kano, A., Maruyama, A., Hadano, M., & Takahara, A. (2008). Thermosensitive transparent semi-interpenetrating polymer networks for wound dressing and cell adhesion control. *Biomacromolecules*, *9*(4), 1313–1321.

Rizzi, R., Centonze, D., Palmisano, F., & Zambonin, P. G. (2000). Simultaneous monitoring of glucose and lactate by an interference and cross-talk free dual electrode amperometric biosensor based on electropolymerized thin films. *Biosensors and Bioelectronics*, *15*(9–10), 531–539.

Rodriguez, J., et al. (2012). Opacification of shape memory polymer foam designed for treatment of intracranial aneurysms (Translated from English). *Annals of Biomedical Engineering*, *40*(4), 883–897 (in English).

Romey, M., et al. (2012). Use of an intravascular fluorescent continuous glucose sensor in subjects with type 1 diabetes mellitus (Translated from English). *Journal of Diabetes Science and Technology*, *6*(6), 1260–1266 (in English).

Roy, D., & Sumerlin, B. S. (2012). Glucose-sensitivity of boronic acid block copolymers at physiological pH. *ACS Macro Letters*, *1*(5), 529–532.

Ruiz, J.-C., et al. (2008). Polypropylene grafted with smart polymers (PNIPAAm/PAAc) for loading and controlled release of vancomycin. *European Journal of Pharmaceutics and Biopharmaceutics*, *70*(2), 467–477.

Sasso, S. V., Pierce, R. J., Walla, R., & Yacynych, A. M. (1990). Electropolymerized 1,2-diaminobenzene as a means to prevent interferences and fouling and to stabilize immobilized enzyme in electrochemical biosensors. *Analytical Chemistry*, *62*(11), 1111–1117.

Schmolka, I. R. (1972). Artificial skin I. Preparation and properties of pluronic F-127 gels for treatment of burns. *Journal of Biomedical Materials Research*, *6*(6), 571–582.
Schultz, J. S., Mansouri, S., & Goldstein, I. J. (1982). Affinity sensor: a new technique for developing implantable sensors for glucose and other metabolites (Translated from English). *Diabetes Care*, *5*(3), 245–253 (in English).
Scott, R. D., II, & Roberts, R. R. (2007). The attributable costs of resistant infections in hospital settings: economic theory and application. In *Antimicrobial resistance: Problem pathogens and clinical countermeasures* (pp. 1–24).
Serrano, M. C., & Ameer, G. A. (2012). Recent insights into the biomedical applications of shape-memory polymers. *Macromolecular Bioscience*, *12*(9), 1156–1171.
Service, R. F. (2002). Can sensors make a home in the body? *Science*, *297*(5583), 962–963.
Shaw, G. W., Claremont, D. J., & Pickup, J. C. (1991). In vitro testing of a simply constructed, highly stable glucose sensor suitable for implantation in diabetic patients. *Biosensors and Bioelectronics*, *6*(5), 401–406.
Shiino, D., et al. (1995). Amine containing phenylboronic acid gel for glucose-responsive insulin release under physiological pH. *Journal of Controlled Release*, *37*(3), 269–276.
Siegrist, J., et al. (2010). Continuous glucose sensor using novel genetically engineered binding polypeptides towards in vivo applications. *Sensors and Actuators B: Chemical*, *149*(1), 51–58.
Sigwart, U., Puel, J., Mirkovitch, V., Joffre, F., & Kappenberger, L. (1987). Intravascular stents to prevent occlusion and re-stenosis after transluminal angioplasty. *New England Journal of Medicine*, *316*(12), 701–706.
Singh, R. P. (2012). Prospects of organic conducting polymer modified electrodes: enzymosensors. *International Journal of Electrochemistry*, *2012*.
Small, W., et al. (2007a). Prototype fabrication and preliminary in vitro testing of a shape memory endovascular thrombectomy device. *IEEE Transactions on Biomedical Engineering*, *54*(9), 1657–1666.
Small, W., et al. (2007b). Shape memory polymer stent with expandable foam: a new concept for endovascular embolization of fusiform aneurysms. *IEEE Transactions on Biomedical Engineering*, *54*(6), 1157–1160.
Small, W., Wilson, T., Benett, W., Loge, J., & Maitland, D. (2005). Laser-activated shape memory polymer intravascular thrombectomy device. *Optics Express*, *13*(20), 8204–8213.
Smith, R. S., et al. (2012). Vascular catheters with a nonleaching poly-sulfobetaine surface modification reduce thrombus formation and microbial attachment. *Science Translational Medicine*, *4*(153), 153ra132.
Striegler, S. (2003). Selective carbohydrate recognition by synthetic receptors in aqueous solution. *Current Organic Chemistry*, *7*(1), 81–102.
Stuart, D. A., et al. (2006). Vivo glucose measurement by surface-enhanced raman spectroscopy. *Analytical Chemistry*, *78*(20), 7211–7215.
Tamai, H., et al. (2000). Initial and 6-month results of biodegradable poly-l-lactic acid coronary stents in humans. *Circulation*, *102*(4), 399–404.
Traitel, T., Cohen, Y., & Kost, J. (2000). Characterization of glucose-sensitive insulin release systems in simulated in vivo conditions. *Biomaterials*, *21*(16), 1679–1687.
Vaddiraju, S., Singh, H., Burgess, D. J., Jain, F. C., & Papadimitrakopoulos, F. (2009). Enhanced glucose sensor linearity using poly(vinyl alcohol) hydrogels. *Journal of diabetes science and technology*, *3*(4), 863–874.
Venkatraman, S. S., Tan, L. P., Joso, J. F. D., Boey, Y. C. F., & Wang, X. (2006). Biodegradable stents with elastic memory. *Biomaterials*, *27*(8), 1573–1578.

Verestiuc, L., Ivanov, C., Barbu, E., & Tsibouklis, J. (2004). Dual-stimuli-responsive hydrogels based on poly(N-isopropylacrylamide)/chitosan semi-interpenetrating networks. *International Journal of Pharmaceutics*, *269*(1), 185–194.

Wache, H. M., Tartakowska, D. J., Hentrich, A., & Wagner, M. H. (2003). Development of a polymer stent with shape memory effect as a drug delivery system. *Journal of Materials Science: Materials in Medicine*, *14*(2), 109–112.

Wang, J. (2007). Electrochemical glucose biosensors. *Chemical Reviews*, *108*(2), 814–825.

Wang, J., Chen, L., Zhao, Y., Guo, G., & Zhang, R. (2009). Cell adhesion and accelerated detachment on the surface of temperature-sensitive chitosan and poly(N-isopropylacrylamide) hydrogels (Translated from English). *Journal of Materials Science: Materials in Medicine*, *20*(2), 583–590 (in English).

Wang, L. S., Chow, P. Y., Phan, T. T., Lim, I. J., & Yang, Y. Y. (2006). Fabrication and characterization of nanostructured and thermosensitive polymer membranes for wound healing and cell grafting. *Advanced Functional Materials*, *16*(9), 1171–1178.

Wang, L. S., Chow, P. Y., Tan, D. C. W., Zhang, W. D., & Yang, Y. Y. (2004). Nanostructured and transparent polymer membranes with thermosensitivity for wound dressing and cell grafting. *Advanced Materials*, *16*(20), 1790–1794.

Wang, C., et al. (2008). Synthesis and performance of novel hydrogels coatings for implantable glucose sensors. *Biomacromolecules*, *9*(2), 561–567.

Williams, D. F. (1987). Definitions in biomaterials. In *Proceedings of a consensus conference of the European society for biomaterials, Chester, England, March 3–5, 1986* (Vol. 4). New York: Elsevier.

Wilson, G. S., & Gifford, R. (2005). Biosensors for real-time in vivo measurements. *Biosensors and Bioelectronics*, *20*(12), 2388–2403.

Wilson, T. S., Small Iv, W., Benett, W. J., Bearinger, J. P., & Maitland, D. J. (2005). *Shape memory polymer therapeutic devices for stroke*, pp. 60070R-60070R-60078.

Wimmer-Greinecker, G., et al. (2011). Randomized clinical trial comparing a thermosensitive polymer (LeGoo) with conventional vessel loops for temporary coronary artery occlusion during off-pump coronary artery bypass surgery. *Annals of Thoracic Surgery*, *92*(6), 2177–2183.

Wischke, C., Neffe, A. T., Steuer, S., & Lendlein, A. (2009). Evaluation of a degradable shape-memory polymer network as matrix for controlled drug release. *Journal of Controlled Release*, *138*(3), 243–250.

Wu, Q., Wang, L., Yu, H., Wang, J., & Chen, Z. (2011). Organization of glucose-responsive systems and their properties. *Chemical Reviews*, *111*(12), 7855–7875.

Xue, L., Dai, S., & Li, Z. (2010). Biodegradable shape-memory block co-polymers for fast self-expandable stents. *Biomaterials*, *31*(32), 8132–8140.

Yakacki, C. M., et al. (2007). Unconstrained recovery characterization of shape-memory polymer networks for cardiovascular applications. *Biomaterials*, *28*(14), 2255–2263.

Yang, S., Atanasov, P., & Wilkins, E. (1995). Glucose biosensors based on oxygen electrode with sandwich-type membranes (Translated from English). *Annals of Biomedical Engineering*, *23*(6), 833–839 (in English).

Yang, B., Huang, W. M., Li, C., Lee, C. M., & Li, L. (2004). On the effects of moisture in a polyurethane shape memory polymer. *Smart Materials and Structures*, *13*(1), 191–195.

Yang, J. M., & Lin, H. T. (2004). Properties of chitosan containing PP-g-AA-g-NIPAAm bigraft nonwoven fabric for wound dressing. *Journal of Membrane Science*, *243*(1–2), 1–7.

Yang, W., Xue, H., Carr, L. R., Wang, J., & Jiang, S. (2011). Zwitterionic poly(carboxybetaine) hydrogels for glucose biosensors in complex media. *Biosensors and Bioelectronics*, *26*(5), 2454–2459.

Yao, H., Shum, A. J., Cowan, M., Lähdesmäki, I., & Parviz, B. A. (2011). A contact lens with embedded sensor for monitoring tear glucose level. *Biosensors and Bioelectronics*, *26*(7), 3290–3296.

Yonzon, C. R., Haynes, C. L., Zhang, X., Walsh, J. T., & Van Duyne, R. P. (2003). A glucose biosensor based on surface-enhanced Raman scattering: improved partition layer, temporal stability, reversibility, and resistance to serum protein interference. *Analytical Chemistry*, *76*(1), 78–85.

Yu, B., et al. (2008). Use of hydrogel coating to improve the performance of implanted glucose sensors. *Biosensors and Bioelectronics*, *23*(8), 1278–1284.

Yu, B., Long, N., Moussy, Y., & Moussy, F. (2006). A long-term flexible minimally-invasive implantable glucose biosensor based on an epoxy-enhanced polyurethane membrane (Translated from English). *Biosens Bioelectron*, *21*(12), 2275–2282 (in English).

Zhang, Y., et al. (1994). Elimination of the acetaminophen interference in an implantable glucose sensor. *Analytical Chemistry*, *66*(7), 1183–1188.

Zhang, Z., et al. (2009). Zwitterionic hydrogels: an in vivo implantation study. *Journal of Biomaterials Science-Polymer Edition*, *20*(13), 1845–1859.

Zhang, H., Wang, H., Zhong, W., & Du, Q. (2009). A novel type of shape memory polymer blend and the shape memory mechanism. *Polymer*, *50*(6), 1596–1601.

Zheng, X., Zhou, S., Li, X., & Weng, J. (2006). Shape memory properties of poly(D,L-lactide)/hydroxyapatite composites. *Biomaterials*, *27*(24), 4288–4295.

Index

Note: Page numbers followed by "f" and "t" indicate figures and tables, respectively.

A

Affinity-based biosensors
boronic acid–saccharide interaction, 285–287
ConA ratiometric fluorescence resonance energy transfer sensor, 284–285
label-free surface affinity-based biosensors, 287
synthetic responsive polymers, 285–287
two glucose-responsive polymers, 285–287, 286f
viscometric-affinity ConA sensor, 285
Aneurysm occlusion devices, 265–266
Aqueous two phase system (ATPS), 20–21

B

Betaine monomer subunits, 46–47, 46f
Biocatalysts, 108, 108f
Biomolecule-sensitive polymers, 244–245
Biosensors, 108–109, 109f
affinity-based biosensors. *See* Affinity-based biosensors
enzyme electrode-based biosensors. *See* Enzyme electrode-based biosensors
long-term implantation, 281
material–body responses, 287–288
pH responsive polymer-based biosensors, 283–284
Block copolymer, 123–124, 124f, 134–135, 135f–137f
Branched polymer brushes, 123–124, 124f
BST-CarGel®, 239
Bulk hydrogels, 22

C

Cardiovascular devices
advantage and disadvantage, 263
aneurysm occlusion devices, 265–266
clot extraction devices, 264, 265f
drug-eluting capability, 266–267
in vivo tests, 267
minimally invasive surgery, 263–264
pre-deformed polyurethane SMP stent retraction, 267–268, 268f
SMPs, 262
temporary endovascular occlusion devices, 268–269
vascular stents, 266
Cell sheet–based regenerative medicine
applications, 223–225, 225f
cardiomyocyte sheet transplantation, 222
cell sheet–based tissue engineering, 220, 221f
corneal epithelial cell sheets, 220
double-layered 3D cocultured cell sheets, 222–223
mice hepatic-layered cell sheets, 222
neonatal rat cardiomyocyte sheets, 222
oral mucosal cell sheets, 220–221
periodontal ligament tissue, 221
rat islet cell sheets, 223
suturing and stapling techniques, 221–222
Cell-surface thiols, 243
Chitosan thiomer, 242–243
Clot extraction devices, 264, 265f
Corneal epithelial cell sheets, 220

D

Dental and orthopedic devices, 274–276
Diblock-copolymers, 119–120
Double-layered 3D cocultured cell sheets, 222–223
Drop-casting method, 216
Dual responsive coatings, 152–153, 153f

E

Elastin-like oligo- and polypeptides polymers (ELP), 18–19, 73–74

Electric-field-sensitive polymers, 250–251, 250f
Electron beam (EB) irradiation method, 203, 204f
  and plasma irradiation methods
    adhesion-promoting layer, 209
    cell–cell junction and ECM components, 207
    cell detachment mechanism, 207
    cell-repellent character, 208
    fibronectin recovery, 209
    graft polymer thickness dependency, 208, 208f
    plasma polymerization method, 209
Electro-responsive brushes, 130, 130f
Environmentally responsive polyelectrolytes and zwitterionic polymers
  advantages, 58
  antipolyelectrolyte effect, 50–51, 51f–52f
  betaine monomer subunits, 46–47, 46f
  industrial applications, 45–46
  natural biological macromolecules, 45
  polyampholyte monomer subunits, 47–48, 47t
  polyelectrolyte, 47–50, 47t, 50f
  polymerization and coating approach, 48–49
  solution-based properties, 49
  stimuli-induced changes
    biomedical community, 52
    in vitro applications, 57–58
    pH-induced responsiveness, 53–55, 54f
    salt concentration induced responsiveness, 55–56, 56f
    temperature-induced responsiveness, 57
Enzyme electrode-based biosensors
  conductive polymers, 283
  dialysis membranes, 282
  electron-transfer process, 281–282
  glucose–enzyme reaction, 281
  glucose sensor, 282
  implantable glucose electrode, 281–282, 282f
  polyHEMA, 283
Enzyme-filled brushes, 133–134
Enzyme-triggered polymers, 243–244
Extracellular membrane (ECM) components, 193

G

Gastrointestinal devices, 279
Gold-based SAMs, 69, 70f
Grafting-from methods, 121f, 122
  ATRP/RAFT polymerization method, 213
  cell adhesion character, 214–216
  graft PIPAAm density, 214, 215t
  graft polymer thickness and polymer graft density, 213–214
  PEGMA component, 213
  polyethylene glycol–based temperature-responsive polymer, 216
  self-assembled monolayer, 214–216
Grafting-to methods, 121–122, 121f, 212
  advantage and disadvantage, 212
  surface-initiated living radical polymerization technique, 212–213

H

Homopolymer brushes
  biotin-attached PAA brushes, 131
  decorate nanoparticles, 133–134, 134f
  enzyme-filled brushes, 133–134
  pH-responsive polymers, 131, 133f
  PNIPAM brushes, grafting density, 131, 132f
  reversible pH-controlled transformation, 131–133, 133f

I

Implantable medical devices
  biocompatibility
    capsulation and thrombus formation, 261–262
    extracellular polysaccharide, 260–261
    foreign body response, 260, 261f
    mitigation, 262
    thrombin, 260
  biological structures, 259
  cardiovascular devices. *See* Cardiovascular devices
  dental and orthopedic devices, 274–276
  drug delivery/infection control coatings and devices
    bacterial and fungal infections, 270
    biofilms, 270
    catheter-based drug release coatings, 270–271

catheters, 269
SMPs, 271
zwitterionic polymers, 270
gastrointestinal devices, 279
insulin delivery devices, diabetes management, 280
ophthalmic devices, 276–278
respiratory devices, 278
surgical devices
sutures, 271–272, 272f
wound dressings, 272–274
thermo-responsive polymers, 259
urogenital devices/urinary devices, 279–280
In situ–forming hydrogel, 240
Insulin delivery devices, diabetes management, 280
Ionic strength responsive brushes, 126–128, 129f
Ion-sensitive polymers, 240–242, 241f

L

Label-free surface affinity-based biosensors, 287
Light-induced brushes, 130
Light-sensitive polymers, 249–250, 250f

M

Macroscopic hydrogels and microgels
crosslinked polymeric porous network, 22
de-swelling behavior, 22–23
on–off drug release systems, 23–24
polymeric gel networks, 22–23, 23f
receptor-mediated endocytosis, 24–25
semi-interpenetrating hydrogel networks, 24
switchable swelling behavior, 24
volume–phase transitions, 24–25
Magnetic-field-sensitive polymers, 247f, 248–249
Mesoporous silica nanoparticles (MSNPs), 109–111, 110f
Mice hepatic-layered cell sheets, 222
Mixed brushes, 136–138, 138f
Mixed polymer brushes, 119–120, 123–124, 124f

N

Naturally inspired peptides, 73f
antimicrobial polymer, 74
ELPs, 73–74
extracellular matrix protein, 73
heterogeneous protein surface, 74
Kemptide, 74–75
layer-by-layer deposition technique, 74
short peptides, 74–75
Near-infrared-light-sensitive polymers, 246–248, 247f
Negative temperature–sensitive polymers, 4
Neonatal rat cardiomyocyte sheets, 222

O

OncoGel™, 239–240
Ophthalmic devices, 276–278
Optobioelectronic devices, 106, 106f–107f
Oral mucosal cell sheets, 220–221

P

Peptide-based switchable and responsive surfaces
adsorption/deposition, 72, 72f
advantages and disadvantages, 67–68, 68t
applications
adhesion events control, 88
biosensor fabrication, 88
bioseparation, 89
artificial peptides, 72–73
attach/release peptides, 77–78, 78f
bioactive surfaces, definition, 67
3D dimensional surfaces, 89–90
designed peptides
amine function, 75
artificial polypeptide surface, 75–76
circular dichroism, 76–77
photosensitizers, 77
pH-responsive peptide surfaces, 76–77
quartz crystal microbalance measurements, 76–77
response mechanisms, 75, 76f
fabrication strategies, 68
glass surfaces
amine-terminated silanes, 70–71
“grafting from” approach, 71–72
silanization, 70, 71f
gold-based SAMs, 69, 70f
hide/reveal peptides, 77–78, 79f
immobilization, 69, 78–80
man-made biomaterials, 66

Peptide-based switchable and responsive surfaces (*Continued*)
  material–biology interface, 66
  naturally inspired peptides, 73f
    antimicrobial polymer, 74
    ELPs, 73–74
    extracellular matrix protein, 73
    heterogeneous protein surface, 74
    Kemptide, 74–75
    layer-by-layer deposition technique, 74
    short peptides, 74–75
  natural stimulus-responsive materials, 66, 67t
  non–peptide-based response mechanisms, 77–78
  physical properties
    benefits and limitations, 83
    electrochemical potential, 84
    enzymes, 84–85
    light, 84
    pH, 83–84
    response types, 85
    temperature, 83
    trigger responses, 83
  protein mimic, 72–73
  protein surfaces
    design, 81, 82f
    protein release, 83
    stimulus-responsive proteins, 82–83
    types, 81
  release of, 78–80
  response characterization, 86–87
  stimulus responsiveness, 65–66
  substrate surface, 69
  surface characterization, 85–86
  triggered display, 80–81
  UV-based turbidity measurements and calorimetry, 85
Photonic sensitive switchable materials
  biocatalysts, 108, 108f
  biomolecular-based photonic sensitive materials, 93
  biosensors, 108–109, 109f
  bR biomaterial, 111
  chemical and physical properties, 93
  drug delivery system, 109–111, 110f
  inorganic complex, 93
  inorganic compound, 94, 94f
  MSNPs, 109–111, 110f
  optical switches
    gating channels, 103–104
    light and pH dual-driven ionic gate, 103–104, 105f
    light-gated nanofluidic diode system, 104, 105f
    light-responsive synthetic solid-state nanochannels, 103–104
    steady and reversible light controlled nanochannel, 103–104, 104f
  optobioelectronic devices, 106, 106f–107f
  organic compound
    aobenzene, 95
    light-induced isomerisation and photochemical reactions, 94, 95f
    liquid-crystal monomer and crosslinker, 95–96, 96f
    photoirradiation, 94
    potochromic diarylethenes, 95
    spirobenzopyran derivatives, 95
    tiphenylmethane leuco derivatives, 95
  photoswitchable biomaterials, 98–99, 98f
  photosynthetic pigment, 97, 97f
  reversible photo-responsive surfaces, 111
  reversibly photostimulate biomaterials, 99, 100f
  semiconductor-biomolecule QDs hybrids, 97–98, 98f
  switchable wettability, photoresponsive surfaces
    flat zinc oxide substrate, 99–101, 101f
    malachite green–terminated alkanethiol, 102, 103f
    photo-switched azobenzene monolayer, 102, 103f
    tin oxide nanorod film, 101–102, 102f
    titanium dioxide surfaces, 99–101
    UV-induced wettability conversion, 99–101
    UV-responsive superhydrophobic/superhydrophilic switches, 102
pH-responsive brushes, 126, 128f
pH responsive polymer-based biosensors, 283–284
pH-responsive polymers, 131, 133f
pH-sensitive polymers
  cationic polymers, 236–237
  nanoparticular polyplexes, 238–239
  PEIs, 238

polymers and polymeric drug delivery systems, 237–238, 238f
polyvinylacetal diethylaminoacetate, 237
Planar polyelectrolyte brushes, 176–177, 177f
Poloxamers/pluronics, 13–15
Poly(acrylic acid) (PAA) brushes, 126
Poly(ethylene glycol)-poly lactic acid-co-glycolic acid (PEG-PLGA) block copolymers, 15–18
Poly(ethylene glycol)-poly L-lactic acid (PEG-PLLA), 15–18
Polyethyleneimines (PEIs), 238
Poly(ethylene oxide)–poly(propylene oxide)–poly(ethylene oxide) (PEO–PPO–PEO) block copolymers, 13–15
Poly(*N*-isopropylacrylamide) (PNIPAAm), 6–8
Polymer coating method
advantages, 216
contamination, 218
dissolution-induced cell detachment character, 217
drop-casting method, 216
electrostatically polymer coating approach, 219–220
molecular dynamics simulation, 217–218
polystyrene surfaces, 217
spin-coating method, 218, 219f
triblock copolymer, 217–218
Poly(methacrylic acid) (PMAA) brushes, 126
Poly(*N*-vinylamides) (PVP), 11–13
Poly(*N*-vinylcaprolactam) (PVCL), 11–13
Positive temperature–sensitive polymers, 4
Protective brush layer, 139–140, 140f
Protein–surface interactions
electric potential, 178–179, 178f
generic modes, 168, 168f
molecular simulation, 182
pH and/ionic strength-responsive surfaces
electrostatic interaction, 174
PAA brushes, 176–177, 177f
polyelectrolyte-modified surfaces, 174
SPBs, 174–176, 175f
photo-induced supermolecular interaction, 179–180, 179f
protein properties, 168–169
selective physical adsorption/covalent immobilization, 167
smart surfaces, 167–168
solution properties, 168–169
surface properties, 168–169
surface wettability, 182
synergistic effect, surface chemistry and nanostructures, 180–181, 181f
thermo-responsive surfaces
aqueous chromatographic methods, 172–173
chain conformation and hydrophobicity, 171
fluorescence-labeling assay, 169–170
graft density, 171–172
Halperin's model, 172
immobilized proteins/enzymes, 173
LCST behavior, 169
nonspecific and biospecific protein adsorption, 173, 174f
PNIPAAm, 169
radiolabeling assay, 170
"smart" mobile affinity matrix, 172–173, 173f
temperature-dependent adhesion force, 170
time-of-flight secondary ion mass spectrometry, 169–170

Q

Quartz crystal microbalance-dissipation monitoring (QCM-D)
coated quartz crystals, 159
phase transition temperature, 159–160, 160f
PNiPAAm-based copolymer film, 159, 160f
quartz oscillation, 158
viscoelastic properties, 158–159

R

Rat islet cell sheets, 223
Redox-sensitive polymers, 242–243
Respiratory devices, 278
Responsive polymer brushes
adsorption, 121, 121f
advantages and disadvantages, 120
biomedical applications, 122–123, 123f
block copolymer, 123–124, 124f, 134–135, 135f–137f

Responsive polymer brushes (*Continued*)
branched polymer brushes, 123–124, 124f
brush regime, 119
colloidal particles, 140
density effect, 119, 120f
diblock-copolymers, 119–120
electro-responsive brushes, 130, 130f
"grafting from" approach, 121f, 122
"grafting to" approach, 121–122, 121f
homopolymers, 123–124, 124f. *See also* Homopolymer brushes
ionic strength responsive brushes, 126–128, 129f
light-induced brushes, 130
mixed brushes, 136–138, 138f
mixed polymer brushes, 119–120, 123–124, 124f
mushroom regime, 119
neutral polymer block, 139, 140f
pH-responsive brushes, 126, 128f
protective brush layer, 139–140, 140f
protein molecules, 119
adsorption and nonadsorption, 139–140
solvent-responsive brushes, 128–129, 129f
SPBs, 139
temperature-responsive brushes, 126, 127f
types, 124–126, 125t
Responsive/switchable surface interaction
cell-material properties and interactions, 189
cytokines, 198
interfacial properties
binding affinity, 191
ECM components, 193
elasticity and curvature, 193
stem cell scaffolding parameters, 193, 194f
surface energy and dynamic contact angle measurements, 192, 192t
wettability, 192–193
physical behavior
adhesion, 195–196
cell:matrix and cell:cell signaling, 194, 195f
differentiation, 197–198
integrin signaling, 194
migration, 196–197
stem cell behavior
adhesion and adsorption, 190, 191f
spreading and migration, 190–191
stem cell fate and phenotypic expression, 190
Reverse thermo-gelation property system, 19–20

S

Semiconductor-biomolecule quantum dots hybrids, 97–98, 98f
Shape memory polymers (SMPs), 262
Solvent-responsive brushes, 128–129, 129f
Spherical polyelectrolyte brushes (SPBs), 174–176, 175f
Spin-coating method, 218, 219f
Stimuli-sensitive polymers
advantages, 252
external stimuli, 235
electric-field-sensitive polymers, 250–251, 250f
light-sensitive polymers, 249–250, 250f
magnetic-field-sensitive polymers, 247f, 248–249
near-infrared-light-sensitive polymers, 246–248, 247f
temperature-sensitive polymers, 245–246, 247f
ultrasound-sensitive polymers, 248
internal stimuli, 235
biomolecule-sensitive polymers, 244–245
enzyme-triggered polymers, 243–244
ion-sensitive polymers, 240–242, 241f
pH-sensitive polymers. *See* pH-sensitive polymers
redox-sensitive polymers, 242–243
temperature-sensitive polymers, 239–240
pH and ionic concentration, 235, 236t
polymeric systems, 252
targeting drug delivery systems, 251
Surgical devices
sutures, 271–272, 272f
wound dressings, 272–274
Switchable copolymers, 150, 150f
anchorage mechanism, 147–148
biomolecular functionalisation, 152
contact angle measurements, axisymmetric drop shape analysis, 153, 154f

cross-linking, 148–149
dual responsive coatings, 152–153, 153f
electrokinetic measurements, 155–157, 161
NiPAAm-based thermo-responsive copolymers, 149
physicochemical mechanism, 147
QCM-D, 161
coated quartz crystals, 159
phase transition temperature, 159–160, 160f
PNiPAAm-based copolymer film, 159, 160f
quartz oscillation, 158
viscoelastic properties, 158–159
scanning force microscopy
bovine serum albumin–coated probes, 157–158
force *vs.* displacement curves, 157, 158f
temperature-dependent elastic modulus, 157, 159f
spectroscopic ellipsometry
film thickness *vs.* temperature, 154–155, 155f
interaction mechanisms, 155
surface-immobilised swellable hydrogel film, 154
temperature-dependent swelling behaviour, 155, 156f
temperature-dependent water contact angle, 149–150, 149f
thermo-responsive materials, 147
tuning cell adhesion and detachment, 150–152, 151f
Synthetic responsive polymers, 285–287
Synthetic thermo-responsive polymers
ELP, 18–19
PEG-PLGA block copolymers, 15–18
PEG–PLLA, 15–18
PEO–PPO–PEO block copolymers, 13–15
PVCL, 11–13
N-substituted acrylamide polymers
co-monomers, 8–9
crumped coil and molten globule intermediate states, 9
inter-molecular H_2O–H_2O hydrogen bond network, 9
intra-molecular segment aggregation, 9
NIPAAm-acrylic acid copolymer, 10–11
phase separation temperatures, 7–8
PNIPAAm, 6–8
polymer–water hydrogen bonds, 10
shorter and longer lifetime, 9–10
side chain lengths and properties, 7–8, 8t

T

Temperature-responsive brushes, 126, 127f
Temperature-responsive cell culture surface (TRCS)
cell sheet–based regenerative medicine. *See* Cell sheet-based regenerative medicine
covalent polymer bonding surface modification
electron beam and plasma irradiation methods. *See* Electron beam and plasma irradiation methods
irradiation procedure, grafting polymer, 206–207
photo-initiator/photo-cross- linker–induced polymer grafting method, 209–211
preactivated surface methods, 211
EB irradiation method, 203, 204f
electrostatically polymer-coated surface, 204, 206t
grafting-from methods. *See* Grafting-from methods
grafting-to methods, 212
advantage and disadvantage, 212
surface-initiated living radical polymerization technique, 212–213
PIPAAm and conventional TRCS, characteristics, 203, 204f
polymer bonding surface modifications, 204, 205t
polymer chains, 211
polymer-coated surface modifications, 204, 206t
polymer coating method. *See* Polymer coating method
reversible hydration and dehydration, 203
Temperature-sensitive polymers, 239–240, 245–246, 247f
Temporary endovascular occlusion devices, 268–269

Thermo-responsive polymers, 259
  anti-inflammatory therapy, 28
  biocompatibility and biodegradability, 30–31
  block copolymer, 26–28, 27f
  characteristics
    chains-to-globules-to-aggregate transition, 4–5, 5f
    dissolution, 5
    enthalpy and entropy, 5–6
    Gibbs free energy, 5
    lower critical solution temperature, 4, 4f
    polymer–water hydrogen bonding interactions, 4–5
    sol–gel transition, 4
    temperature-induced phase transitions, 4–5, 5f
    upper critical solution temperature, 4, 4f
  classification, 6
  critical micelle concentration, 26–28
  crosslinked polymeric network, 20
  drug delivery vehicle, 26–28, 27f
  drug solubilization capacity, 28
  free polymer chains
    affinity precipitation, 20–22
    ATPS, 20–21
    EOPO, 21
  gold nanoparticles, 28–29
  hydrolysis/enzymatic degradation–triggered drug release systems, 28
  hydrophilic and hydrophobic blocks, 6, 7f
  macroscopic hydrogels and microgels. *See* Macroscopic hydrogels and microgels
  nano-range size distribution, 28–29
  natural thermo-responsive polymers, 19–20
  polymersomes, 29–30, 29f
  properties, 3
  responsive behavior, 3
  smart/stimuli-responsive materials, 3
  surface-grafted polymers, 25–26
  synthetic thermo-responsive polymers, 6, 7f. *See also* Synthetic thermo-responsive polymers
  temperature sensitivity, 30–31
  thermosensitive polymer, 20
Thiolated gelatin, 242
Thiolated poly(methacrylic acid) polymer, 243
TRCS. *See* Temperature-responsive cell culture surface (TRCS)
Two glucose-responsive polymers, 285–287, 286f

U

Ultrasound-sensitive polymers, 248
Urogenital devices/urinary devices, 279–280

V

Viscometric-affinity ConA sensor, 285

Edwards Brothers Malloy
Ann Arbor MI. USA
June 26, 2015